Applied Instrumentation in the Process Industries

Volume IV

Control Systems: Theory, Troubleshooting, and Design

Applied Instrumentation in the Process Industries

Volume IV

Control Systems: Theory, Troubleshooting, and Design

Leslie M. Zoss

Leslie M. Zoss

Gulf Publishing Company
Book Division
Houston, London, Paris, Tokyo

Applied Instrumentation in the Process Industries, Volume IV
Control Systems: Theory, Troubleshooting, and Design

Library of Congress Catalog Card Number 72-94067
ISBN 0-87201-391-X

First Printing, February 1979
Second Printing, August 1982

Contents

6

7

8

9

10

11

12

13

14

Preface

Over the past 25 years, I have had the opportunity to present control system programs for many educational, technical, and industrial organizations. During this period, there has been a rapid technological change in both control equipment and methods of analysis. In all this change, however, a number of stable concepts has come to my attention. First, regardless of analytical technique, control system theory has not changed. Second, each year, more individuals from all walks of life are requesting help in understanding control system theory and application. Third, there is a large number of practitioners who find it difficult to relate theory to practice, or practice to theory. The layout and planning for this text has developed around these three concepts.

Since theory has not changed, it does not seem advisable or necessary, in a beginning text, to cover all possible techniques of analysis. Thus, I have selected one technique, namely frequency response, and used it throughout the text. For the reader interested only in a basic understanding, one technique should be adequate. The serious student, however, will want to examine other texts to determine the advantages and disadvantages of all available techniques. If this student has comprehended the frequency response technique and the material presented in this text, continued study should be much less difficult.

While the self-study approach of this text is designed to aid all readers, it should be particularly helpful to the "business reader," i.e., any individual who requires an understanding of control systems but has primary responsibility in another area. Examples of business readers include process designers, plant managers, chief engineers, vice presidents, company presidents, purchasing agents, advertising agents, and magazine editors. A combination of the self-study text and an instructor having a knowledge of control systems should prove to be very beneficial to both industrial and technical groups. This combination could be used to provide training programs for business readers as well as practitioners.

The sequence of theory, troubleshooting, and design may seem to be out of order to many readers. Most will suggest that design should precede troubleshooting. I cannot argue with this logic; however, experience indicates that it does not work out this way. Many systems are designed by people with theoretical knowledge, practical experience, or both. In most instances these systems work. When they do not function properly or when changing process conditions cause malfunctions, troubleshooting becomes the important role. In fact, the information learned from troubleshooting often becomes the basis for theory and design. To start with troubleshooting and ignore available theory, however, would be senseless. Thus, I have chosen the sequence of theory, troubleshooting, and design, which has proven to provide an effective learning procedure. The introduction of troubleshooting techniques immediately after theory provides concrete examples for both the theoretical and practical student. Pedagogically, it brings the students together and provides the basis for relating theory and practice in the design process.

The text does not try to cover examples in all process areas. Rather, it is an attempt to present a program for study that has been developed during a 25-year apprenticeship in teaching control. I am hopeful that this program of study will prove to be beneficial to the reader and if required, provide a basis for continued study.

Leslie M. Zoss
January 1979

Acknowledgments

I extend thanks to H.S. Wilson of Moore Products Company for his encouragement to continue with this work and for his review of early chapters. Appreciation is also extended to Professor W. Shewan of Valparaiso University and H.W. Stoll, consulting engineer, for reviewing chapters of this manuscript.

Thanks are also extended to Professors K. Mortimer, M.C. Vocke, and R.J. Bohlmann of Valparaiso University for their review of specific material in Chapters 11, 12, and 14.

I am particularly indebted to Mrs. Joyce DelMastro for her patience and attention to detail when typing the original manuscript.

In addition, I wish to acknowledge Valparaiso University and my many students for providing the facilities and environment in which to produce and test this work.

To my wife Carolyn and my family,
for their encouragement and understanding

1 The Closed Loop Control System

This chapter is devoted to defining terminology, block diagram symbolism, and fundamental objectives of a simple single closed loop control system.

1.1. Terminology of Closed Loop Control

In any control system the prime objective is to maintain a particular variable at a desired condition. The particular variable to be regulated is termed the "controlled variable" and is assigned the notation c. The desired condition of the controlled variable is the "reference" or "set point" and is assigned the notation r. In the ideal system the controlled variable c would be regulated or controlled to be equal to the reference or set point r at all times.

While the basic idea is simple, there is a number of problems that occur. A shell and tube heat exchanger (Figure 1-1) is typical of many systems and will be used as an example. In this example, product outlet temperature is the controlled variable.

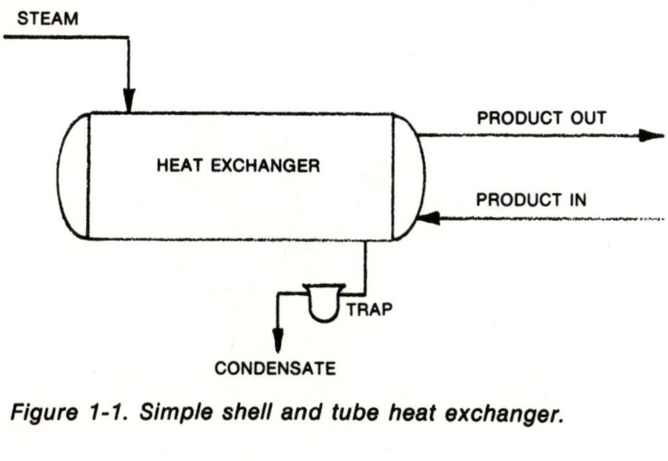

Figure 1-1. Simple shell and tube heat exchanger.

A reference or set point of 160°F will be assumed. The objective is to maintain the controlled variable, namely, product outlet temperature, at the reference or set point temperature of 160°F. The first problem is to measure the controlled variable, product outlet temperature. Although there are numerous ways to accomplish temperature measurement, the majority of industrial applications will be accomplished using resistance thermometers, thermocouples, or filled tube systems. Examples of commercially available items are shown in Figures 1-2, 1-3, and 1-4.

In some installations the signal from the measuring element (or "measured variable") will be transmitted directly to an indicating or recording controller as shown in Figure 1-5. As a rule, however, this type of system is only used with locally mounted controllers with short sensing line lengths (approximately 25 ft or less, although longer distances can be obtained). In most modern applications where indicating or recording controllers are mounted in a unit control room, a transmitter is normally close coupled to the measuring element in the field as shown in Figure 1-6. The transmitter serves the dual function of a power amplifier and transducer. The term "power amplifier" in this sense is general and not limited to electrical systems. In pneumatic transmitters the power amplifier is commonly called the relay valve.

A transmitter is needed because measuring elements in general are low-energy devices, whose output signals are a function of the range zero and the range span chosen for the controlled variable. In the example of Figures 1-1, 1-5, and 1-6, a range span of 100°F could be selected with a range of zero 100°F. The range of measurement would then be from 100 to 200°F. The transmitter serves the function of converting or transducing the measuring element signal into a form that is compatible with the reference or set point in the controller. The transmitter also supplies the necessary power for transmitting the measured variable to the control room. In either case the value of the measured variable is assigned the notation b.

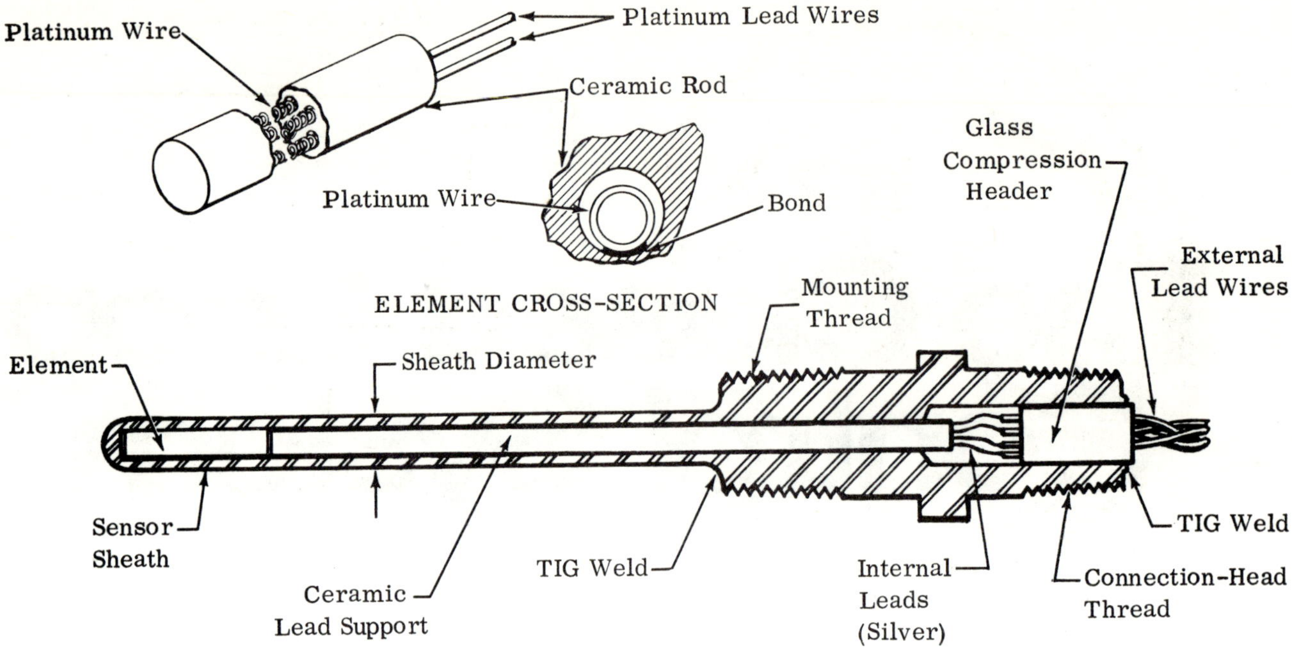

Figure 1-2. A typical industrial resistance thermometer bulb. Courtesy of Rosemount, Inc., Minneapolis, Minnesota.

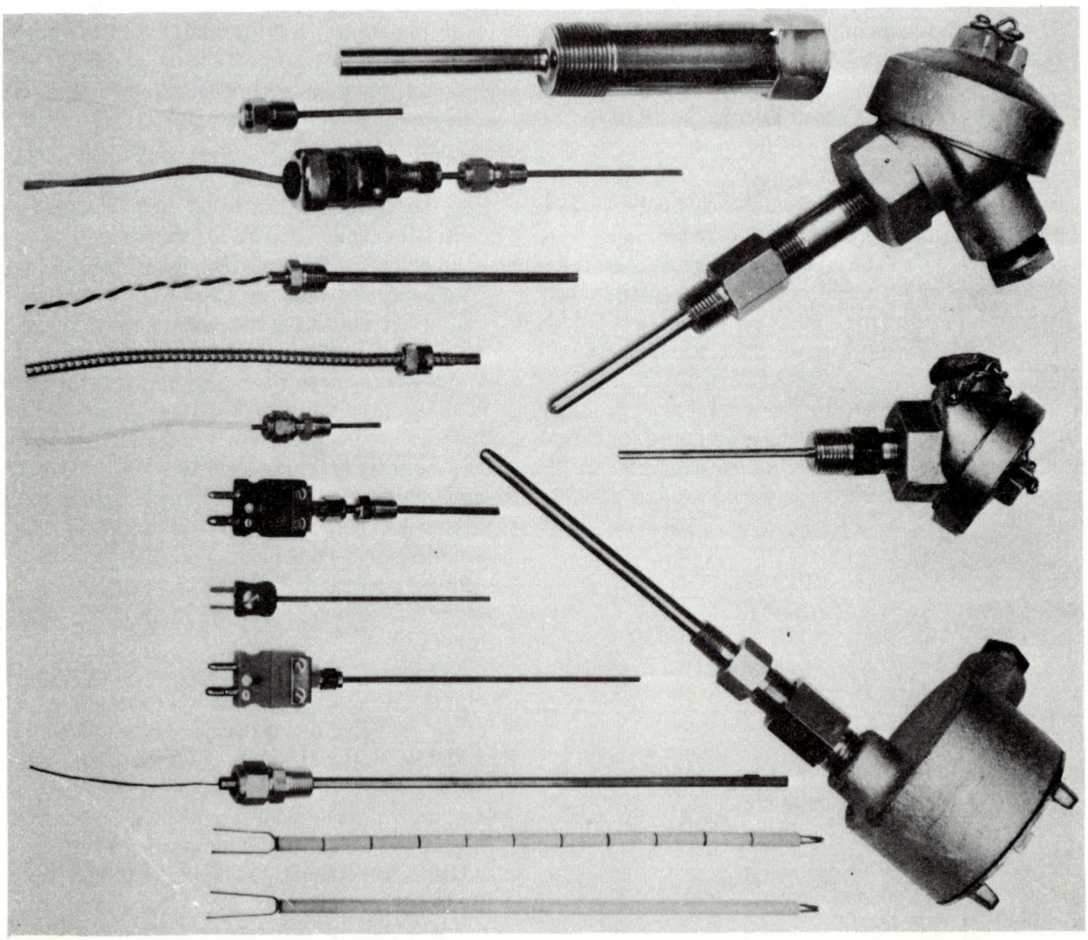

Figure 1-3. Industrial thermocouples and mountings. Courtesy of Hy-Cal Engineering, Santa Fe Springs, California.

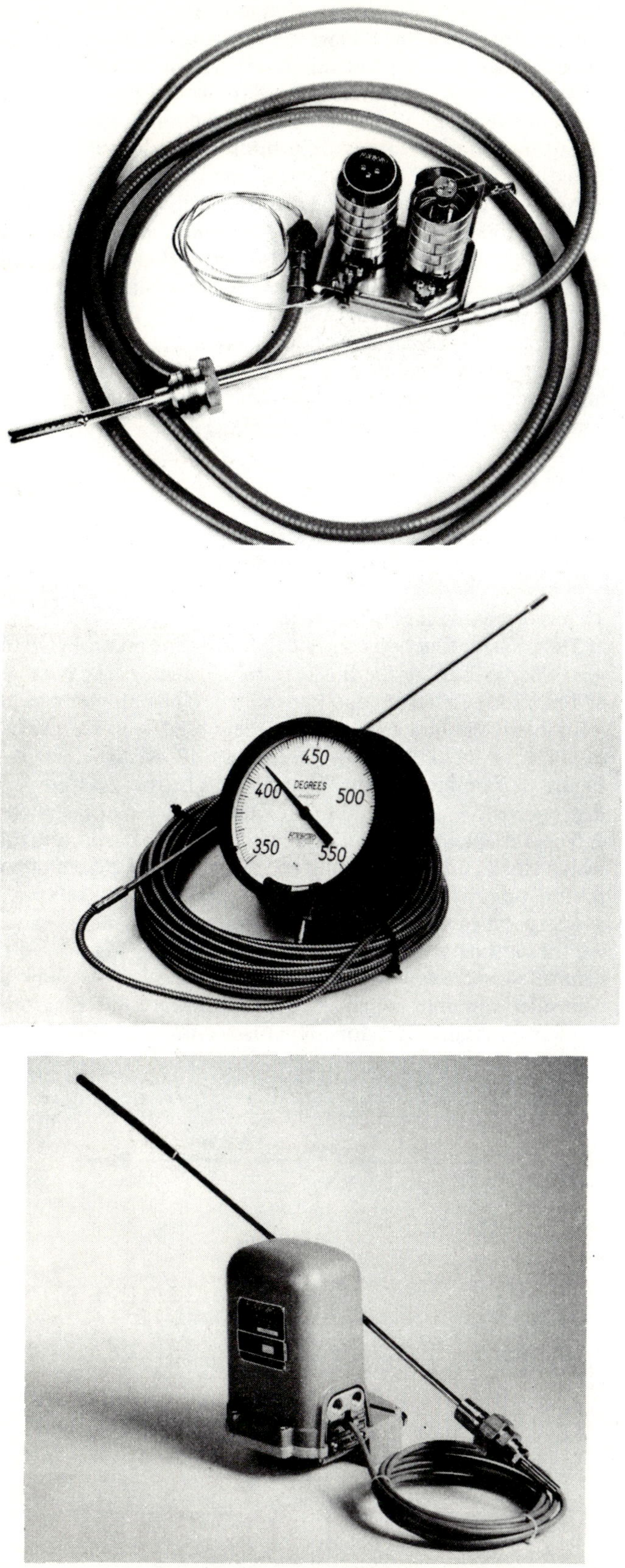

Figure 1-4. Industrial filled tube systems. Courtesy of The Foxboro Co., Foxboro, Massachusetts.

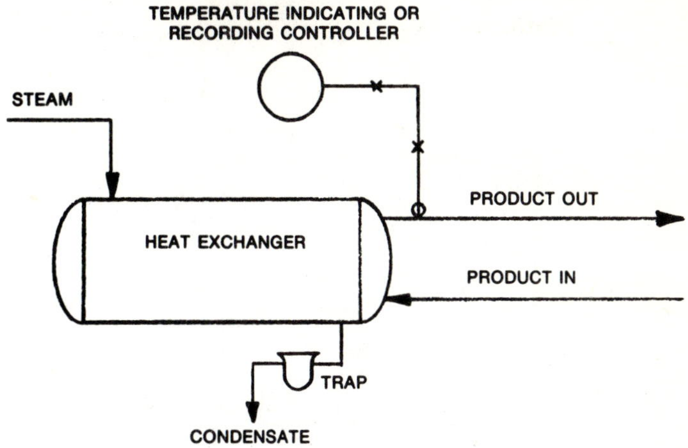

Figure 1-5. Controller locally mounted.

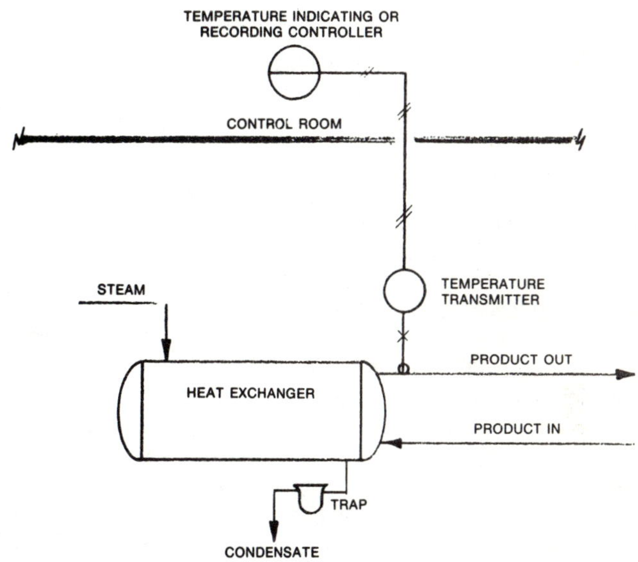

Figure 1-6. Controller mounted in control room.

In the controller the difference between the reference or set point and measured variable is determined. This difference is the error or actuating signal and is assigned the notation e. In equation form this is

$$e = r - b, \tag{1-1}$$

where e = error or actuating signal,
r = reference or set point,
b = measured variable.

The controller acts on the error signal to produce an output that will bring the controlled variable to the reference or set point value. Here, again, a problem develops. In acting on the error signal the controller assumes that the measured variable b is equal to the controlled variable c. This, of course, is unrealistic,

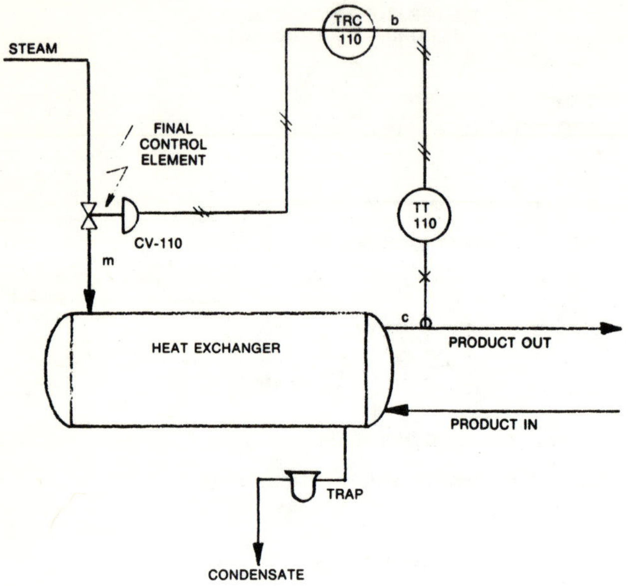

Figure 1-7. Closed loop control system.

since measuring elements do not instantaneously respond to changes and time is required to transmit signals. Thus, the time required or the dynamics of measuring and transmitting elements can become a problem. This will be considered in more detail in Chapter 2.

The particular action taken by the controller can take many forms and these will also be discussed later. For now, it is sufficient to note that the controller produces an output which is some function of the error signal. The output actuates the "final control element" which in turn regulates the "manipulated variable," m. In the example shown, a steam control valve is the final control element, and steam flow rate is the manipulated variable. With all elements combined the closed loop for the heat exchanger example is as shown in Figure 1-7. In the heat exchanger shown in Figure 1-7, a filled-tube measuring element has been used along with a pneumatic transmitter and controller. Schematic symbols used are standard for industry.[1,2,3] Notation of defined variables has been used where possible in Figure 1-7.

1.2. Block Diagram Concepts

The type of diagram shown in Figure 1-7 is similar to what the reader will find on industrial plant drawings. This type of drawing is sufficient for operating purposes, assuming that experienced operators are available. From the viewpoint of training and analysis, however, it does have certain limitations. For this reason it is common practice to use block diagrams, which show in detail the relations existing between components and signals in a closed loop control system.

One of the basic elements in block diagram symbolism is the summing point. In the summing point, two or more signals are added or subtracted according to the specifications of the defining equation. To explain this in more detail, assume the simple heat exchanger system referred to in Section 1.1 and, more specifically, Equation (1-1) which states that the error is equal to the reference or set point minus the measured variable. The summing point for Equation (1-1) is shown in Figure 1-8. More explicitly, the summing point is a pictorial representation of an equation which involves only addition and subtraction of signals.

As noted in Section 1.1, the controller acts on the error signal to produce an output that will bring the controlled variable to the reference or set point value. As also noted, this action can take many forms. In the general case the relation can be defined by the term "transfer function." A transfer function is "a mathematical, graphical, or tabular statement of the influence which a system or element has on a signal or action compared at input and output terminals."[2] Assuming a notation of g_1 for the controller of the heat exchanger example gives

$$p = g_1 e, \tag{1-2}$$

where p = controller output signal,
g_1 = controller transfer function.

This is represented graphically in Figure 1-9.

The transfer function concept used to define controller action may also be used for the final control element, heat exchanger, and measuring element and transmitter. When all elements are assembled, the block diagram of the closed loop for the heat exchanger system is as shown in Figure 1-10. All the elements in Figure 1-7 are included in Figure 1-10. The basic difference is that Figure 1-7 is pictorial to provide assistance in installation and operation, where Figure 1-10 is a mathematical presentation which can be used for training and analysis. The element transfer functions assigned the g notation are commonly referred to as "forward elements." By convention, these elements exist in the forward path between the error or actuating signal and the controlled variable. In Figure 1-10, it is easy to see how the controlled variable signal is fed back through the measuring element and transmitter for comparison with the reference or set

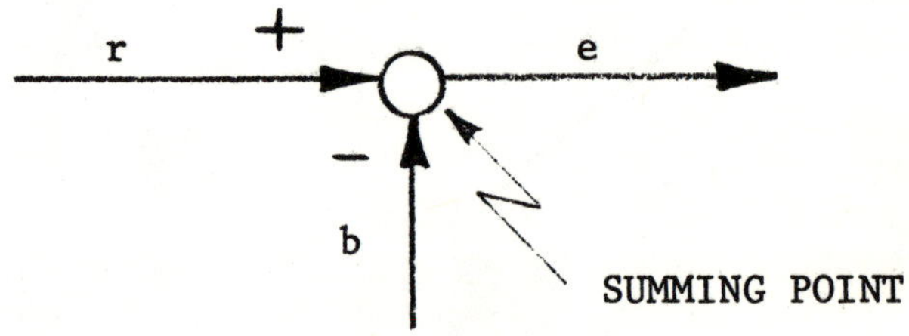

Figure 1-8. Block diagram summing point.

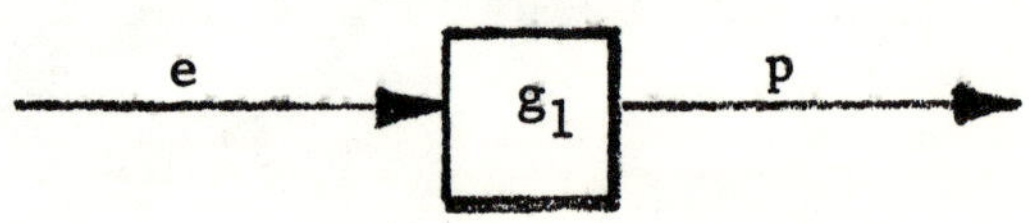

Figure 1-9. Block diagram transfer function.

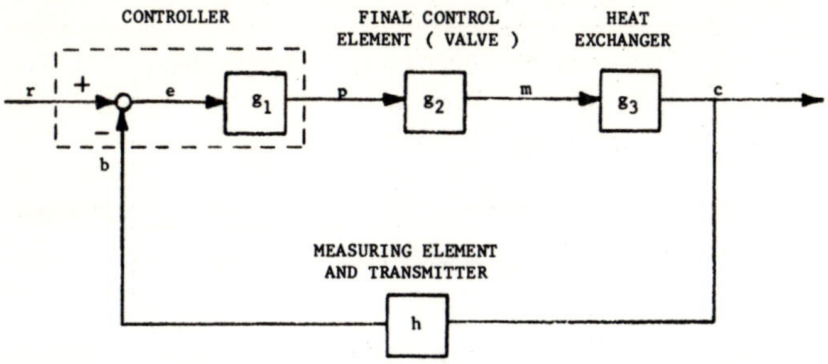

Figure 1-10. Block diagram closed loop.

point. This, of course, is the basis for the term "feedback control." The h notation is commonly used for feedback elements.

As noted in Section 1.1, time can be a problem in the dynamics of the measuring element and transmitter. This is true of all the element transfer functions in a closed loop. In the heat exchanger, for example, the time problem is readily apparent from an inspection of the transfer function equation

$$c = g_3 m, \tag{1-3}$$

where g_3 = heat exchanger transfer function.

Equation (1-3) states that the controlled variable, namely product outlet temperature, is equal to the product of the heat exchanger transfer function and manipulated variable, which is steam flow rate. Anyone who has waited for water to boil in a teapot will recognize that this exchange of energy takes time. Thus, in order to predict what will happen in a given control system, both magnitude and time effects of each transfer function must be known.

1.3. Fundamental Objectives

Unless the reader is very cautious in the solution of Problem 16 at the end of this chapter, the load flow term will probably be neglected in the first attempt at solving the problem. Tank pressure is a function of the capacitance (or tank volume) and the number of pounds of air in the tank. Thus, the tank transfer function relating pressure (controlled variable) to flow rate is not a simple function of inlet flow (manipulated variable); it is also a function of outlet flow or system load flow. The positive sign on m indicates an increase in c as m increases. The negative sign on l indicates a decrease in c with an increase in l. The logic is simple: increased inlet flow causes an increase in tank pressure, and increased outlet flow results in decreasing tank pressure.

For simplicity, the load or disturbance problem was intentionally by-passed in the heat exchanger example. In fact, several unstated assumptions were made:

—constant product flow rate,
—constant product inlet temperature,
—constant steam supply pressure and temperature, or quality, and
—constant steam pressure in the exchanger.

In actual practice such ideal conditions will not exist. If such were the case, the controller and control valve could be replaced by a hand valve.

Effects of supply and tank pressure were also neglected in the tank pressure control system. These could have been handled by analyzing pressure effects on the final control element or valve. For example, the manipulated variable or air flow is not only a function of valve opening but is also a function of supply or upstream pressure and tank or downstream pressure. In equation form this is

$$M = g_2 p + g_4 P_s - h_1 P_t, \tag{1-4}$$

where M = total change in inlet air flow to tank,
p = controller output signal,
g_2 = final control element transfer function based on input signal from controller,
P_s = upstream or supply pressure,
g_4 = final control element transfer function based on upstream or supply pressure variations,
P_t = downstream or tank pressure (this is also the controlled variable c),
h_1 = final control element transfer function based on downstream or tank pressure.

The resultant air flow to the tank, which produces tank pressure changes, is now equal to the difference between the total change in inlet flow M and load flow l. In equation form this becomes

$$Q = M - l, \tag{1-5}$$

where Q = resultant air flow to tank.

If the complete block diagram is now assembled, it appears as shown in Figure 1-11. The result is a combination of Problems 11 and 16 and Equation (1-5). Even now, the picture is not complete. For such a system, load resistance must also be considered. For example, In Problem 8, the load is a number of

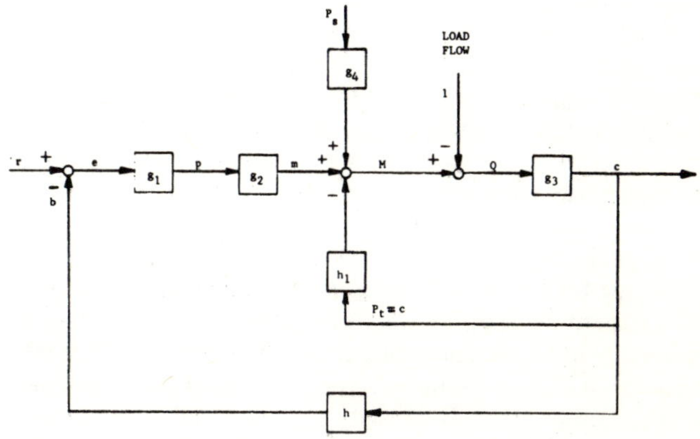

Figure 1-11. Tank pressure control system with load disturbances.

paint spray booths. Thus, the load is variable over a range of some minimum flow resistance when all booths are in operation to an infinite resistance when all booths are off. For any given load, however, load flow is a function of tank pressure and exhaust pressure, which will normally be at atmospheric pressure. If exhaust pressure is assumed constant, the load flow l will vary as a function of load resistance and tank pressure P_t. Since tank pressure is also the controlled variable c, it is obvious there is another feedback function.

The functional relation in this case is somewhat more complicated and will be delayed to Chapter 8. For now, note the influence of external disturbances such as supply pressure and load flow variations. Such external changes, along with reference or set point changes, are the disturbances that the controller must compensate for in any system to maintain the controlled variable at the desired value.

Determining how a system responds to disturbances is the fundamental problem of control system analysis and design. In fact, the remainder of this book is devoted to this problem. However, before starting on the problem, it is necessary to reexamine objectives and recognize basic limitations. In both Sections 1.1 and 1.2, the problem of time or dynamics was noted and can now be explained in more detail. The time required for a given element to produce an output signal proportional to a change in input signal depends on the physical characteristics of the element. In practice, time can vary from a few nanoseconds in electronic equipment to a few hours in heat exchange equipment. These time effects (as previously noted) are included in the individual element transfer functions. A familiar example is the nurse waiting for a thermometer to respond to body temperature. Similar delays occur in all physical equipment. If this fact is accepted, it becomes readily apparent that no controller can accomplish perfect regulation of the controlled variable. The question, then, is what can be expected from a given system?

To answer this question, refer to the simple heat exchanger in Figures 1-7 and 1-10, where external disturbances of steam supply pressure, product flow rate, and inlet product temperature have been assumed constant. Consider a reference or set point change of 10°F (from 160 to 170°F). Possible temperature-time or transient response records are shown in Figures 1-12 a-d. Figure 1-12a shows the theoretical step change in the reference or set point r, while Figures 1-12b, c, and d show possible responses of the controlled variable c. While the responses in Figures 1-12b and c may not be ideal or optimum, they do indicate that the controller does achieve its objective of regulating the controlled variable at the reference or set point. The response of Figure 1-12d, however, is totally unacceptable and indicates an unstable system. The potential for an unstable response exists in any closed loop system and will be considered in more detail in Chapter 5. The important objective at this time for any closed loop system is the requirement for a stable system such as shown in Figures 1-12b or c. Similar responses will be obtained for external disturbances. The only difference will be that the reference or set point r will not change.

As a general rule the transient response shown in Figure 1-12c is a more desirable result than that of Figure 1-12b. The basis of this decision will be discussed in Chapter 5. For the present it is

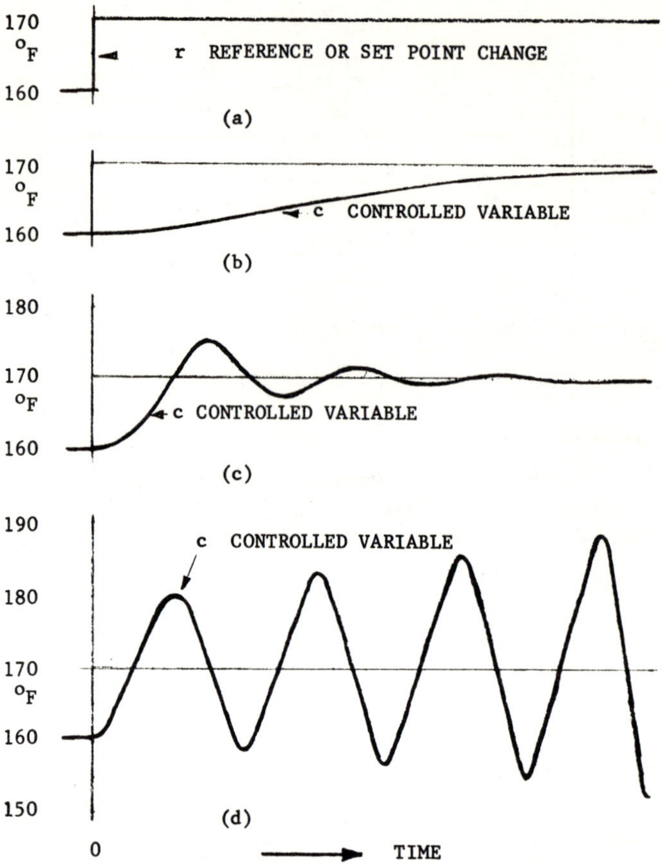

Figure 1-12. Possible transient control responses.

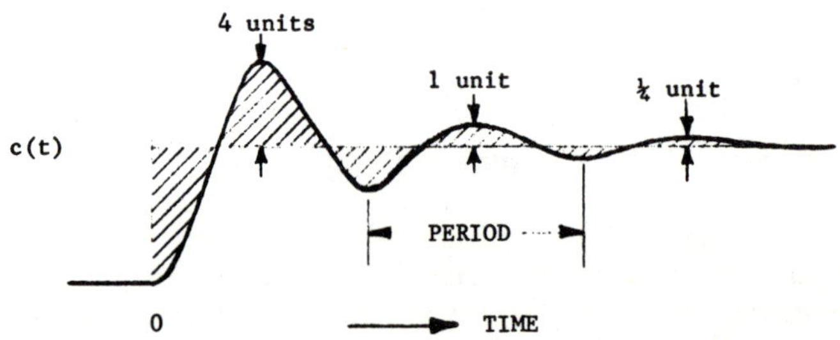

Figure 1-13. Time-error product area and 1:4 decay ratio.

sufficient to assume that this type of response will produce minimum waste product. Waste product in the mathematical sense is shown by the shaded area in Figure 1-13 and is defined as the time-error product. In an actual system there are certain tolerance limits. However, until an actual system is selected and limits are defined, any comparison is meaningless. Thus, for training purposes the total time-error product area will be used. A practical approach to minimum area is achieved with a quarter decay ratio. A quarter decay ratio is defined as one in which each successive peak in the decaying sine wave has an amplitude equal to 1/4 the prior peak. This concept is also shown in Figure 1-13.

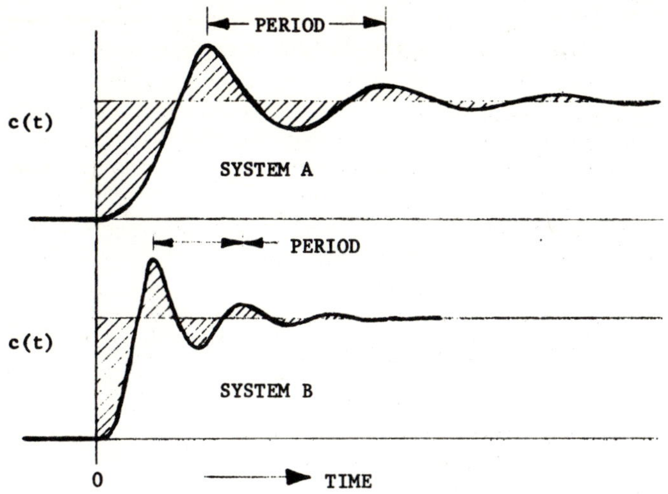

Figure 1-14. Time-error product area as a function of period.

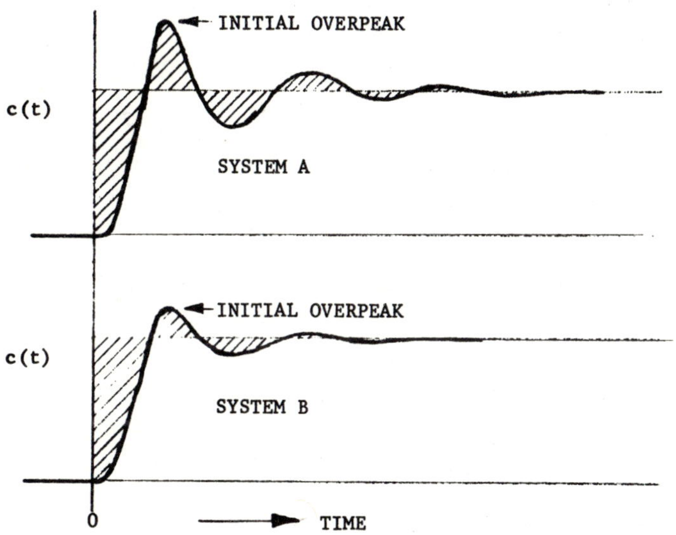

Figure 1-15. Time-error product area as a function of initial overpeak.

Once a stable system with approximately quarter decay is obtained, the analysis problem changes to consideration of the period of the decaying response. Period is the time required for one cycle and is also shown in Figure 1-13. Transient responses for two systems both having a quarter decay ratio are shown in Figure 1-14. It is obvious that system B with the shorter period has a smaller time-error product area, which means less waste product. Thus, a second requirement is the desirability of minimum period.

The third requirement concerns the initial overpeak in the transient response. Assume two systems with quarter decay stability and equal periods as shown in Figure 1-15. Here again, it is obvious that system B with less initial overpeak has a smaller time-error product, or less waste product. Thus, the third requirement or control objective is minimum initial overpeak.

Stability, period, and initial overpeak are dependent on the magnitude and time (or dynamic) characteristics of each element of the system and their relationship to reference or set point changes and external disturbances. Thus, the next step is to develop the transfer function or dynamic characteristics of typical control and process elements.

Chapter 1—Study Questions and Problems

(Answers appear on pages 151-152)

1. The prime objective of any control system is to maintain the __________ variable at the reference or set point.

2. The signal received at the controller from the measuring element (or measuring element and transmitter combination) is termed the __________ variable.

3. Is the measured variable added to or subtracted from the reference or set point to produce the error signal?

4. The controller acts on the __________ signal and produces an output signal which actuates the __________ element.

5. In attempting to maintain the controlled variable at the reference or set point, the controller is placed at a disadvantage. This problem results from the fact that the controller only sees the values of the __________ variable and reference or set point.

6. The final control element regulates the __________ variable.

7. Give alphabetical notations for controlled variable, measured variable, reference or set point, error signal, and manipulated variable, respectively.

8. Make a schematic sketch similar to Figure 1-7 for a tank pressure control system. Assume that the fluid is air and is used in paint spraying. Control tank pressure by regulating inlet air from an available supply. Label all elements and use alphabetical notation where possible to define variables.

9. Sketch the summing point block diagram for $m = x + y - z$.

10. Sketch the transfer function block diagram for $u = g_x a$.

11. Sketch a combined block diagram for $m = g_2 p$ and $M = m + g_4 P_s - h_1 P_t$.

12. A __________ is actually a graphical representation of the equations representing a control system.

13. A transfer function defines the influence which a system or element has on a signal compared at the __________ and __________ terminals.

14. Forward elements in a closed loop control system are by convention located in the forward path between the __________ or __________ signal and the controlled variable.

15. To predict how a closed loop control system will operate, it is necessary to know what two effects of each transfer function?

16. Sketch a closed loop block diagram for the tank pressure control system of Problem 8.

17. In the heat exchanger example of Figure 1-10, external disturbances of product flow rate and inlet temperature were neglected. Using the following equation, redraw the block diagram of Figure 1-10 showing influence of external disturbances.

$$c = g_3 m - g_4 Q + g_5 \Theta_i,$$

where Q = product flow rate,
Θ_i = product inlet temperature,
g_4 = heat exchanger transfer function based on product flow rate,
g_5 = heat exchanger transfer function based on product inlet temperature.

18. Explain the physical significance of the minus and plus signs on the terms $g_4 Q$ and $g_5 \Theta_i$ in Problem 17.

19. The fundamental problem in analyzing a given control system is determining how the controlled variable will respond to changes in the reference or set point and __________.

20. The time required for any control or process element to produce an output signal proportional to a given input signal depends on the __________ of the given element.

21. What is the first requirement of any closed loop system?

22. A practical approach to a minimum time-error product (minimum waste) is achieved by a system having a decay ratio of approximately __________.

23. To minimize the waste or time-error product, a system should be stable and should also produce a controlled variable response having a minimum period and __________.

24. Stability, period, and initial overpeak of any system are dependent on the physical characteristics of each element of the system and their relationship to reference or set point changes and __________.

References

1. *Instrumentation Symbols and Identification,* I.S.A.—S5.1, Instrument Society of America, March 1968.
2. *Terminology for Automatic Controls,* A.S.A.—C85, The American Society of Mechanical Engineers, April 1961.
3. Glossary of Analog and Digital Process Control Terms, Pub. 124-10M, The Foxboro Co., Foxboro, Massachusetts, August 1967.

2 Introduction to System Dynamics

To predict stability, period, and initial overpeak of control loops as noted in Chapter 1, the dynamics of the system must be known. This chapter is devoted to defining two basic building blocks—the exponential lag and dead time lag.

2.1. Similarity of Systems

It is no secret that the use of dynamics involves energy changes, and in the case of control analysis, energy changes as a function of time. This implies the use of differential or time-dependent equations, a procedure which induces withdrawal pains in most practicing engineers. In an effort to minimize pain and negative reaction to a necessary evil, similarity of systems will be established. The objective, of course, will be a common solution for the general equation.

In the heating or cooling of solids, liquids, or gases, it is a well-known physical fact that time is required to transfer energy from the source to the material being heated. Since many industrial processes include heating or cooling operations, this presents a logical starting point. As a more specific example, assume one of the temperature measuring elements of Chapter 1. All these elements produce an output signal (resistance, voltage, volume, or pressure change) proportional to changes in thermal energy storage. Thus, to determine sensing element dynamics, it is necessary to determine the equations for heat transfer and storage. These equations are readily available in any text on heat transfer and need only be properly applied. As a start, assume a brass rod (1/4 in. diam.) inserted through a suitable seal into a pipe in which water is flowing. A detailed sketch is shown in Figure 2-1.

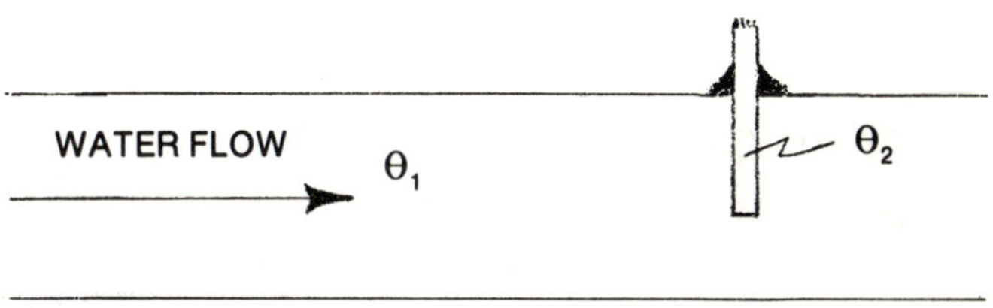

Figure 2-1. Pseudo temperature-measuring element.

As shown in Figure 2-1, water temperature is Θ_1 and rod temperature is Θ_2. If loses to the atmosphere through the rod are neglected and equilibrium conditions are assumed, then $\Theta_1 = \Theta_2$. The problem is to determine how Θ_2 will change in response to a change in Θ_1, i.e., to determine the transfer function for this element. When water temperature Θ_1 changes, energy transfer to the rod may be defined by the following equation.[1]

$$q = hA\,(\Theta_1 - \Theta_2), \tag{2-1}$$

where q = heat transferred, B/hr,
h = coefficient of heat transfer, B/(hr-ft^2 − °F),
A = area, ft^2,
Θ_1 = water temperature, °F,
Θ_2 = rod temperature, °F.

From this equation it is easy to see that rate of energy transfer q is a function of driving force or temperature difference, available area, and the coefficient of heat transfer. The temperature difference is, of course, what has to be determined. Area and the heat transfer coefficient are physical characteristics of the system and will be discussed in more detail as the problem develops.

The energy transferred from the water to the rod must be stored in the rod. This storage rate of energy is what effects a change in rod temperature and may be defined by the following equation:

$$q = M_r C_p \, (d\Theta_2/dt), \tag{2-2}$$

where M_r = mass of rod, lb,
C_p = specific heat, B/(lb − °F),
$d\Theta_2/dt$ = rate of change of Θ_2 as a function of time, °F/hr.

In this case, rate of energy transfer q is a function of the rod mass, specific heat, and the rate of change of rod temperature with time. Here, again, mass and specific heat are physical characteristics of the system, and temperature is a force or potential measurement.

Since the energy leaving the water must be equal to the energy stored in the rod, Equations (2-1) and (2-2) may be equated. This has the effect of eliminating the heat transfer (or flow term q) and results in an equation relating temperature to physical characteristics of the system. The result is:

$$hA \, (\Theta_1 - \Theta_2) = q = M_r C_p \, (d\Theta_2/dt)$$

$$d\Theta_2/dt = (\Theta_1 - \Theta_2)/\left[(M_r C_p)/(hA)\right] \tag{2-3}$$

The result is a differential equation that explains the rate at which Θ_2 changes with time, based on the difference between Θ_1 and Θ_2 and physical properties of the system. The particular format used for Equation (2-3) will become more reasonable as other systems are studied. A second example will now be examined in an effort to demonstrate similarity between systems.

For the second example assume air flowing through a valve into a tank as shown in Figure 2-2. Numerous equations may be obtained for defining flow through the valve. These are special equations which are selected based on type of flow, Reynolds number, and experimentally determined correction factors.[2,3] The basic equation, however, shows that valve flow is proportional to the square root of the pressure drop:

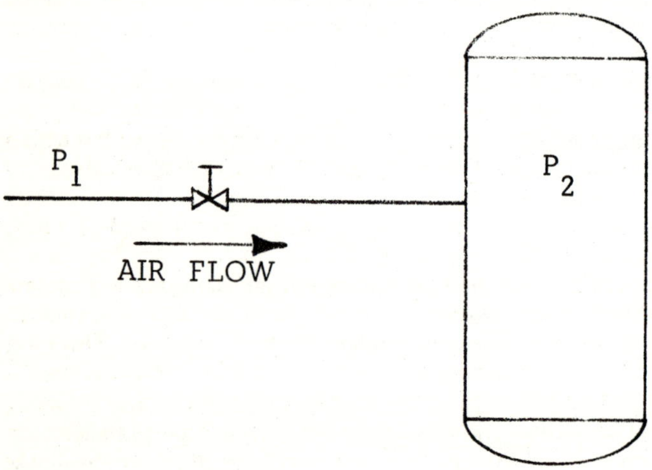

Figure 2-2. Air flow process.

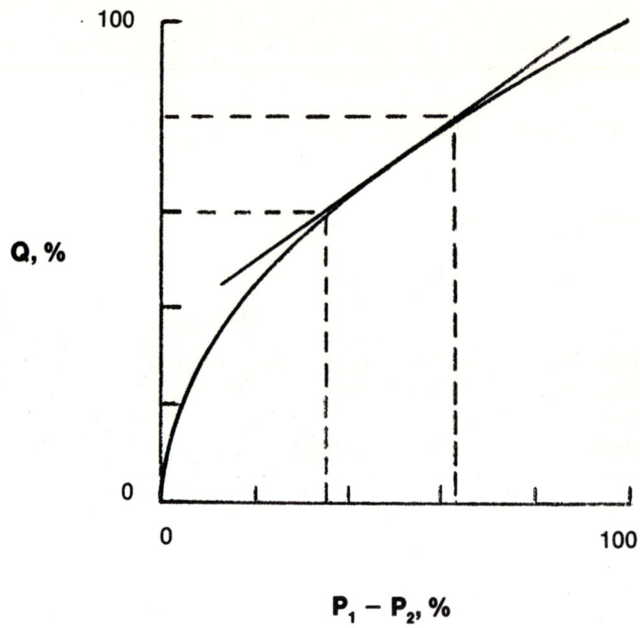

Figure 2-3. Linearization of square-root function.

$$Q = K_1 (P_1 - P_2)^{1/2} \tag{2-4}$$

where Q = flow rate, cfm,
P_1 = upstream pressure, psia,
P_2 = downstream pressure, psia,
K_1 = constant, cfm/(psi)$^{1/2}$.

A plot of Equation (2-4) is shown in Figure 2-3. Although this is a square root function, it is evident that any small portion of the curve may be closely approximated by a straight line. This practice is very common and is referred to as linearization. Since most systems are nonlinear to some extent, linearization is often used without consideration. Application to such an obviously nonlinear function as the square root, however, causes some concern for most readers. In this case an understanding of the problem will usually quiet the nervous system. If the problem is start-up from zero to maximum flow, the approximation is obviously invalid. On the other hand, if the process is running and changes around an operating point are being considered, valid results can be obtained using the linearization technique. Based on this latter case, Equation (2-4) may be restated:

$$Q = K_2 (P_1 - P_2), \tag{2-5}$$

where K_2 = constant (slope at operating point), cfm/psi.

Flow rate Q, as in the thermal example, is again equal to the product of the driving force or pressure difference and the physical characteristics of the system included in K_2.

The quantity or mass flow through the valve must be stored in the tank. Here again, it is this storage that will produce a change in tank pressure. This result in equation form is

$$Q = (V/P^*)(dP_2/dt), \tag{2-6}$$

where V = tank volume, cu ft,
P^* = atmospheric pressure, psia,
dP_2/dt = rate of change of P_2 as a function of time, psia/min.

Thus, the flow rate Q is a function of tank volume, atmospheric pressure, and the rate of change of tank pressure with time. As in the thermal example, tank volume and atmospheric pressure are physical characteristics of the system, and pressure is a force or potential measurement. The ratio of V/P^* represents the cubic feet of free air required to raise tank pressure 1 psi. This is based on the assumption of isothermal flow and the application of Boyle's law, which is a reasonable procedure for analyzing changes around a given operating point.[4]

Since the mass flow passing through the valve must equal the quantity entering the tank, Equations (2-5) and (2-6) may be equated. Again, the effect is to eliminate the flow rate term, giving an equation relating pressure to physical characteristics of the system. The result is

$$K_2(P_1 - P_2) = Q = (V/P^*)(dP_2/dt)$$

$$dP_2/dt = (P_1 - P_2)/\left[(V/P^*)/K_2\right] \tag{2-7}$$

The result is again a differential equation showing the rate at which P_2 changes with time as a function of the difference between P_1 and P_2 and physical characteristics of the system. Equation (2-7) results from a volume or mass flow balance, whereas Equation (2-3) resulted from an energy balance. Similarity of the resultant equations is obvious and is the reason for selecting the format of Equation (2-3). However, before attempting a solution, one more example will be examined.

The third and final example is an electrical system shown in Figure 2-4. This is the classical RC circuit, which by definition must be included in every text on control. Assuming that resistance does not vary with temperature, Ohm's law may be stated as follows:

$$i = (1/R)(E_1 - E_2), \tag{2-8}$$

where i = current, amps,
R = resistance, ohms,
E_1 = input voltage, volts,
E_2 = output voltage, volts.

Although the format of Equation (2-8) is similar to Equations (2-1) and (2-5), unit definition produces an aura of mystery. This can be easily overcome by using fundamental definitions for current and resistance. In the current case an ampere is equivalent to a coulomb per second. Current in these units is readily recognized as a flow, which is similar to the previous examples.

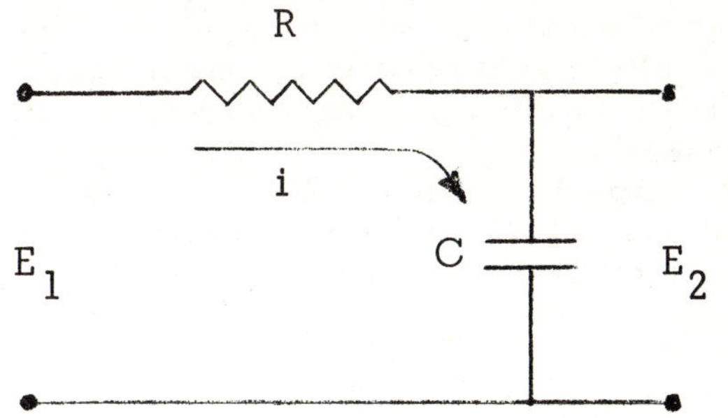

Figure 2-4. Electrical RC circuit.

Using these units, resistance in ohms can be replaced by volts per coulomb per second. As in the previous examples, flow rate is a function of the driving force (or potential difference $E_1 - E_2$) and the resistance which is dependent on the physical characteristics of the system.

As in previous examples, flow through the resistance must be stored in the capacitance. This storage in turn produces the voltage change across the capacitor. The resulting equation is

$$i = C(dE_2/dt), \tag{2-9}$$

where C = capacitance, farads,
dE_2/dt = rate of change of E_2 as a function of time, volts/sec.

In this case the farad unit on capacitance provides the confusion. If the farad is replaced by its equivalent of coulomb per volt, the storage concept is more easily explained. Stated in these units, capacitance is the units of charge (coulombs) required to increase the potential across the capacitor by 1 volt. As before, the flow rate is a function of physical characteristics of the system and the rate of change of the driving force or potential E_2 as a function of time.

If the current or flow rate through the resistance is now equated to storage in the capacitor, the common differential equation is obtained:

$$(1/R)(E_1 - E_2) = i = C(dE_2/dt)$$

$$dE_2/dt = (E_1 - E_2)/(RC) \tag{2-10}$$

As in previous examples, the rate of change of E_2 with time is a function of the difference between E_1 and E_2 and physical characteristics of the system.

By this time it should be evident that Equations (2-3), (2-7), and (2-10) are identical in format and should, therefore, have a common solution. It should also be evident that any simple system having a resistance to flow and a single storage element will also have this same form of differential equation. This form is commonly referred to as a "first-order linear equation." "First-order" refers to the exponent on the rate term $d^1(\quad)/dt^1$, which is one. "Linear" refers to the assumption of linearity made in all the equations. Although the need for the linear assumption is not always as obvious as in the case of the square

Table 2-1
Units for Resistive and Capacitive Terms of Several Physical Systems

	Thermal System	Pneumatic System	Electrical System
Resistance	$\frac{1}{hA}\ \frac{\text{hr} - °\text{F}}{\text{B}}$	$\frac{1}{K_2}\ \frac{\text{min-psi}}{\text{cu ft}}$	$R\ \frac{\text{volts-sec}}{\text{coulomb}}$ (ohms)
Capacitance	$M_rC_p\ \frac{\text{B}}{°\text{F}}$	$\frac{V}{P^*}\ \frac{\text{cu ft}}{\text{psi}}$	$C\ \frac{\text{coulomb}}{\text{volt}}$ (farad)
Time Constant	$T = \frac{M_rC_p}{hA}$ hr	$T = \frac{V}{P^*K_2}$ min	$T = RC$ sec

root term, it is normally assumed. For example in the thermal and electrical systems, electrical resistance, the heat transfer coefficient, and specific heat terms are assumed as constants. All these coefficients, however, vary with temperature and are nonlinear. Although the nonlinearities are normally not as evident as in the square root case, they do exist, and the linear term is included to jog the user's memory. At the present, however, the problem is to obtain a solution for this general equation.

2.2. First-Order or Exponential Lag Solution

Before proceeding with the solution of the first-order linear differential equation, it is necessary to examine the denominators of Equations (2-3), (2-7), and (2-10). As noted in Section 2.1, the format of the resulting equation for each example was selected with a specific purpose in mind. The denominators of the right-hand side of each equation include only system physical properties, which in the range of analysis are considered as constants. These constants are in fact the resistive and capacitive terms of each system. A breakdown of each example is shown in Table 2-1.

If the resistive and capacitive units are combined, the result in all cases is time as shown in Table 2-1. The reader should recognize that this time quantity is a constant (based on linear assumptions) and a function of the physical characteristics of the system. It is now possible to reduce each equation to a more general form:

$$d\Theta_2/dt = (\Theta_1 - \Theta_2)/T \qquad (2\text{-}3a)$$

$$dP_2/dt = (P_1 - P_2)/T \qquad (2\text{-}7a)$$

$$dE_2/dt = (E_1 - E_2)/T, \qquad (2\text{-}10a)$$

where T = system "time constant," units of time.

The equations may now be stated in the following manner—The rate of change of the output variable as a function of time is equal to the driving force difference at a given time divided by the time constant of the system. A general solution can now be determined.

Since the equations are similar, only one example is necessary. The first example of a thermal system will be used. A solution for the first-order linear differential equation may be found in almost any book on differential equations or control systems.[5] A graphical intuitive solution may be obtained by recognizing that the general equation is

$$\text{slope} = \text{rise/run} \qquad (2\text{-}11)$$

Assuming a step input change in Θ_1 at time zero, Equations (2-3a) and (2-11) may be depicted graphically as shown in Figure 2-5. After some time, however, it is evident that the driving force or rise $\Theta_1 - \Theta_2$ will decrease due to the change in Θ_2. The run or time constant T, however, remains fixed since this is fixed by the physical properties of the system. Thus, if some small value of time (namely Δt) is selected, the process of Figure 2-5 may be repeated. The result after several steps is as shown in Figure 2-6. It is easy to see from Figure 2-6 that the solution for Θ_2 must be in the form of an exponential. In addition, Θ_2 must equal the original value of Θ_1 at time zero and equal the value of

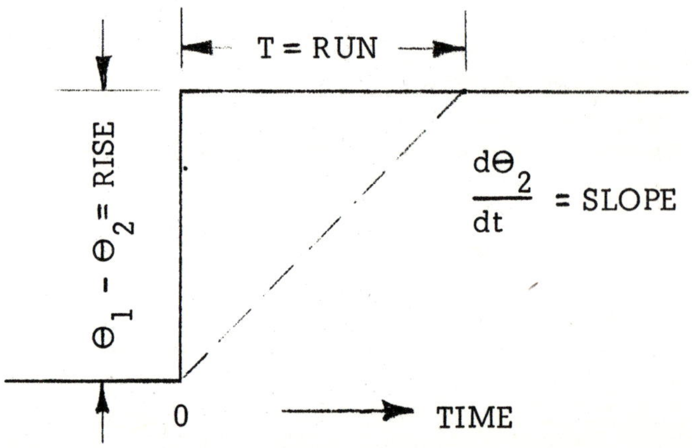

Figure 2-5. Graphical representation of first-order equation.

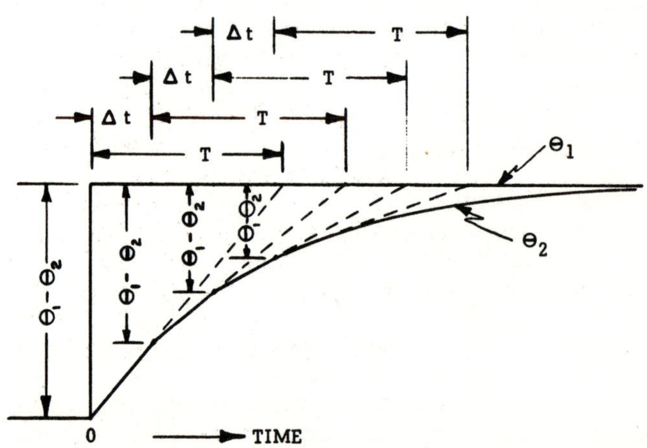

Figure 2-6. Graphical solution of first-order equation.

Θ_1 after the step change at infinite time. This results in an equation as follows:

$$\Theta_2(t) = \Theta_1(0^-) + \Theta_1(0^+)\left[1 - e^{-t/T}\right], \tag{2-12}$$

where $\Theta_2(t)$ = value of Θ_2 at any time,
$\Theta_1(0^-)$ = initial value of Θ_1 before step change,
$\Theta_1(0^+)$ = final value of Θ_1 after step change,
e = exponential factor or base of natural logarithms, 2.718...,
t = time,
T = time constant.

Since in most control system studies only changes around an operating point are of interest, the initial value of the input variable is often neglected. In this case, only the magnitude of the step change is used, and the equation is

$$\Theta_2(t) = A\left[1 - e^{-t/T}\right], \tag{2-13}$$

where $A = \Theta_1(0^+) - \Theta_1(0^-)$ = step change magnitude.

Note that this form of the solution is valid only for a step change in input. For other input disturbances different forms of the exponential solution will result. Some of these will be discussed in Chapters 3 and 11.

There are several interesting facts associated with Equation (2-13) that are worth noting; the output variable, which in this particular case is Θ_2, can be calculated at any time if the magnitude of the input step and the value of the time constant are known. Values for the input step magnitude will, of course, be readily known from analytical or experimental design. Calculating time constants, however, may not be as simple as implied thus far in the text. This will be especially true as systems become more complex. For this reason experimental testing is often used to determine system dynamics. While experimentation on a simple system is basically of academic interest, it is a necessary step in development of techniques for handling more complex systems.

For variety assume the experiment shown in Figure 2-7. Assume that the hand-operated reducing valves in the air supply

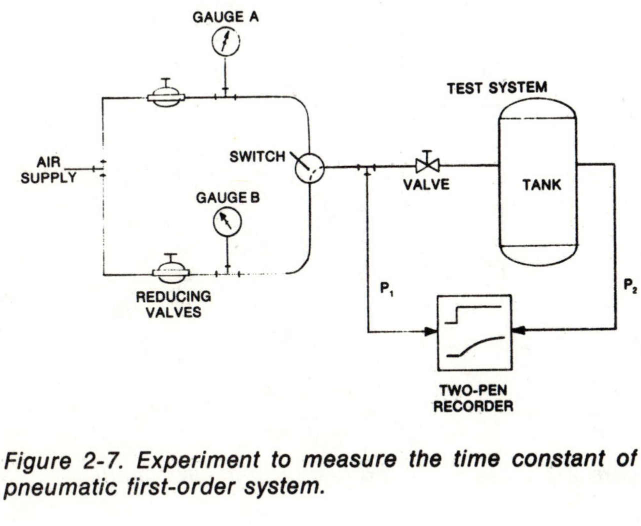

Figure 2-7. Experiment to measure the time constant of a pneumatic first-order system.

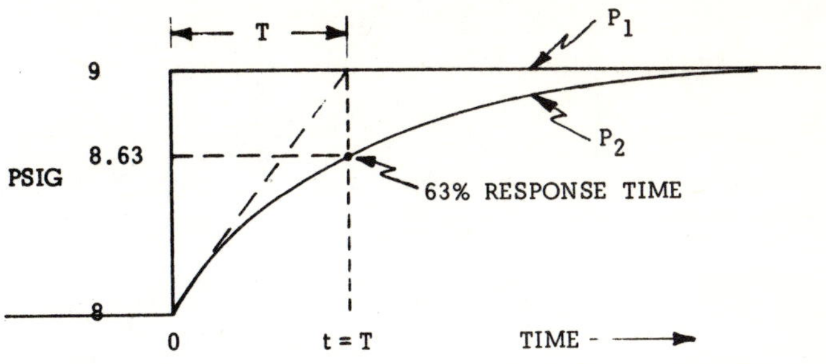

Figure 2-8. Experimental step response for a pneumatic first-order system.

lines are set such that test gauges A and B read 9 and 8 psig, respectively. As shown, the system is in equilibrium with the 8-psig supply shown on test gauge B. If the two-way valve is changed to the 9-psig supply a 1-psi step change will be imposed on the system under test. Using a two-pen recorder to obtain records of P_1 and P_2 will produce the result shown in Figure 2-8. Experimental evaluation of the time constant may be obtained using the slope concept of Equation (2-11) (also see Equation (2-7a)). In this method the time constant is equal to the time at which the initial slope intersects the final value of P_1. While this is a very simple method, problems often result in determining the initial slope. This is particularly true if noise exists in the system. A second method which may be used in addition to the slope technique is the classical 63% response time.

If Equation (2-13) is rewritten as the solution of the pneumatic example, the result is

$$P_2(t) = A(1 - e^{-t/T}) \tag{2-14}$$

The easiest point of evaluation for this equation is when $t = T$. Under this condition, Equation (2-14) reduces as follows

$$P_2(T) = A(1 - e^{-1}) = A(1 - [1/e])$$

using $e = 2.718$ gives

$$P_2(T) = A(0.632) \simeq 0.63A \tag{2-15}$$

Equation (2-15) states that the time at which P_2 is equal to $0.63A$, or 63% of the input step, is the time constant of the system. In the present example, $A = 1$ psi; thus, the time at which P_2 is equal to 0.63 psi (or 8.63 psig) should be the time constant. This result is also shown in Figure 2-8. This is often referred to as the 63% response time. The reader should note that the concept has little mathematical value or meaning except for the first-order or exponential lag.

Another tool that is convenient in the analysis of more complex systems is the incomplete response technique.[4] Here, again, the application to a first-order or exponential lag system is trivial but necessary for the educational process. If both sides of Equation (2-14) are divided by A, the equation is said to be normalized such that the output signal will vary from 0 to 1. If each side is then subtracted from 1 as shown below, a value is obtained which is defined as the incomplete response or IR.

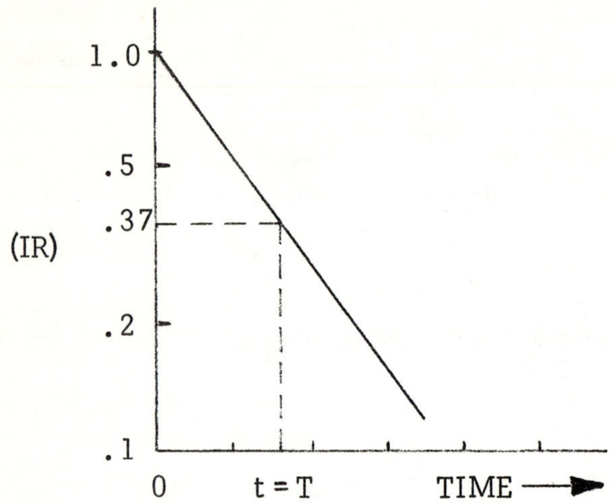

Figure 2-9. Incomplete response plot of a first-order system.

$$IR = 1 - \left[P_2(t)/A\right] = 1 - (1 - e^{-t/T})$$

$$IR = e^{-t/T} \qquad (2\text{-}16)$$

Taking the logarithm of Equation (2-16) gives

$$\log_{10}(IR) = -(0.4343/T)\,t \qquad (2\text{-}17)$$

where $\log_{10} e = 0.4343$.

Equation (2-17) demonstrates that $\log_{10}(IR)$ is a linear function of t. Thus, plotting the incomplete response versus time on semi-log paper will result in a straight line as shown in Figure 2-9. If in the complete response form the time constant can be found at the 63% response time, it is logical to assume that the time constant can be obtained at 37% in the incomplete response. As shown in Figure 2-9, the time constant is evaluated when the output signal is 0.37 on the incomplete response plot. Meaningful application of this technique to more complex systems will be used in Chapter 11.

2.3. Dead Time Lag

The second basic building block that is present in most process control systems is dead time lag. Dead time lag is also referred to as transportation or distance velocity lag. Pure dead time lag may exist in a rolling, textile, or paper mill application. In these cases the dead time is equal to the distance between the action and sensing points divided by the velocity of the sheet. A similar situation can exist in pipe flow mixing or blending. The dead time lag is again equal to the distance between the mixing and sensing points divided by the product flow rate. It is easy to see how the terms "transportation" and "distance velocity lag" came into use.

Most systems exhibit dead time in either a real or pseudo condition. The real or pure dead time as noted above is normally obvious and easy to calculate. The pseudo or aproximate dead time is a different problem. In some cases the problem can be easily resolved with the aid of practical assumptions. A typical example is shown in Figure 2-10, where steam is injected to heat water flowing in a pipe. The shaded area at 1 represents a slug of water at the outlet of the steam nozzle, while 2 represents a point downstream where a sensing element will be located. If thermal sensing elements having negligible time constants are now placed at positions 1 and 2, and temperature of the water slug is instantly changed from 160 to 170°F, the results will be as shown in Figure 2-11. The response shown in Figure 2-11 assumes that the slug of water flows as a plug from position 1 to 2 with little or no mixing (plug flow assumption) and that there is no heat transfer to the atmosphere. Neither of these assumptions is completely true. They do, however, give an adequate approximation of what happens in actual practice. The dead time lag L may again be calculated by dividing the distance d by the velocity V.

In more complex equipment such as heat exchangers, distillation columns, reactors, and furnaces, the pure and even approximate dead time lag may not be evident. As previously noted, however, some dead time lag will usually be present. The dead time may result from flow paths within the equipment producing a result similar to the pipe flow heating example of Figure 2-10. It may also result from a combination of small time constants within the system which produce a pseudo dead time lag. This effect can be seen by combining several exponential lags as

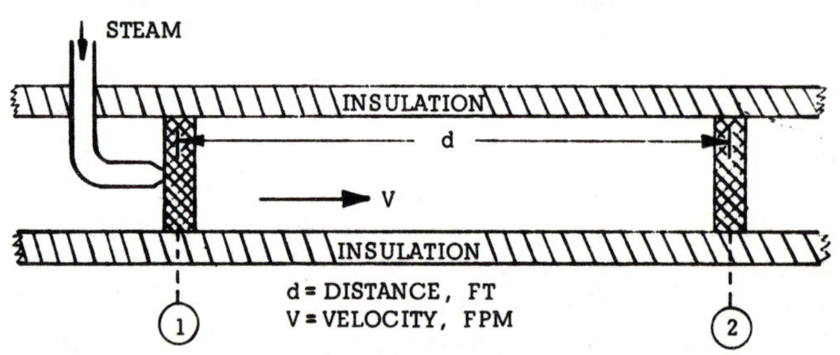

Figure 2-10. Pipe flow heating system.

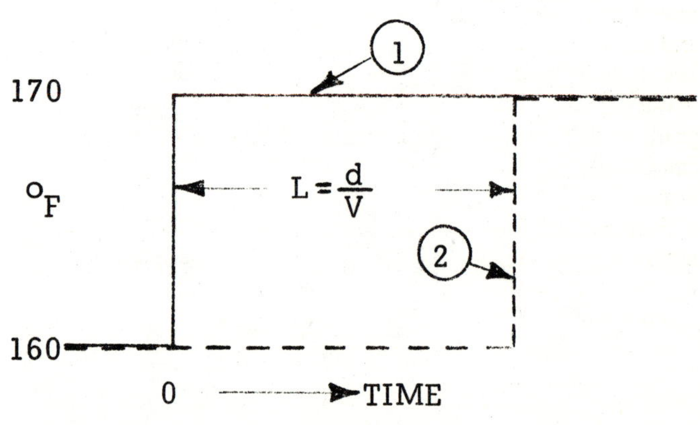

Figure 2-11. Dead time lag response to step input.

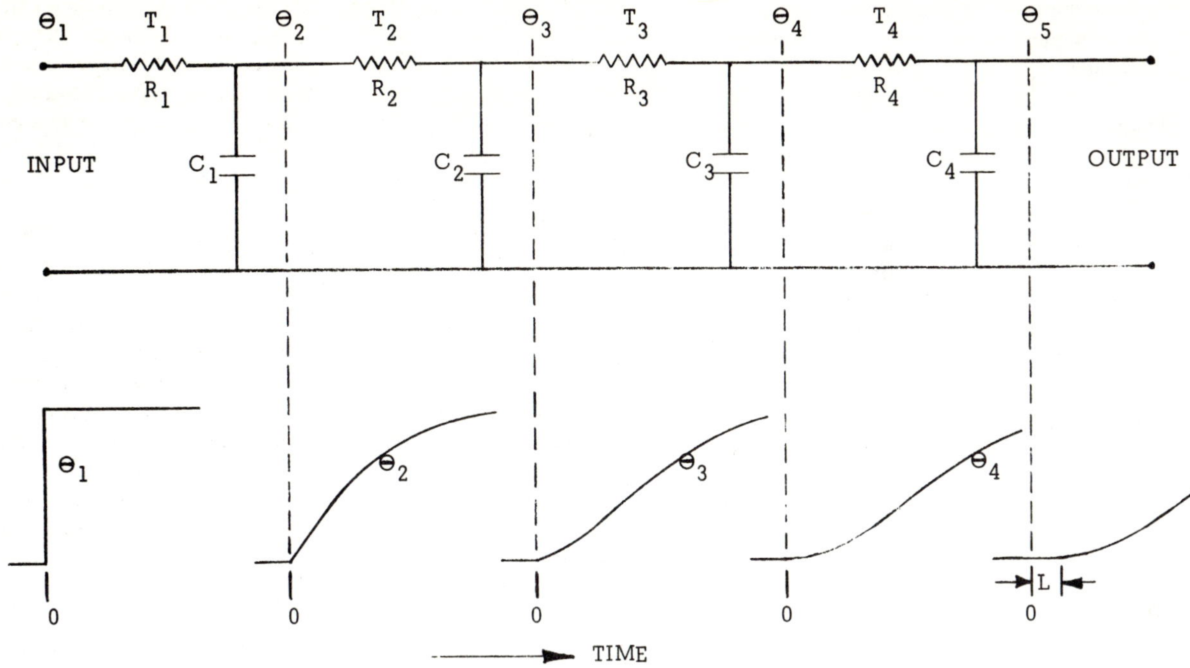

Figure 2-12. Pseudo dead time lag resulting from series-connected exponential lags.

shown in Figure 2-12. An "electrical analog" is used in Figure 2-12 to simulate four RC thermal time constants. Imposing a step change at Θ_1 will produce an exponential output at Θ_2 based on Equation (2-3a). The point to note is that Θ_2 is also the input to the second exponential lag element T_2. Since this input does not initially produce a large difference between Θ_2 and Θ_3, the output Θ_3 will start changing very slowly. This can be seen by revising subscripts on the first-order equation.

$$d\Theta_3/dt = (\Theta_2 - \Theta_3)/T_2 \qquad (2\text{-}18)$$

By using the same pattern of reasoning, it is easy to see that the driving forces to T_3 and T_4, namely, $(\Theta_3 - \Theta_4)$ and $(\Theta_4 - \Theta_5)$, will initially be even smaller then $(\Theta_2 - \Theta_3)$. Thus, the initial rate of change of Θ_4 and Θ_5 with time will be almost negligible, and the output Θ_5 will appear to have dead time lag. This pseudo lag may be determined experimentally or analytically. The methods will be discussed in Chapters 11 and 12. The important point at this time is to recognize the existence of this pseudo dead time lag, since its effect on control is as real as pure dead time.

From a mathematical viewpoint the equations for dead time lag may be written as follows:

$$y(t) = x(t - L) \qquad (2\text{-}19)$$

$$y(t) = x(t_0), \text{ for } t < L, \qquad (2\text{-}20)$$

where $y(t)$ = output signal at a given time,
$x(t - L)$ = input signal at (time − dead time lag),
$x(t_0)$ = initial input signal before change.

Equations for the first-order or exponential lag and the dead time lag are, by themselves, relatively simple and easy to understand. Problems arise for the most part when individual equations for a control system must be combined to obtain the total system equation. The problem is usually twofold: (1) the development of the equation and (2) the solution.

The objective in this text will be, as far as possible, to develop the total system equation using the two building blocks discussed in this chapter. The solution of the total system equation may be handled in numerous ways. In the opinion of the author the technique of frequency response provides the easiest path to an understanding of basic problems and will, therefore, be initially used for problem solving. Other techniques will be noted and introduced as required once the basic theory is established. The next problem is then to transform the time equations of this chapter into the frequency domain.

Chapter 2—Study Questions and Problems

(Answers appear on pages 152-153)

1. In any given system, __________ is required for the transfer of energy or mass.

2. The ratio of output to input change for a given system is referred to as the __________.

3. Most nonlinear systems may be assumed to be linear by assuming that any small portion of the system curve may be approximated by a __________.

4. The linearization technique is not applicable to ________ since the variable in question will usually change through a wide range of values.

5. The linearization technique will apply to many control problems; is the analysis concerned with small or large changes around a given operating point?

6. The differential equations for many systems may be obtained from an ________ or ________ balance on the system.

7. What kind of differential equation is shown in Equations (2-3), (2-7), and (2-10)?

8. The first-order linear system will have a ________ to energy or mass flow and a ________ element.

9. The product of the ________ and ________ terms of any first-order or exponential lag system will result in the ________ constant of the system.

10. The time constant will always be expressed in what unit?

11. The general form of the differential equation for any first-order or exponential lag system may be characterized according to slope, rise, and run. Write the equation using these terms.

12. A thermometer is taken from a bath at 70°F and plunged into a second bath at 90°F. Recorded data are shown below. Plot these data and determine the time constant using the initial slope technique.

Time-Temperature Records for a Thermometer

Time, min	Temperature, °F
0.00	70.0
0.04	76.6
0.08	80.9
0.12	84.0
0.16	85.9
0.20	87.3
0.24	88.2
0.28	88.8
0.32	89.2
0.50	89.9

13. Determine the time constant of Problem 12 using the 63% response technique. Read the value of time from the graph drawn in the solution to Problem 12.

14. Using data from Problem 12, calculate and plot the incomplete response curve.

15. Using the incomplete response plotted in the solution to Problem 14, determine the value of the time constant.

16. The 63% response time has no mathematical or physical meaning except for the ________ or ________ lag.

17. Dead time lag is also commonly referred to by what two other terminologies?

18. In a rolling mill, the sheet travels at a speed of 3000 fpm. If the sheet thickness gauge is 10 ft downstream of the rolls, what is the dead time in seconds?

19. A thermal sensing element is located 12 ft downstream of a heat exchanger outlet. Product flow rate in the pipe is 4 fps. What is the value of dead time from heat exchanger outlet to the thermal sensing element in minutes?

20. In Problem 19, two assumptions were made in approximating pipe flow as dead time. What were the two assumptions?

21. Using Equation (2-11), explain why several series-connected exponential or first-order lags will produce the effect of dead time lag. Assume a step input change.

22. The dead time lag produced by several small exponential or first-order lags within a complex system may be referred to as a ________ dead time lag.

References

1. Giedt, W.H., *Principles of Engineering Heat Transfer,* pp.21-23, Princeton, NJ: Van Nostrand, 1957.
2. *Flow of Fluids Through Valves, Fittings, and Pipe,* Technical Paper N. 410, pp. 1-7 to 1-9, Crane Co., New York, NY, 1969.
3. *Valve Sizing,* Catalog 10, Sec. II, pp.117-119, Fisher Governor Co., Marshalltown, Iowa, 1969.
4. Caldwell, W.I., Coon, G.A., and Zoss, L.M. *Frequency Response for Process Control,* Chap. 4, New York: McGraw-Hill, 1959.
5. Rainville, E.D., *Elementary Differential Equations,* pp.50-52, New York: MacMillan, 1958.

3 System Dynamics in the Frequency Domain

In Chapter 2, a simple exponential or first-order lag and dead time lag were introduced. Under individual or isolated conditions of a simple block, the equations were relatively easy to understand. When these simple blocks must be combined to form a complex system, order of the differential equation increases and the problem becomes much more difficult. This problem of handling higher-order (three or more) differential equations in the time domain has caused more than a few instrument and control engineers to give up on control theory. Although it is no less time-consuming, the frequency response technique presents a simplified method for overcoming this problem. To work in the frequency domain, the problem must be transformed from the time domain. Since the end objective, as noted in Chapter 1, is a time response record of the controlled variable, the frequency domain solution must finally be transformed back to the time domain. To accomplish the transformation, certain mathematical and/or graphical techniques are required. This chapter will be devoted to introducing some of these techniques along with defining the exponential or first-order lag and dead time lag in the frequency domain.

3.1. Complex Numbers

The frequency response technique involves use of a sine wave signal for analysis. A sine wave is the time projection of a vector of given amplitude rotating with constant angular velocity as shown in Figure 3-1. The important concept or key word is "vector." Once the vector concept is accepted, it becomes immediately obvious that both magnitude and direction must be defined. There are several ways in which the vector quantity may be expressed. For example a vector having a magnitude

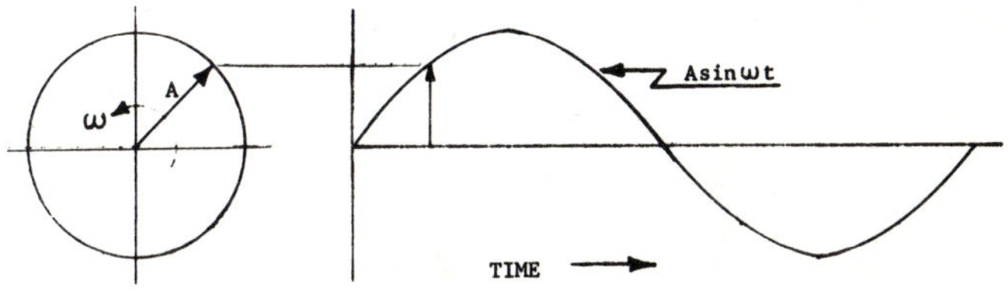

Figure 3-1. Sine-wave development from a rotating vector.

equal to 1.414 ($\sqrt{2}$) at an angle of 45° may be expressed in terms of x and y coordinates:

$$(x, y) = (1, 1) \tag{3-1}$$

It may also be expressed as a magnitude and angle combination:

$$\text{magnitude} \angle \text{angle} = 1.414 \angle 45° \tag{3-2}$$

A more useful mathematical form is provided by the application of complex numbers:[1]

$$a + jb = 1 + j1, \tag{3-3}$$

where a = real or x-axis projection of magnitude,
b = imaginary or y-axis projection of magnitude,
$j = \sqrt{-1}$

In Equation (3-3) the j term implies a ± 90° directional change of the imaginary term and provides the means for solving the

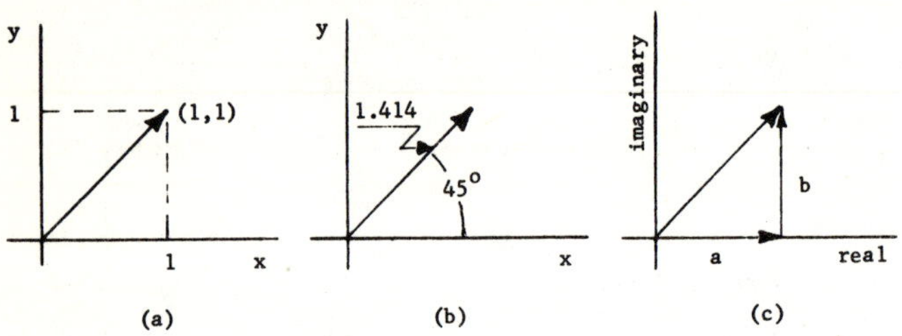

Figure 3-2. Methods of vector representation.

equation. Equations (3-1), (3-2), and (3-3) are shown in Figure 3-2a,b,c. Given Figure 3-2a, the reader would quickly determine vector magnitude and direction using either a graphical parallelogram method or the Pythagorean theorem:

$$\text{magnitude} = \sqrt{(x^2 + y^2)} \tag{3-4}$$

$$\text{angle} = \tan^{-1}(y/x) \tag{3-5}$$

In essence the complex number technique which is shown in Figure 3-2c and Equation (3-3) is nothing more than a mathematical expression of the Pythagorean theorem. Thus, magnitude and direction for Eq. (3-3) are determined in an identical manner:

$$\text{magnitude} = \sqrt{(a^2 + b^2)} \tag{3-6}$$

$$\text{angle} = \tan^{-1}(b/a) \tag{3-7}$$

At this point the reader may question why additional confusion of the complex number technique is required. The reason of course is that the graphical or Pythagorean theorem techniques become very cumbersome as order of the system or number of vectors to be combined increases. The complex number technique, however, is specifically designed to handle these cases with relative ease. As an example assume that the product of two vectors having coordinates $(x_1, y_1) = (2,3)$ and $(x_2, y_2) = (3,5)$ is required. A vector product is obtained by multiplying magnitudes and adding phase angles. Using the Pythagorean theorem, the procedure is

$$M_1 = \sqrt{[(2)^2 + (3)^2]} = \sqrt{13} \cong 3.6 \tag{3-8}$$

$$\Phi_1 = \tan^{-1}(3/2) \cong 56.3°$$

$$M_2 = \sqrt{[(3)^2 + (5)^2]} = \sqrt{34} \cong 5.8$$

$$\Phi_2 = \tan^{-1}(5/3) \cong 59°$$

The resultant product is then

$$M = (3.6)(5.8) \cong 20.9$$

$$\Phi = 56.3° + 59° = 115.3°$$

Using complex numbers, the procedure is

$$\text{Vector product} = (2 + j3)(3 + j5) \tag{3-9}$$

$$\text{Vector product} = 6 + j9 + j10 + j^2 15$$

$$\text{Vector product} = (6 + j^2 15) + j(9 + 10);$$

$$\text{Since } j^2 = (\sqrt{-1})(\sqrt{-1}) = -1,$$

$$\text{Vector product} = -9 + j19.$$

Thus,

$$M = \sqrt{[(-9)^2 + (19)^2]} = \sqrt{442} \cong 21$$

$$\Phi = \tan^{-1}(19/-9) \cong 115.3°$$

For this elementary example the reader may question the advantage of complex number techniques. The answer is that with complex number techniques more difficult systems can be easily handled using conventional mathematical methods.

The main thrust of this text will be devoted to frequency response techniques. For this reason application of complex number techniques will be limited. It was, however, necessary to introduce this method to emphasize vector concepts. If the reader refers back to the block diagrams and transfer functions of Chapter 1, the need for vector products is also easily seen. For example using Figure 1-10, the following equations may be written.

$$c = mg_3 \tag{3-10}$$

$$m = pg_2 \tag{3-11}$$

$$p = eg_1 \tag{3-12}$$

To combine these parts of the system Equations (3-10), (3-11), and (3-12) are manipulated to produce the following:

$$c = eg_1g_2g_3 \text{ or } c/e = g_1g_2g_3 \tag{3-13}$$

The transfer functions g_1, g_2, and g_3 represent vectors in the frequency domain. Thus to combine systems, it is necessary to have a procedure that will provide vector products. Both the complex number and frequency response techniques provide this capability.

3.2. Basic Concepts of the Frequency Response Technique

In Chapter 2 the time response solution of a first-order or exponential lag was determined for a step input using the physical description that slope equaled rise/run. Unfortunately for a sine wave input, no such simple physical description is available. There are, however, certain physical concepts that may be employed to anticipate the solution format. The first of these concepts concerns the assumption of linearity. If the system is

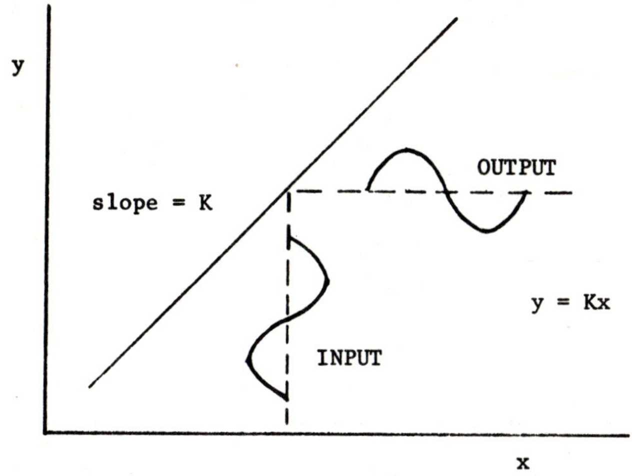

Figure 3-3. Input and output signals for a linear system.

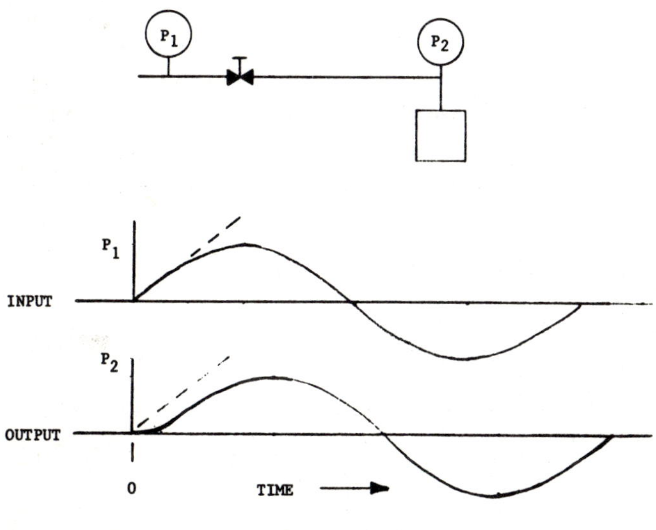

Figure 3-4. Sine-wave input and output responses for a first-order system.

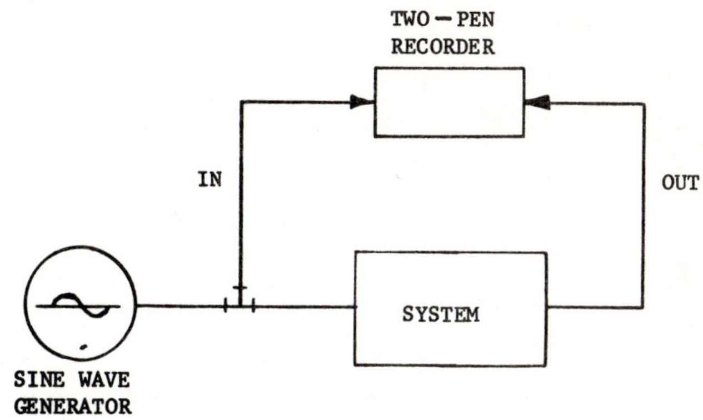

Figure 3-5. Experimental technique for obtaining frequency response data.

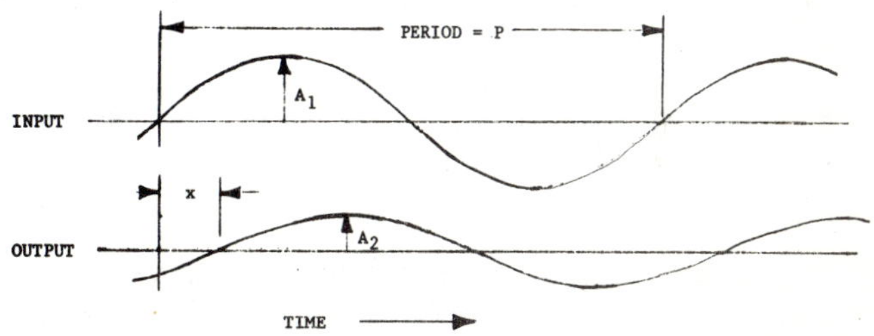

Figure 3-6. Sine-wave records for a system at one given frequency.

linear, the output signal for a sine wave input must be a sine wave having the same frequency as the input. This concept is shown in Figure 3-3.

If the rise over run technique of Chapter 2 is applied, a second concept may be developed. Assume that the input to the pneumatic first-order system of Figure 3-4 is suddenly switched from a constant input to a sine wave input. The initial slope of the sine wave input signal may be approximated by a ramp input or even exponential input as noted in Chapter 2. Since the potential difference ($P_1 - P_2$) will initially be very small, the rate of change or slope of P_2 will be very small. However, after a period of time, P_2 will settle into a periodic cycle, following the input P_1 according to some relationship that is dependent on input signal frequency and the system time constant. The result is an initial transient followed by a steady-state sine wave response. Thus, the solution must have a transient term that decays after some period of time coupled with a steady-state sine term.

If consideration is now given to the steady-state sine term, the third and fourth concepts concerning the solution may be anticipated. If the input signal frequency is very low, period of the sine wave will be very long. Stated in another way, period is very much larger than the system time constant. Under this condition the output will have sufficient time to approach a condition, which is nearly in equilibrium with the input signal. Thus, at low frequencies both magnitude and phase angle of the output signal will be close to magnitude and phase angle of the input signal. However, as the frequency increases, period decreases, and the system does not have sufficient time to achieve an equilibrium status between input and output. Under these conditions output magnitude will decrease, and output phase angle will fall behind or lag the input phase angle. In fact it is easy to visualize a practical system in which input signal changes occur at such a high frequency that output signal response is negligible. Thus, it is evident that any solution must show a decrease in ratio of output-to-input magnitude as frequency increases. By the same reasoning, the solution must show that the difference between input and output phase angles is negative or lagging as frequency increases.

To determine the frequency response of a system experimentally, a variable frequency sine wave generator and a two-pen recorder are connected to a system as shown in Figure 3-5. To obtain frequency response data, the generator is set to a particular frequency and allowed to "drive" the system. Both the input signal to the system and the output signal from the system are recorded. Typical steady-state sine wave records for a particular frequency are shown in Figure 3-6. Magnitude and phase angle at this particular frequency may be calculated as follows:

$$M = A_2/A_1, \tag{3-14}$$

where M = magnitude ratio,
A_2 = output amplitude, chart units,
A_1 = input amplitude, chart units,

$$\Phi = -360°(x/P), \tag{3-15}$$

where Φ = phase angle, degrees,
x = lag measurement, time in chart units,
P = period measurement, time per cycle.

The frequency may be read from the generator dial or calculated from chart measurements of the period and chart speed.

$$f = n/P, \tag{3-16}$$

where P = period measurement, chart units per cycle (in./cyc),
n = chart speed, chart units per unit time (in./sec),
f = frequency, cycles per unit time (cyc/sec).

By repeating this process at a number of different frequencies, the system frequency response can be determined.

Frequency response data are normally plotted in terms of Bode or Nyquist diagrams.[2] In the case of the Bode diagram, which will be used throughout this text, vector magnitude may be plotted on a $\log_{10}$ scale or as a decibel. The decibel is a sound level measurement and results from early work in sound transmission. A decibel is equal to 20 $\log_{10}$M, and thus, may be plotted on rectangular coordinates. At one time most control literature used the decibel unit, and the serious student will need to become familiar with this unit to study other texts and published papers. The $\log_{10}$ magnitude plot, however, is generally accepted and will be used in this text since it is more descriptive of the actual hardware, and it involves one less computation for each data point. A typical frequency response or Bode diagram is shown in Figure 3-7. In the Bode diagram magnitude versus frequency is plotted on log-log coordinates, and phase angle versus frequency on semi-log coordinates. The selection of these scales, while not immediately obvious, is mathematically sound and will be explained in Chapter 4. The selection of this particular method of plotting does not imply that it is the best or easiest. It has been the author's experience, however, that detailed explanation of all methods at this time merely confronts the reader with excess material. On the other hand, once the reader becomes familiar with one technique, the others become fairly obvious.

3.3. Introduction to Operational Calculus

One of the great mysteries of control theory is operational calculus or the Laplace transform. Operational calculus provides a very powerful and timesaving tool for solving time-dependent or differential equations.[3] The technique is a simple procedure of transforming a problem into a format that can be handled algebraically, arriving at a solution and then transforming the solution back to its original format. This later step is often referred to as the inverse transform. While details are slightly more complex and will be explained in steps, a simple analogous example of a similar procedure may help the reader.

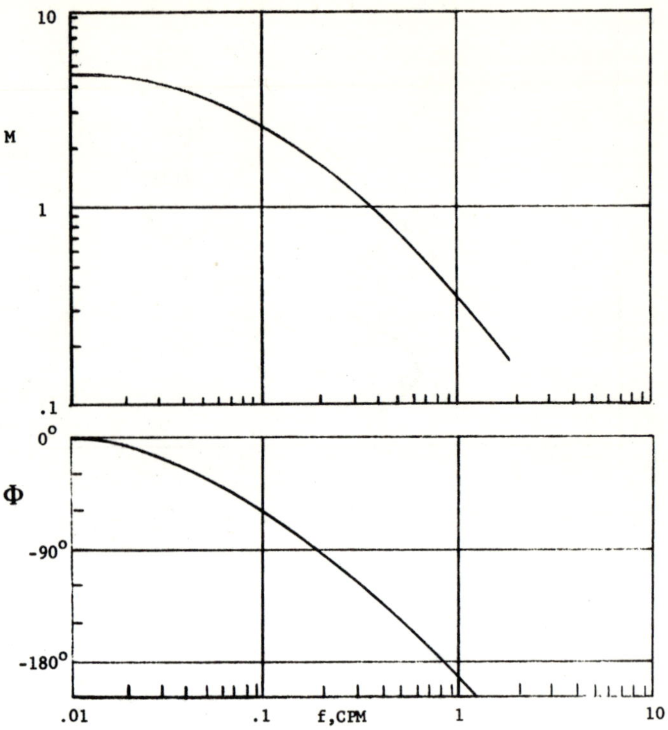

Figure 3-7. System frequency response or Bode diagram.

Assume that the square root of 19 is desired. The answer may be obtained by trial and error, from a table of square root values, or by calculation on a slide rule or calculator. Since trial and error is too time-consuming and a table provides a precalculated solution, the calculation technique is the only method of interest in the present example. The square root of 19 may be expressed as $\sqrt{19}$ or $(19)^{1/2}$. If this is now transformed to 1/2 (log 19), the solution may be easily obtained on a slide rule or calculator by multiplying the logarithm of 19 by 1/2. To obtain the actual answer, however, the antilogarithm or inverse transform of the product must be obtained. When this step is completed, an answer of 4.359 is obtained in the original format. The solution of a differential equation by operational calculus can be as simple as this example.

Before proceeding with the operational technique, however, it is necessary to review some basics of solving equations. Assume Equation (3-17) as follows:

$$X - 10 = 0 \tag{3-17}$$

For Equation (3-17), the solution is $X = 10$. It may also be stated that 10 is the root of the equation. For Equation (3-18), the solutions or roots are $X = -2$ and $X = 4$.

$$X^2 - 2X - 8 = 0 \tag{3-18}$$

The roots for Equation (3-18) were easy to obtain because the equation was factorable into $(X + 2)$ and $(X - 4)$. In the case of

Equation (3-19), however, the procedure is more complex since real number factors are not available:

$$X^2 + 2X + 5 = 0 \qquad (3\text{-}19)$$

To determine the roots in this case requires the aid of the quadratic factor shown in Equation (3-20):

$$X = \left[-b \pm \sqrt{(b^2 - 4ac)}\right]/2a \qquad (3\text{-}20)$$

Equation (3-20) of course gives roots for any equation of second-order as noted in Equation (3-21):

$$aX^2 + bX + c = 0 \qquad (3\text{-}21)$$

By comparing Equations (3-19) and (3-21) it is apparent that $a = 1, b = 2$, and $c = 5$. Substituting these values in Equation (3-20) yields the following roots for Equation (3-19):

$$X = \left[-2 \pm \sqrt{(4 - 20)}\right]/2 \qquad (3\text{-}22)$$

or

$$X = -1 + \sqrt{-4} \qquad (3\text{-}23)$$

and

$$X = -1 - \sqrt{-4} \qquad (3\text{-}24)$$

The negative signs within the radical terms of Equations (3-23) and (3-24) indicate complex roots. By using the complex number notation explained in Section 3.1, these roots may be expressed as follows:

$$X = -1 + j2 \text{ and } X = -1 - j2 \qquad (3\text{-}25)$$

There are several observations that may be gleaned from this simple refresher on solving equations. First, the solution of an equation is a root. Second, the number of roots depends on the order of the equation. In Equation (3-17), the order of X is 1, thus one root. In Equations (3-18) and (3-19) the order of X is 2 (the largest exponent), thus two roots. Third, roots may be positive or negative real numbers, they may be equal (none shown), or they may be complex numbers. Fourth, in the case of complex roots they must occur in pairs as can be seen from Equation (3-20). Based on this introduction and review, the method of operational calculus will be applied to the solution of the first-order differential equation or exponential lag.

The thermal system of Chapter 2 will again be taken for the general case. In its time-constant format the equation is

$$d\Theta_2/dt = (\Theta_1 - \Theta_2)/T \qquad (2\text{-}3a)$$

Equation (2-3a) may be transformed into an algebraic equation by making the operational substitution or transformation shown in Equation (3-26):

$$p = d/dt, \qquad (3\text{-}26)$$

where p = Heaviside operator, time.$^{-1}$

Substituting Equation (3-26) into Equation (2-3a) gives

$$p\Theta_2 = (\Theta_1 - \Theta_2)/T \qquad (3\text{-}27)$$

At this point it is necessary to recall the objective, which is to obtain a transfer function that will describe how the output signal Θ_2 reacts in response to a change in the input signal Θ_1. This transfer function, or ratio of output to input Θ_2/Θ_1, can now be determined from Equation (3-27) by algebraic manipulation. The solution is shown in Equation (3-28):

$$\Theta_2/\Theta_1 = 1/(1 + Tp) \qquad (3\text{-}28)$$

Equation (3-28) is the general solution in operational form of the first-order differential equation or exponential lag. Actually it shows two solutions. If it is remembered that $p = d/dt$, then it is easy to see that, as t approaches infinity, p approaches zero, and a steady-state solution results. The initial or transient response, however, is dependent on p and is characterized by the denominator $(1 + Tp)$. For this reason the $(1 + Tp)$ term is referred to as the characteristic solution. It is also important at this time to note that the form of the input signal Θ_1 has not been specified. It could take any form such as an impulse, a step as in Chapter 2, a sine wave which is the immediate interest of this chapter, or a random variation.

By setting the characteristic equation equal to zero, it is possible to obtain the root of the system.

$$1 + Tp = 0 \text{ or } p = -(1/T) \qquad (3\text{-}29)$$

Since the time constant is a function of physical parameters as shown in Chapter 2, it is obvious that the root in this case is real and negative. It is also evident that for this first-order equation there is only one root. For a second-order equation there would be two roots. Likewise for an nth-order equation there would be n roots.

Using this simple review of equation solving as a model, it seems fairly clear that a general expression could be written to define all roots. Such an expression would have a positive and negative real part, and a positive and negative imaginary part. In the case of differential equations, however, there must also be a frequency term included in the root. This conclusion is intuitive based on the observed control responses of Figure 1-12c and d. Thus, a general expression for roots may be written as follows:

$$p = \pm\sigma \pm j\omega, \qquad (3\text{-}30)$$

where σ = real part of root,
ω = imaginary part of root, angular frequency, rad/time,
$= 2\pi f$

Using the general root expression of Equation (3-30) it is possible to establish a so-called "p-plane" or plot which shows all possible roots. If the root of Equation (3-29) for the first-

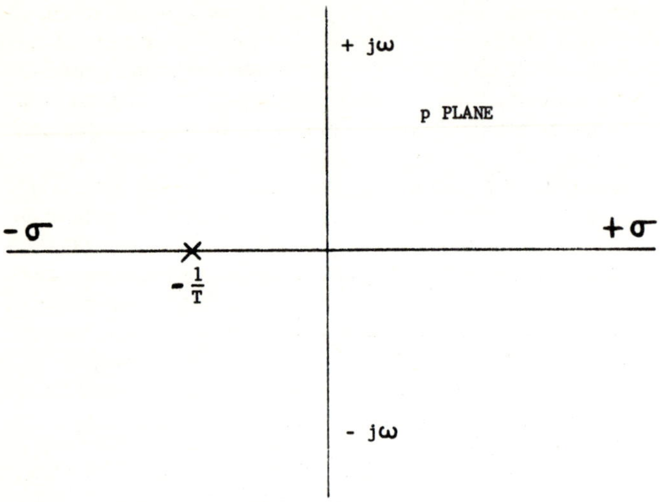

Figure 3-8. p-Plane showing a first-order root.

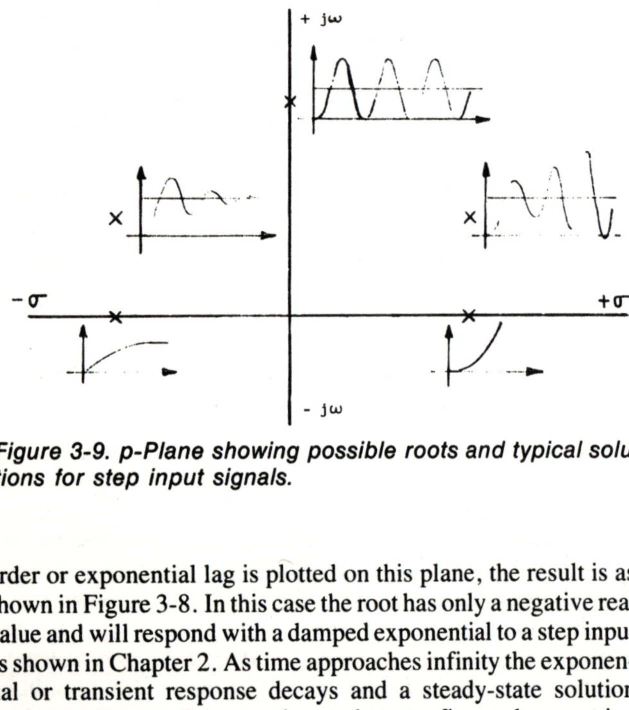

Figure 3-9. p-Plane showing possible roots and typical solutions for step input signals.

order or exponential lag is plotted on this plane, the result is as shown in Figure 3-8. In this case the root has only a negative real value and will respond with a damped exponential to a step input as shown in Chapter 2. As time approaches infinity the exponential or transient response decays and a steady-state solution exists. In Chapter 2 it was shown that any first-order equation would have this same exponential response to a step input and could be characterized by the time constant. From the p-plane and Equation (3-29) it can also be seen that the root is characterized or located by the time constant. Thus, it may be concluded that for any real negative root the solution of the general equation for a step input will be in the form of a decaying exponential. By a similar analysis it also is possible to show that for positive real roots the solution to a step input is an undamped or increasing exponential. These results are shown in Figure 3-9.

If the order of the system increases for example to 2, then several conditions are possible. If the roots are real, unequal, and negative, there is no frequency term and an overdamped exponential response to a step input exists as shown in Figure 2-12 (see Θ_3). In the case of real, unequal, and positive roots, a similar undamped or increasing exponential response results. When the system roots include both real and imaginary terms, the response must then include not only an exponential but also a sine or cosine function. Typical results using step inputs for these cases are also shown in Figure 3-9. If the roots have only imaginary terms (no real parts), the solution will have only sine or cosine functions with no decay. This result is also shown in Figure 3-9. The interesting point is that all the responses shown in Figure 3-9 are similar to the possible control responses shown in Figure 1-12. Thus, two conclusions become immediately evident. First, if the roots of the system equation are known, stability of the system can be determined. Second, if any roots occur on the positive or right-hand half of the p-plane, an unstable system will occur.

The technique of transforms is of course the classical approach to control theory. The remaining step is to inverse transform the general solution of Equation (3-28) back to the time domain for a particular input signal. Unfortunately, this step can become very complicated in complex systems, and for this reason will not be used in this text. For a complete explanation of the method the reader is referred to the references.[4,5] The preceding detailed explanation, however, was necessary for several reasons. Much present literature concerning control theory uses this approach. Thus, the reader will need an understanding of root concepts to relate this text to other literature. An understanding of roots and their potential time domain formats helps explain how potential solutions may be assumed as is done in many texts. The most important reason of course is the substitution of $j\omega$ for p or s (the Laplace) operator to obtain the frequency response, which will be discussed in the next section.

3.4. First-Order or Exponential Lag Solution

If the reader has previous exposure to control literature, use of the Heaviside or p operator rather than the Laplace or s operator may be causing some confusion at this point. The difference is minor but is often overlooked. Unfortunately, in the author's opinion this frequently leads to problems for the reader. The Heaviside operator p assumes that the initial conditions on all functions are zero. For example take the responses of Figure 1-12, where the initial condition is 160°F and the step input change is 10°F. If the Heaviside method is used, a solution results which shows only the change in either the controlled or measured variable. To get an absolute answer the bias or initial value of 160° must be added to each value. If used properly the Laplace operator s will account for the initial condition, and the solution will show actual temperature for each time value. The equivalent transforms for a derivative are shown in the following equations to demonstrate this concept:

$$P\{df/dt\} = pf \tag{3-31}$$

$$L\{df/dt\} = sf - f(0) \tag{3-32}$$

The capitals P and L in Equations (3-31) and (3-32) imply a transformation of the bracketed function. The actual transformation is on the right-hand side of the equal sign. As can be seen, the only difference is the value of the function at $t = 0$, ($f(0)$), which is included in the Laplace transformation. With the assumption of zero initial conditions, the Heaviside and Laplace operators are identical, and since most texts and papers use the Laplace operator s, this author will now convert to s for the sake of consistency. Equation (3-28) will then take on the conventional appearance:

$$\Theta_2/\Theta_1 = 1/(1 + Ts), \tag{3-28a}$$

where s = Laplace operator, time^{-1} = d/dt

The problem now is to determine the frequency response of the first-order differential equation. Experimentally this is done as shown in Figure 3-5. From a mathematical viewpoint the input signal is $\Theta_1 = A \sin \omega t$ or $A \sin 2\pi ft$, where A = amplitude and t = time. If $A \sin 2\pi ft$ is substituted for Θ_1 in transform notation, the result is

$$L\{\Theta_1\} = L\{A \sin 2\pi ft\} = A2\pi f/[s^2 + (2\pi f)^2] \tag{3-33}$$

Equation (3-28a) may then be written in operational form for a sine input as follows:

$$\Theta_2 = \{A2\pi f/[s^2 + (2\pi f)^2]\}[1/(1 + Ts)] \tag{3-34}$$

Taking the inverse transform[4] of Equation (3-34) gives a solution of Θ_2 in the time domain as follows:

$$\Theta_2 = [2\pi fA/(1 + (2\pi fT)^2)]\, e^{-t/T} + \{A/[1 + (2\pi fT)^2]^{1/2}\} \sin(2\pi ft - \tan^{-1} 2\pi fT) \tag{3-35}$$

Equation (3-35) clearly shows the combination of initial transient and steady-state response terms that were previously noted and shown in Figure 3-4. Once the transient term of Equation (3-35) decays, input and output signals may be expressed as follows:

$$\text{Input} = \Theta_1 = A \sin 2\pi ft \tag{3-36}$$

$$\text{Output} = \Theta_2 = (A/[1 + (2\pi fT)^2]^{1/2}) \times \sin(2\pi ft - \tan^{-1} 2\pi fT) \tag{3-37}$$

In a manner similar to the experimental method of Equation (3-14), the magnitude ratio as a function of frequency may be determined from input and output amplitude terms of Equations (3-36) and (3-37).

$$M = \text{OUTPUT/INPUT} = A/\{[1 + (2\pi fT)^2]^{1/2}A\} = 1/[1 + (2\pi fT)^2]^{1/2} \tag{3-38}$$

The phase angle may be obtained by subtracting the input angle from the output angle. This procedure results from the procedure of dividing vectors. To divide vectors, the magnitudes are divided (as in Equation (3-38)) and phase angles are subtracted.

$$\Phi = (2\pi ft - \tan^{-1} 2\pi fT) - 2\pi ft = -\tan^{-1} 2\pi fT \tag{3-39}$$

Using Equations (3-38) and (3-39) it is now possible to get an exact plot of magnitude and phase angle as a function of frequency, which would duplicate the curves shown in Figure 3-7.

While this technique of transforming and inverse transforming is a very powerful tool, it can require a degree of mathematical sophistication in complex systems that many readers will not have experienced. The results of Equations (3-38) and (3-39), however, can be obtained using a much simpler technique. In the experimental method sine waves are imposed on the system at particular frequencies with magnitudes and phase angles measured from recorded results. An inspection of the p-plane in Figure 3-9, or s-plane as it will now be called, shows that sine waves occur only along the $j\omega$-axis. Thus, by forcing s to take on values of $j\omega$ or $j2\pi f$, a mathematical result is obtained that is similar to the experimental frequency test. This effect can be seen by substituting $j2\pi f$ for s in Equation (3-28a).

$$\Theta_2/\Theta_1 = 1/(1 + j2\pi fT) \tag{3-40}$$

Equation (3-40) is now in the complex format described in Section 3.1, where the j term in the numerator is zero. The magnitude and phase angle of the numerator and denominator can now be expressed according to Equations (3-6) and(3-7). Here again, vectors must be divided, which means dividing magnitudes and subtracting angles. Results are

$$M = 1/[1 + (2\pi fT)^2]^{1/2} \tag{3-41}$$

$$\Phi = 0 - \tan^{-1} 2\pi fT = -\tan^{-1} 2\pi fT \tag{3-42}$$

The result is the same as in Equations (3-38) and (3-39). Also, the procedure of letting $s = j2\pi f$ to obtain the frequency response terms requires much less work. Note, however, that this does not provide a time response equation for Θ_2. This problem will be discussed in Chapter 7.

A detailed plot of Equations (3-40) and (3-41) for a time constant $T = 1/(2\pi)$ min is shown in Figure 3-10. As in the case of the time response curve of Chapter 2, the frequency response curves may be characterized by the time constant. At low frequencies the magnitude approaches the steady-state or zero frequency magnitude of 1, and the phase approaches zero. At high frequencies the magnitude slope approaches -1, and the phase angle approaches $-90°$. If high and low frequency asymptotes are sketched on the magnitude curve, they will intersect at the so-called corner frequency. The corner frequency is defined as

$$f_c = 1/(2\pi T), \tag{3-43}$$

where f_c = corner frequency, cyc/time.

For $T = 1/(2\pi)$ min as in Figure 3-10, $f_c = 1.0$ cpm. At this frequency the magnitude is 0.707 of the zero frequency magnitude, and the phase angle is $-45°$. Proof of this can be shown using a vector diagram of Equation (3-40) for $T = 1/(2\pi)$ and $f = f_c = 1$. Making these substitutions gives

$$\Theta_2/\Theta_1 = \text{OUTPUT/INPUT} = 1/(1 + j1) \tag{3-44}$$

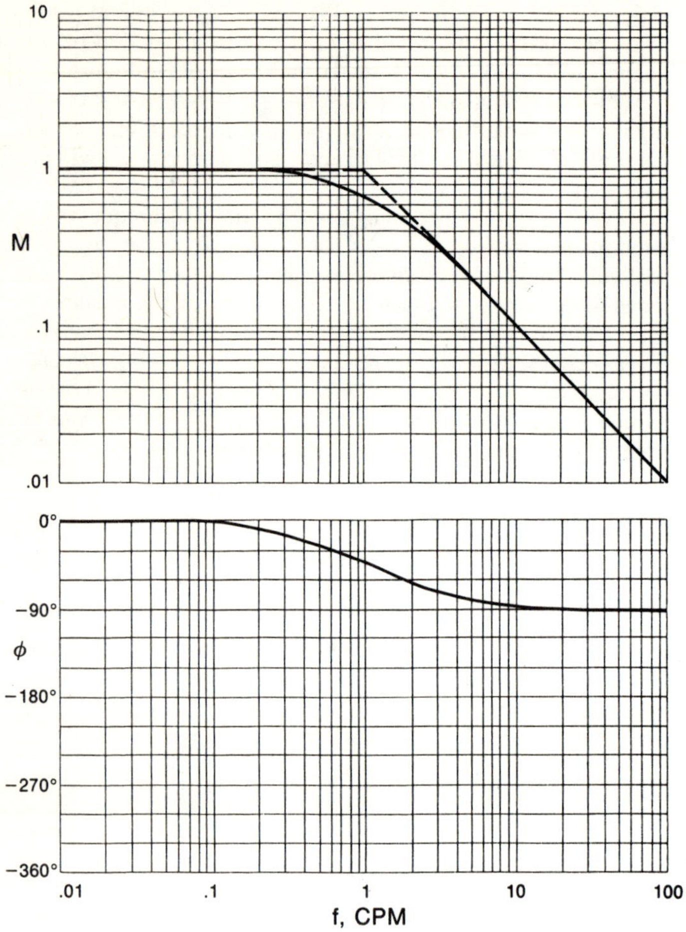

Figure 3-10. Frequency response for an exponential lag with T = 1/(2π) min.

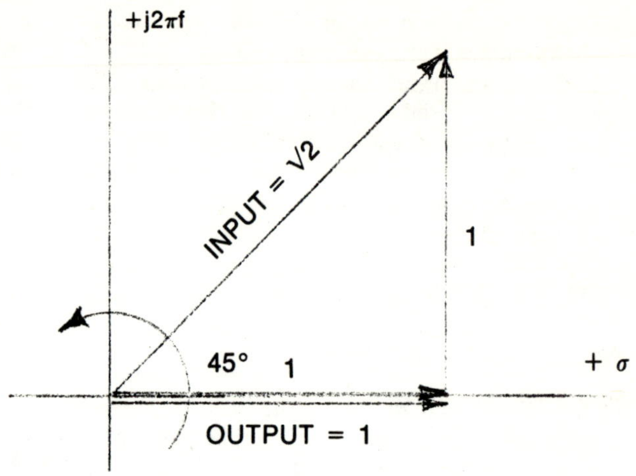

Figure 3-11. Vector diagram for the corner frequency.

Assuming the numerator 1 to be the output and the denominator $1 + j1$ to be the input, a vector diagram can be sketched as shown in Figure 3-11. As can be seen in Figure 3-11, magnitude of output/input is $1/\sqrt{2}$ or 0.707, and output lags the input by 45°. Vectors are offset from the axis for clarity.

Although it may not be immediately obvious, the curves of Figure 3-10 are "constant" in the sense that their shapes remain constant. Since these curves are characterized by the time constant through the corner frequency, it is possible to make a template. By indexing the corner frequency on the template, the frequency response can be quickly sketched on a Bode plot for any time constant by calculating the corner frequency, locating the template index at this frequency on the diagram, and tracing the template. If it is assumed, for example, that g_1, g_2, and g_3 of Equation (3-13) are all simple exponential lags, each curve could be quickly sketched. To obtain the product of these vectors, magnitudes must be multiplied and phase angles added. Although the detailed procedure will be left to the next chapter, it is now easy to see the basis for scale selection on the Bode diagram. Use of the log scale on magnitudes allows multiplication of vectors by addition of distances. Since phase angles are directly additive, a linear scale may be used. Although it may be obvious, it is worth noting that the template procedure will only work when template and paper scales are matched.

3.5. Dead Time Lag

The pipe flow heating example of Figure 2-10 will be used in developing the frequency response for the dead time or distance velocity lag. The conditions of plug flow and negligible heat loss are assumed as was the case in Chapter 2. If heat energy is added in the form of a sine function at point 1, temperature recordings at points 1 and 2 will result as shown in Figure 3-12a. The magnitude, phase angle, and frequency may be found using Equations (3-14) through (3-16). It is evident from Figure 3-12a that with the noted assumptions $A_1 = A_2$ and $M = 1.0$. The phase angle may be found using Equation (3-15) by substituting the lag measurement L and period P_1 in Equation (3-15).

$$\Phi_1 = -360\,(L/P_1) \tag{3-45}$$

Equation (3-45) may also be written in terms of frequency by making the substitution $f_1 = 1/P_1$.

$$\Phi_1 = -360 f_1 L \tag{3-46}$$

If the frequency of the input signal is doubled, the period is halved. Temperature recordings at points 1 and 2 will then appear as shown in Figure 3-12b. As previously experienced, the amplitudes A_1 and A_2 will be equal, resulting in a magnitude equal to 1. In case of the phase angle, however, there is an interesting change. Although the period was halved, the value of L remains constant. This results from the fact that L is not a function of the frequency of heat energy variation but is dependent only on the distance between measuring points and flow velocity. Since $f_2 = 2f_1$ and L is constant, the phase angle Φ_2 must be equal to 2 Φ_1. Based on these two tests, general ex-

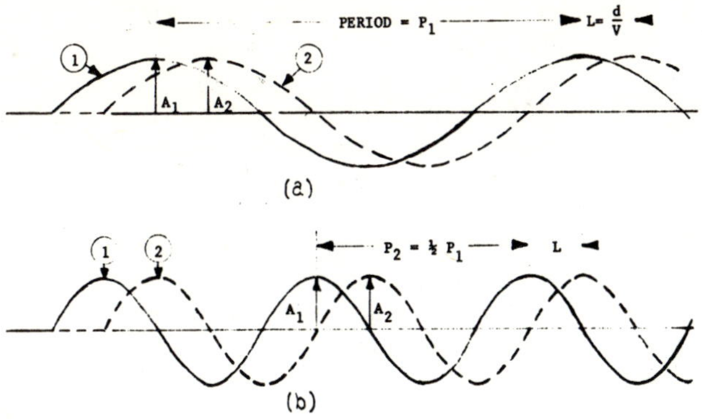

Figure 3-12. Sine-wave records for a dead time lag.

pressions for magnitude and phase angle of the dead time lag may be expressed as follows:

$$M = 1.0 \text{ (for dead time only)} \tag{3-47}$$

$$\Phi = -360 fL \text{ (for dead time only)} \tag{3-48}$$

Using Equations (3-47) and (3-48) the frequency response for a dead time lag of 0.5 min is shown in Figure 3-13.

The operational or Laplace transform for dead time lag is[5]

$$\Theta_2/\Theta_1 = e^{-Ls} \tag{3-49}$$

Equation (3-49) is the general transfer function for which the form of the input signal is not specified. The frequency response may be obtained by again substituting $s = j2\pi f$. This is the same procedure that was used for the exponential lag in Equation (3-40).

$$\Theta_2/\Theta_1 = e^{-j2\pi fL} \tag{3-50}$$

Equation (3-50) may be rewritten by using the trigonometric identity $e^{-jx} = \cos x - j \sin x$, which gives

$$\Theta_2/\Theta_1 = \cos(2\pi fL) - j \sin(2\pi fL) \tag{3-51}$$

An easy way to examine Equation (3-51) is to divide the right-hand side of the equation by 1. This does not change the value of the equation. It does, however, again allow the assumption of output being equal to numerator and input to denominator.

$$\begin{aligned}\Theta_2/\Theta_1 &= \text{OUTPUT/INPUT} \\ &= \left[\cos(2\pi fL) - j \sin(2\pi fL)\right]/1\end{aligned} \tag{3-52}$$

A vector diagram similar to Figure 3-11 may now be sketched as shown in Figure 3-14. The magnitude of the output-to-input ratio can now be obtained using either complex number methods or by observation of Figure 3-14.

$$\begin{aligned}M &= \text{OUTPUT/INPUT} \\ &= \sqrt{\left[\cos^2(2\pi fL) + \sin^2(2\pi fL)\right]}/1 = 1\end{aligned} \tag{3-53}$$

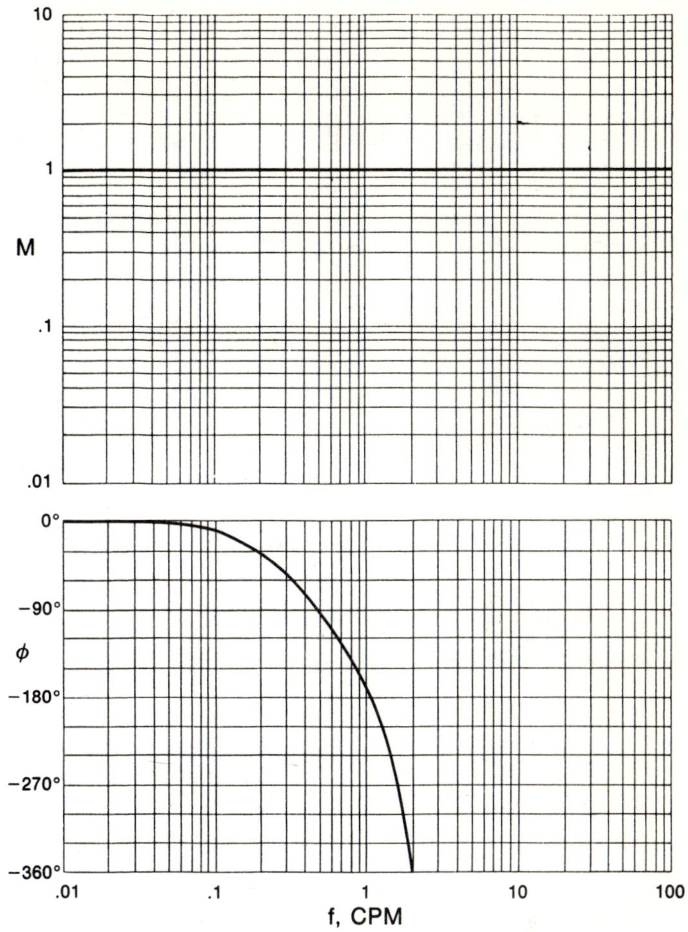

Figure 3-13. Frequency response for a dead time lag with L = 0.5 min.

The result of Equation (3-53) may be checked by the doubters or accepted by recalling that the identity $\cos^2 x + \sin^2 x = 1$. The phase angle is again the angle of the numerator minus the angle of the denominator. In this case, however, the numerator angle is negative since the j term is negative. This is also shown in Figure 3-14. Since the j term of denominator is zero (Equation (3-52)), the result is as shown in Equation (3-54).

$$\begin{aligned}\Phi &= -\tan^{-1}\left[\sin(2\pi fL)/\cos(2\pi fL)\right] - 0 \\ &= -\tan^{-1}\left[\tan(2\pi fL)\right]\end{aligned} \tag{3-54}$$

In Equation (3-54) sine over cosine has been replaced by the tangent, another trigonometric identity. The tongue-teasing result is that "the angle whose tangent is the tangent is the angle." Thus, Equation (3-54) may be reduced to

$$\Phi = -2\pi fL \tag{3-55}$$

The resulting angles in Equation (3-55) will be in radians. To convert to degrees, use the conversion factor $360° = 2\pi$ radians.

$$\Phi = -(2\pi fL)(360/2\pi) = -360° fL \tag{3-56}$$

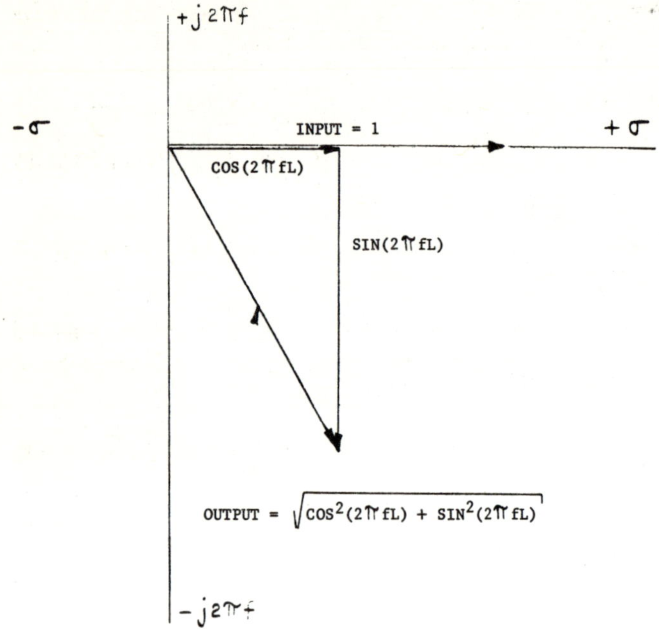

Figure 3-14. Vector diagram for a dead time lag.

The results obtained in Equations (3-53) and (3-56) are the same as results of Equations (3-47) and (3-48). Although it is possible to construct a template for dead time lag, its value is often questioned due to the ease of constructing the magnitude and phase angle curves as a function of frequency.

Using the exponential and dead time lag equations and assumptions concerning practical hardware, it is a simple matter to construct a frequency response curve for a control loop. This procedure will be considered in Chapter 4.

Chapter 3—Study Questions and Problems

(Answers appear on pages 153-155)

1. A vector quantity has both magnitude and__________.

2. A vector may be expressed by what three methods? In mathematical format a vector is called a__________.

3. Express the vector having coordinates $(x, y) = (3, 4)$ in the form of a complex number.

4. What are the magnitude and angle values of the vector $(3 + j4)$?

5. The product of vectors is obtained by multiplying the__________ and adding the__________.

6. Find the product of the following vectors: $(3 + j4)\ (5 + j2)$.

7. In a linear system the output signal for a sine wave input signal gives what kind of wave? Does it have the same frequency as the input signal?

8. The initial response at the output of a system to a sine wave input signal will be __________ in nature.

9. Following the initial transient response the system output signal (assuming a sine wave input signal) will show what kind of wave response?

10. When the period of a sine wave input is much larger than the system time constant, the output signal magnitude will be almost (equal to, 0, or greater than) the input signal magnitude, and the phase angle lag of the output signal will be almost __________.

11. As the frequency of the input signal increases, that is, period decreases, will the output signal magnitude increase or decrease? And will the phase angle lag of the system normally increase or decrease?

12. The magnitude ratio is defined as the amplitude of the __________ signal divided by __________ signal amplitude.

13. The phase angle lag for a given system may be determined from experimental data by multiplying the __________ measurement by 360° and dividing by the __________ measurement.

14. In the Bode plot for frequency response data, __________ is plotted against frequency on log-log coordinates, while __________ is plotted on semi-log coordinates.

15. In a simple equation such as $X + 5 = 0$, the solution is $X = -5$. This solution is also known as the __________of the equation.

16. In a second-order equation, there are how many roots?

17. In a case where roots cannot be factored, how can they be obtained?

18. What is the equation for the quadratic factor?

19. Determine the roots for $X^2 + 4X + 20 = 0$, and express the roots in complex notation.

20. In operational calculus, what operator is used to replace the differential d/dt to transform the differential equation into an algebraic equation?

21. The general solution of a first-order differential equation in operations form is OUTPUT/INPUT $= 1/(1 + 10p)$. For this case determine the characteristic equation, the time constant, and the root of the equation.

22. All roots of differential equations may be expressed in what common format?

23. A difference between the Heaviside and Laplace operators exists when considering the __________.

24. If initial conditions can be assumed to be zero, a differential such as dP_2/dt may be transformed in operational notation to ________.

25. The units on the Laplace operator s are ________.

26. The substitution of $j2\pi f$ for s is a mathematical technique for obtaining the ________ response. This is similar to using what kind of wave input signal on an actual system?

27. Using Equations (3-41) and (3-42), calculate the frequency response for $T = 1/(2\pi)$ min. Make the calculation over the frequency range of 0.1 to 10 cpm.

28. Plot the data obtained in the answer to Problem 27 and construct a template. *Note*: Although any standard log-log and semi-log paper may be used, a special combined paper for a plot of this type is available.[6] Availability of supply will normally dictate the paper choice. The template may be constructed by cementing the curve to a sheet of cardboard and cutting along the curve. An alternate technique is to make a plastic transparency of the plot such as used for an overhead projector.[7] By cutting the plastic along the curves, a transparent template is obtained. Indexing of the template should of course be completed before making the transparency. Some shrinkage is possible in the transparency technique and this should be checked.

29. To sketch magnitude and phase curves for an exponential lag of $T = 0.5$ min, the index of the template should be set at a frequency of ________ cpm.

30. In considering pipe flow heating as a dead time lag, what two basic assumptions must be made?

31. The magnitude of a dead time lag is always ________.

32. The phase angle for a dead time lag may be found from the equation $\Phi =$ ________.

33. Give the general expression for the dead time lag transfer function in operational form.

34. A thermal sensing element is located 9 ft downstream from the outlet of a heat exchanger. If product velocity is 3 fps, what is the dead time lag in minutes?

35. Plot the frequency response for a dead time lag of 0.05 min.

References

1. Spiegel, M. R., *Theory and Problems of Complex Variables*, New York: Schaum, 1964.
2. Harrison, H. L., and Bollinger, J. G., *Introduction to Automatic Controls*, Chapters 9, 10, and 11, Scranton, PA: International Textbook Co., 1963.
3. Kells, L. M., *Elementary Differential Equations*, Chapter 6, New York: McGraw-Hill, 1954.
4. Nixon, F. E., *Principles of Automatic Controls*, Englewood Cliffs, NJ: Prentice-Hall, 1960.
5. Buckley, P. S., *Techniques of Process Control*, New York: John Wiley, 1964.
6. Form 97219, Frequency Response Graph Paper, available from Taylor Instrument Companies, 95 Ames St., Rochester, N.Y. 14601.
7. Ostroot, G. F., personal communication, General Mills, Inc., 9000 Plymouth Ave. N, Minneapolis, MN 55427.

4 Process Combinations in the Frequency Domain

Having established the exponential and dead time lags in the frequency domain, it is now possible to describe individual components and combine their responses.

4.1. Components of the Process

A simple heating system as shown in Figure 4-1 will be used to demonstrate procedures. The system may be described by four components as shown in the block diagram in Figure 4-2. Although the result will not be exact, the transmission line and valve actuator can normally be expressed as an exponential lag as follows:

$$\Delta P_2/\Delta P_1 = g_2 = k_2/(1 + T_2 s), \tag{4-1}$$

where P_1 = controller output signal, psi,
P_2 = valve actuator pressure, psi,
g_2 = transmission line and valve actuator transfer function,
k_2 = zero frequency gain, psi/psi,
T_2 = time constant, min.

The value of k_2 is the steady-state or zero frequency gain. For this component $k_2 = 1$. This can be deduced by noting that any change in P_1 will after some time be duplicated at P_2. Thus, in

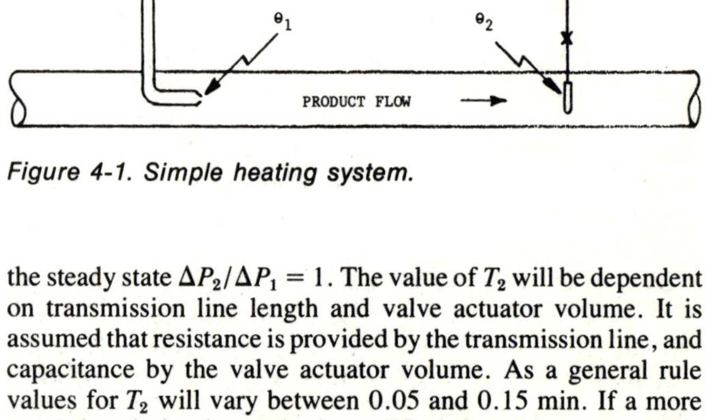

Figure 4-1. Simple heating system.

the steady state $\Delta P_2/\Delta P_1 = 1$. The value of T_2 will be dependent on transmission line length and valve actuator volume. It is assumed that resistance is provided by the transmission line, and capacitance by the valve actuator volume. As a general rule values for T_2 will vary between 0.05 and 0.15 min. If a more exact description is required, specific information can usually be obtained from suppliers or their literature.[1] For the example at hand a value of $T_2 = 0.1$ min will be assumed.

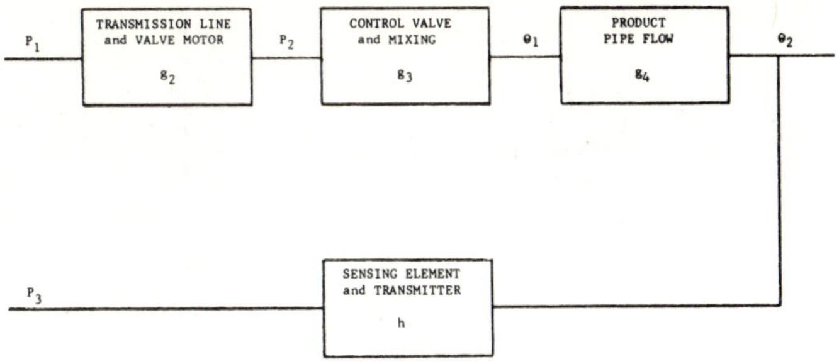

Figure 4-2. Block diagram of heating system shown in Figure 4-1.

The second component in the system represents the valve and mixing process. Steam flow through the valve is a function of pressure drop and valve opening. Any change in steam supply pressure or downstream pressure will cause a steam flow change even though valve opening (or position) remains constant. This type of change produces the disturbances or load changes noted in Chapter 1. For the present, however, these changes will be neglected. This requires the assumption of constant pressure drop across the valve. Under this condition, steam flow is proportional to stem lift or valve actuator pressure. If steam storage in the nozzle and mixing time are neglected (assumed to be small in comparison to other system lags), the valve and steam nozzle may be treated as a gain term. If this is true, a change at P_2 will cause an instantaneous change in product temperature at Θ_1. The transfer function may be expressed as follows:

$$\Delta \Theta_1/\Delta P_2 = g_3 = k_3, \tag{4-2}$$

where Θ_1 = product temperature at steam nozzle, °F,
g_3 = control valve and mixing transfer function,
k_3 = zero frequency gain or sensitivity, °F/psi.

The value of k_3 for a given product flow rate and inlet temperature may be determined from valve characteristics and an energy balance. Here again, potential disturbances or load changes such as product flow rate and inlet temperature will be temporarily neglected. A typical plot of Θ_1 versus P_2 is shown in Figure 4-3. Using Figure 4-3 and selecting a reference or set point of 150° results in

$$k_3 = \Delta\Theta_1/\Delta P_2 = (200 - 90)/(14 - 3) = 110/11 = 10°\text{F/psi} \tag{4-3}$$

Product pipe flow between the steam nozzle and temperature sensing element is the next system component. Following the discussion of Chapters 2 and 3, it is evident that g_4 in Figure 4-2 may be represented as a dead time lag. In arriving at this decision, plug flow and negligible heat loss assumptions have been made. The transfer function may be expressed as follows:

$$\Delta\Theta_2/\Delta\Theta_1 = g_4 = e^{-Ls}, \tag{4-4}$$

where Θ_2 = product temperature at sensing element, °F,
g_4 = product pipe flow transfer function,
L = dead time lag, min.

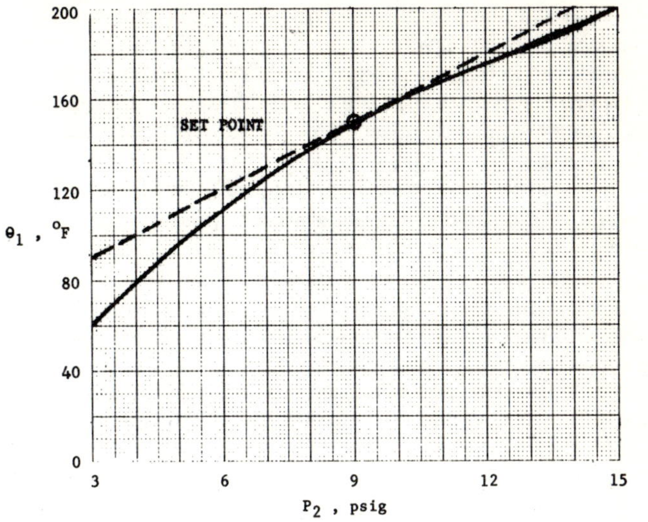

Figure 4-3. Control valve and mixing curve for a constant valve pressure drop, product flow, and product inlet temperature.

Dead time lag may be calculated knowing product velocity and distance from steam nozzle to sensing element. Assuming a distance of 27 ft between the steam nozzle and sensing element, and a velocity of 3 fps gives

$$L = (27\text{ ft})/\left[(3\text{ ft/sec})(60\text{ sec/min})\right] = 0.15\text{ min} \tag{4-5}$$

The remaining component shown in Figures 4-1 and 4-2 is the temperature transmitter with its sensing element. Here again, a simple exponential lag will be assumed. This type of an approximation is certainly open to question, particularly in the case of the example system. Due to the lack of any large time constants or dead time lags in the process equipment, the time constant of the exponential lag representing the temperature sensor and transmitter will be of the same order of magnitude as other system components. When this happens it may be necessary to make a more detailed analysis of this component.[2] In most industrial applications, however, this will be a reasonable assumption, and a detailed study will not be required. The transfer function is then

$$\Delta P_3/\Delta\Theta_2 = h = k_5/(1 + T_5 s), \tag{4-6}$$

where P_3 = transmitter output signal, psi,
h = sensing element and transmitter transfer function,
k_5 = zero frequency gain or sensitivity, psi/°F,
T_5 = time constant, min.

A value for k_5 may be determined from transmitter input and output ranges. For example, assume an input temperature range of 100°F within limits of 100 to 200°F and an output pressure range of 12 psi for the conventional 3 to 15 psig limits. The value of k_5 is then

$$k_5 = 12\text{ psi}/100°\text{F} = 0.12\text{ psi/°F} \tag{4-7}$$

At this point note that the zero frequency gain or sensitivity of the system is not always 1. In Chapter 3 this point was bypassed, and in Equation (4-1), $k_2 = 1$. When the zero frequency gain or sensitivity term is something other than 1, the template of Chapter 3 is merely raised or lowered on the magnitude scale. As noted before, the shape of the curve is constant. Its position on the frequency spectrum is determined by the time constant, and its position on the magnitude scale by the gain magnitude or sensitivity.

The value of T_5 will be a function of the type of thermal sensing element selected and how it is installed, namely, with or without a well. For the example at hand, T_5 will be assumed to be 3 sec or 0.05 min.

The combined transfer function may now be obtained as a product of the individual component transfer functions.

$$\Delta P_3/\Delta P_1 = g_2 g_3 g_4 h \qquad (4\text{-}8)$$
$$= (\Delta P_2/\Delta P_1)(\Delta \Theta_1/\Delta P_2)(\Delta \Theta_2/\Delta \Theta_1)(\Delta P_3/\Delta \Theta_2)$$

The transfer function in Equation (4-8) will be defined as the process transfer function. By definition the process transfer function will include every component in the control loop except the controller. This causes some confusion with process engineers since they are not in the habit of including valves and measuring components as parts of the process. From a control viewpoint, however, the distinction between "controller" and "process" is an ideal division. If the transfer functions of Equation (4-8) are replaced by their particular values the result is

$$\Delta P_3/\Delta P_1 = \left[1/(1 + 0.1s)\right]\left[10\right] \qquad (4\text{-}9)$$
$$\times \left[e^{-0.15s}\right]\left[0.12/(1 + 0.05s)\right]$$

The problem now is to combine these system components. To do this the frequency response of each component must be plotted. The process transfer function plot may then be obtained by multiplying magnitudes and adding phase angles.

4.2. Vector Multiplication

Development of the process transfer function requires a technique for vector multiplication which allows individual component responses to be combined. To demonstrate how this technique works on a Bode plot, assume two transfer functions as shown in Figure 4-4. To combine these two transfer functions requires multiplying magnitudes and adding phase angles. One method of multiplication uses logarithms. In this method the product of x and y is equal to $\log x + \log y$. This technique may now be applied to the Bode plot and suggests why the log scale was selected for the magnitude plot. By plotting magnitude on log-coordinates, the log of the gain or sensitivity is automatically taken and represented by a distance on the plot. Since the logarithm of 1 is zero, the distance in either direction from a magnitude or gain of 1 represents the number given on the ordinate scale. Thus, to obtain the product of two numbers, the distances are added. This procedure is identical to the slide rule technique for multiplication or division.

To demonstrate, the gain product will be obtained at frequencies of 0.1, 1.0 and 10 cpm using data from Figure 4-4. Figure

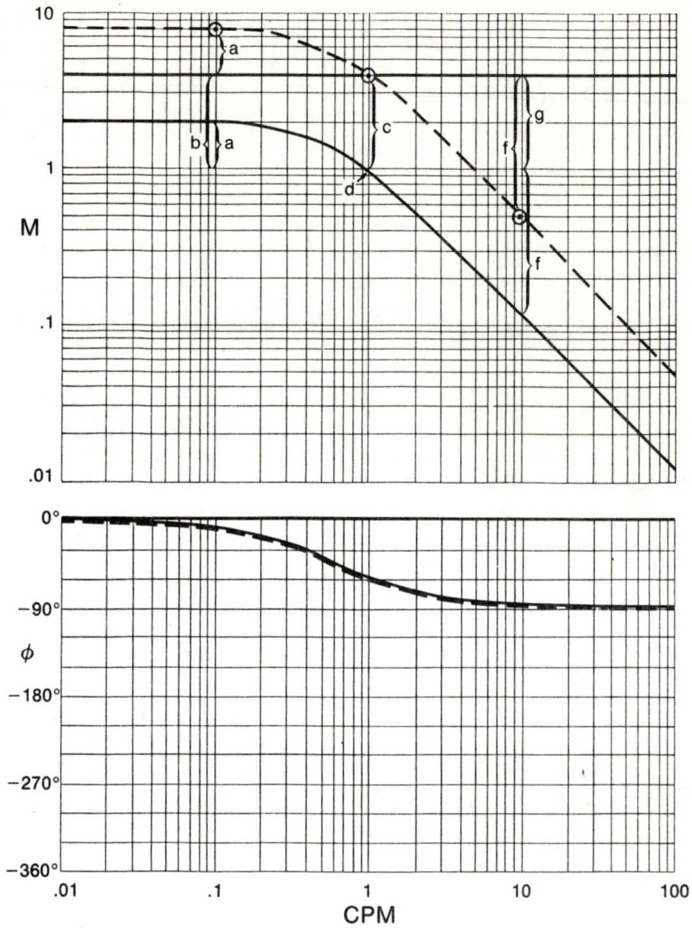

Figure 4-4. A demonstration of vector magnitude multiplication on a Bode plot.

4-4 shows two transfer function plots—one being an exponential lag, the other a constant gain term. At 0.1 cpm, gain magnitude of the individual components are $a = 2$ and $b = 4$. The product of these two vector magnitudes is of course 8. The same result can be obtained as follows:

$$M = a \times b \text{ or } \log M = \log a + \log b \qquad (4\text{-}10)$$

The distance from a magnitude of 1 to the component curve on the Bode plot is actually the logarithm. Thus, the sum of the two distances a and b gives the resultant vector product M.[3]

At 1.0 cpm in Figure 4-4, component magnitudes are $c = 4$ and $d = 1$, giving a product of 4. Since the log 1 = 0, there is no distance to add and the result is simply the distance c. Values at a frequency of 10 cpm are $g = 4$ and $f = 0.125$, giving a product of 0.5. In this case the f value is less than 1, which results in a negative logarithm. The negative sign implies a difference and thus, log f must be subtracted from log g to obtain the result shown in Figure 4-4. From the three examples a pattern is evident: component magnitudes greater than 1 are added; component magnitudes equal to 1 are neglected since 1 serves as the reference point; and component magnitudes less than 1 are subtracted.

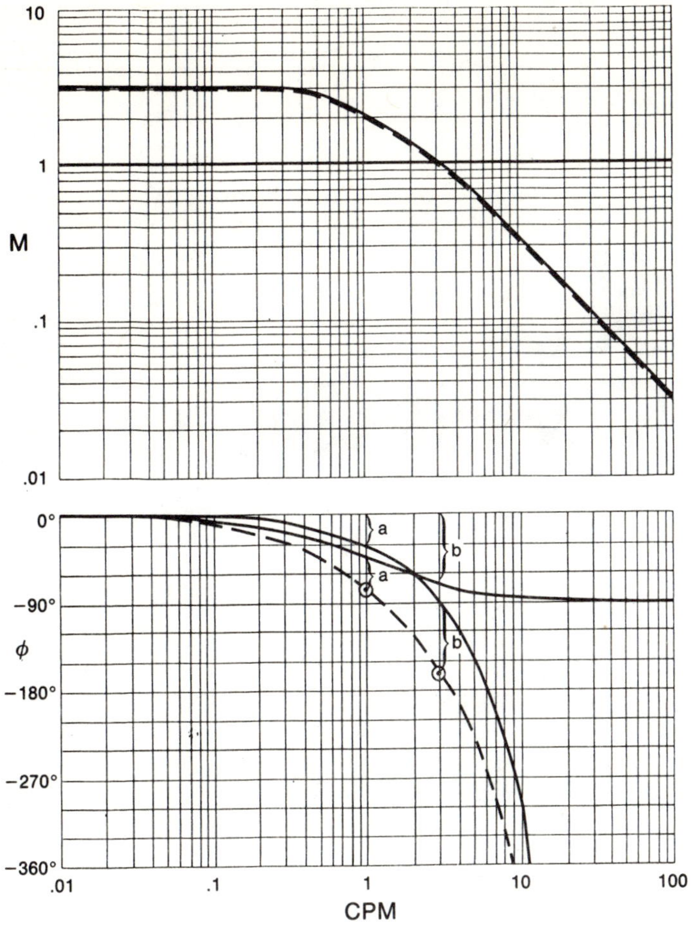

Figure 4-5. A demonstration of vector phase angle summation on a Bode plot.

The resultant phase curve for Figure 4-4 is a trivial case since the phase angle for a constant gain term is zero at all frequencies. In vector multiplication, magnitudes are multiplied and phase angles added. The sum in this case is of course equal to the exponential lag curve. When exponential and dead time lag components are combined in Figure 4-5, the condition is reversed. The magnitude curve of Figure 4-5 is of course trivial, since magnitude of a dead time lag is 1 at all frequencies. As shown, the result is the exponential lag plot. Phase angles may be added as shown by the two examples at 1 and 3 cpm. It is also possible to have a positive phase angle as will be shown later in Chapter 9. When this type of component occurs, the resultant phase lag is reduced by the value of the positive angle at that frequency. As long as sign conventions are adhered to, the reader should not have a problem.

The answers to Problems 5, 6, and 7 demonstrate a step-by-step procedure for combining individual components to obtain the process transfer function. As the student becomes more at home with the technique, the procedure can be completed in one step. As a further test of method and technique, the reader should combine $g_2g_3g_4h$ in one step using the Bode plot in answer to Problems 1 through 4. The result should be the same as shown in the answer to Problem 7.

With the process transfer function now available, the problem becomes one of adding the controller and closing the loop. This procedure, as will be shown in the next chapter, will allow determination of stability and period.

Chapter 4—Study Questions and Problems

(Answers appear on pages 155-156)

1. Using the exponential lag template constructed in Chapter 3, construct the Bode plot for g_2.

2. Plot the transfer function g_3 on the same Bode plot used in the answer to Problem 1.

3. Plot the transfer function g_4 on the same Bode plot used in the answer to Problem 1.

4. Plot the transfer function h on the same Bode plot used for the answer to Problem 1.

5. Using the answers to Problems 1 and 2, combine the transfer functions g_2 and g_3.

6. Using the answers to Problems 3 and 4, combine the transfer functions g_4 and h.

7. Using the answers to Problems 5 and 6, combine the transfer functions g_2g_3 and g_4h.

8. The transfer function $g_2g_3g_4h$, which includes all control loop components except the controller, is referred to as the __________ transfer function.

References

1. Lloyd, S.G., and Anderson, G.D., *Industrial Process Control,* Marshalltown, IA: Fisher Controls Co., 1971.
2. Caldwell, W.I., Coon, G.A., and Zoss, L.M., *Frequency Response for Process Control,* New York: McGraw-Hill, 1959.
3. Zoss, L.M., and Delahooke, B.C., *Theory and Application of Industrial Process Control,* Albany, NY: Delmar, 1961.

5 Basic System Stability

One of the first objectives noted in Chapter 1 was the problem of a stable system. The simple heating system or "process" developed in Chapter 4 will now be used to determine stability limits, period, and proportional controller settings.

5.1. Introduction of the Controller

For purpose of analysis, a sine wave generator, temperature recording controller, and three additional recorders will be connected to the "process" as shown in Figure 5-1. The transfer function for the process $\Delta P_3/\Delta P_1$ is equal to $g_2g_3g_4h$ as developed in Chapter 4. The controller TRC-1 (temperature recording controller 1) is a simple proportional controller whose transfer function is

$$\Delta P_4/\Delta P_3 = g_1 = k_1, \tag{5-1}$$

where P_4 = controller output signal, psi,
g_1 = controller transfer function,
k_1 = proportional controller gain, psi/psi.

When valve A is closed and valve B is open, note that the controller output P_4 is actually the "process" input P_1. Although a more detailed analysis of the controller will be provided in Chapter 9 on controllers, it is necessary at this time to note two basic characteristics of controllers.

Once a controller is selected for a process, the "action" of the controller must be determined. The term action refers to response of the controller output as a function of input signal. Using Equation (5-1) as an example, it is easy to see that if the input signal increases ΔP_3, then the output response ΔP_4 must also increase. This result is known as direct action and may be stated as follows—an increase or decrease in the measured variable causes a respective increase or decrease in controller output. The ratio of change for a proportional controller is of course proportional to the controller gain. If the action of the controller is reversed, output will decrease as the input signal increases. Changing the action of a controller is a simple field procedure and is usually either specified on ordering or checked by the installer at start-up. To see why this arrangement is necessary requires an analysis of the process in Figure 5-1.

Safety is one of the main considerations in industrial design and includes both personnel and equipment protection. For the process shown in Figure 5-1 the question that needs answering concerns a failure of air supply to the control valve, and the resulting action of steam flow. If on loss of air supply pressure the control valve opens, then maximum steam flow will enter the product line. If under such conditions a potential hazard to personnel, equipment, or product exists, then an air-to-open valve must be selected. With an air-to-open valve, a loss of air pressure will close the valve and cut-off steam flow. If on the other hand maximum steam flow causes no potential hazard, but a cooling of product will result in equipment or product damage, then an air-to-close valve should be selected. In this case air failure will allow the valve to open, keeping equipment and product heated. Having now established a basis for selecting valve action, an air-to-open valve will be assumed for the process of Figure 5-1.

If the signal to the valve is increased, namely P_1, steam flow will increase. An increase in steam flow will cause an increase in product temperature. An increase in product temperature will cause an increase in transmitter output P_3. The controller of course is placed in the control loop to provide corrective action; thus when P_3 increases, it must take action to reduce P_3 to the desired value. To reduce P_3 it must reduce steam flow. It produces this objective by reducing its output P_4, which under closed loop conditions would be connected to the valve. As the valve pressure is reduced, steam flow is reduced and the system is brought under control. Note that in this case a reverse-acting controller is required to provide corrective action. It is interest-

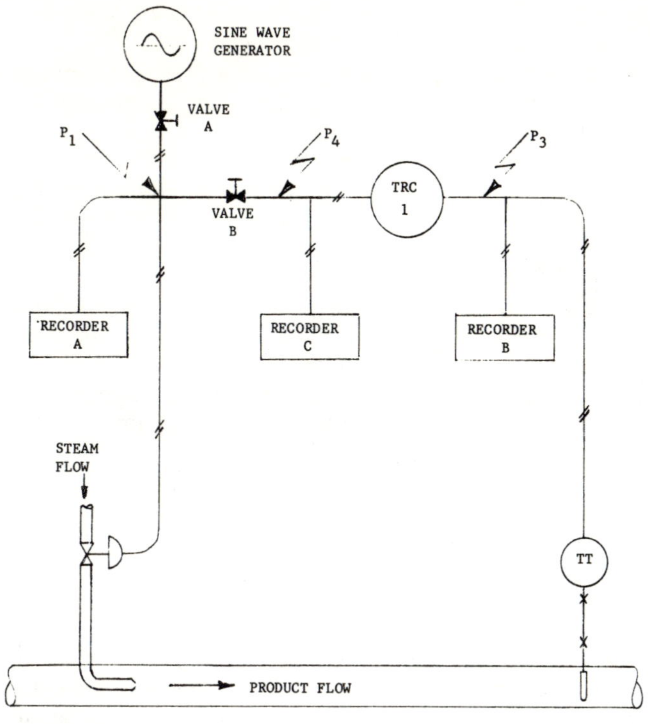

Figure 5-1. Special process and controller combination for stability analysis.

ing to note that all other components in the loop have what may be considered as direct action.

With an air-to-close valve, an increase in P_1 causes a decrease in steam flow, product temperature, and the measured variable P_3. The controller must now decrease its output P_4 to provide corrective action. In this case direct controller action is required. Under these conditions, the equivalent of reverse action is found in the control valve.

The important concept is that corrective action must exist in every control loop if it is to provide control. As shown by the examples, this action can exist in the controller or valve. It can also exist in process equipment, and there can be a combination of actions in one loop. A refrigeration unit operating with an air-to-close valve is a typical example. If valve pressure increases, refrigerant flow decreases. With reduced refrigerant flow, product temperature increases and the measured variable increases. Under these conditions the controller must decrease its output signal to bring the system under control. In this example the valve, refrigeration equipment, and controller all have the equivalent of reverse action.

If an air-to-open valve is used, an increase in valve pressure causes an increase in refrigerant flow and a decrease in product temperature and the measured variable. In this case the controller must decrease its output. Here the controller and valve have direct action, and the process is reverse-acting.

The examples show that corrective action must exist in a control loop if the loop is to work. If it is assumed that direct action is positive and reverse action is negative, all four examples show a net negative result. In both of the heating examples in Figure 5-1 and the refrigeration example with an air-to-open valve, only one reverse-acting component exists. In the refrigeration example with the air-to-close valve, three components have reverse action. Since transfer function values must be multiplied, a net negative action results.

In the process of analysis, action of any particular component is often neglected, or more often assumed positive. The assumed corrective action is introduced by negative feedback and shows up as the negative sign on the measured variable as noted in Equation (1-1) and in Figure 1-10. This is a valid procedure, since if the loop is to control, this result must be obtained. It also accounts for the terminology of "negative feedback." The effect of corrective action is to produce a phase reversal of 180° in the control loop. Physically this may be explained by noting that as the controlled variable increases, some action must be taken in the opposite direction (180° out of phase) to provide a correction. The application of this concept will be used in the stability analysis of the process shown in Figure 5-1.

The second controller characteristic which requires discussion at this time is the frequency response. The proportional controller as represented by Equation (5-1) shows a gain term k_1, that for any set value is constant at all frequencies. This is an ideal situation, since all physical components will have some limiting value on their ability to respond to change. This fact is well known by hi-fi enthusiasts who are conscious of fidelity reproduction of low and high frequencies. In process control applications, however, the assumption of a constant gain term is usually valid since most controllers will have a cut-off frequency well beyond the area of process interest. Thus, as shown in Figure 5-2, the controller can be assumed to have a constant gain and phase angle or "flat" characteristics in the frequency spectrum of interest.[1]

With the introduction of the controller to the process, the problem of stability may now be examined.

5.2. Stability Limits

To investigate system stability, the specialized test arrangement of Figure 5-1 will now be used. To start, assume an air-to-

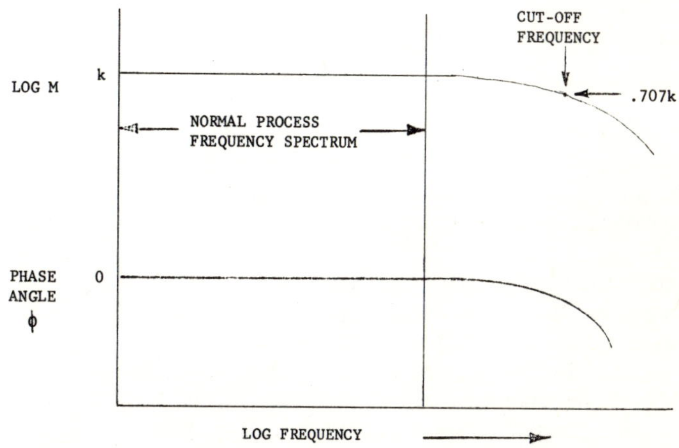

Figure 5-2. Proportional controller response.

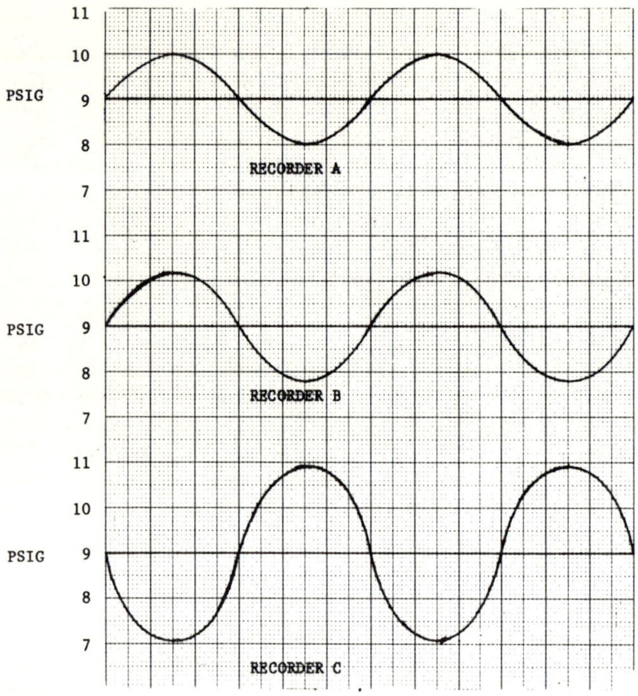

Figure 5-3. Chart records for 0.01 cpm.

open control valve and that valve A is open and B closed. Under these conditions sine wave signals from the sine wave generator may be recorded as they enter the "process" by recorder A, as they leave the "process" and enter the controller by recorder B, and as they leave the controller by recorder C. The actual chart on TRC-1 could also be used in place of recorder B.

To start the analysis assume that a sine wave having an amplitude of 1 psi and a frequency of 0.01 cpm is generated and fed into the "process." The record of recorder A will appear as shown in Figure 5-3. Using the "process" frequency response, $g_2g_3g_4h$ (as shown in the answer to Probelm 7, Chapter 4), it can be seen that at 0.01 cpm the magnitude ratio is 1.2 and the phase angle is zero. Thus, at this frequency the sine wave in traveling through the "process" should be amplified by 1.2 and remain in phase with the input. This result is shown by the record of recorder B in Figure 5-3. It is now necessary to see what happens as this signal proceeds through the controller.

From the discussion in Section 5.1, it may concluded that with an air-to-open valve the controller must have reverse action to provide control. Thus, the first conclusion is that the controller will have an output that is 180° out of phase with its input or the signal at recorder B. The magnitude of the output will of course depend on the controller gain k_1. Like the weightless string and the frictionless pulley, the controller gain will now be assumed equal to 1.6. The reason for setting $k_1 = 1.6$ will shortly become obvious. With an input amplitude of 1.2 psi and a gain of 1.6, the controller output amplitude will be 1.92 psi. The resulting output is shown by the record of recorder C in Figure 5-3. The bias signal of 9 psig is selected only because it is the midpoint of the standard commercial instrument range of 3 to 15 psig. Note that a similar analysis could be made with electrical signals using a conventional 4-20 ma range.

While it may not be immediately obvious, the reader should remember that we are interested in determining process stability limits. Thus, although Figure 5-3 does not prove anything, it was a necessary step to this end. The next step is to readjust the sine wave generator to a frequency of 0.84 cpm. Referring to the answer to Problem 7, Chapter 4, shows that at 0.84 cpm the "process" has a magnitude of approximately 1.05 and a phase angle of $-90°$. The choice of 0.84 cpm results from the fact that at this frequency the phase angle is $-90°$. This has no technical significance; it merely notes the author's reluctance to draw sine waves at intervals other than multiples of 90°.

If a sine wave of 1 psi amplitude and 0.84 cpm frequency is now fed to the process as shown by recorder A in Figure 5-4, the resulting "process" output will be as shown by recorder B in Figure 5-4. The amplitude will be 1.05 psi and will lag the input by 90°. As in Figure 5-3 at 0.01 cpm, the controller will again amplify its input by 1.6 and produce a phase reversal of 180°. This result is shown by the record of recorder C in Figure 5-4. The amplitude is 1.05 times 1.6 or 1.68 psi.

One more example is required to determine the stability limits. The sine wave generator is now adjusted to give a sine wave having an amplitude of 1 psi at a frequency of 2 cpm. Again, from the answer to Problem 7, Chapter 4, the "process" magnitude is read as approximately 0.625 and the phase angle as $-180°$. The implied accuracy is beyond graphical techniques presented and is not required in practical applications. It is used here only to establish procedure. With these conditions, records of recorders A and B will appear as shown in Figure 5-5. The amplitude as shown on recorder B will be 0.625 psi, and the record will be 180° out of phase with the input signal. The controller will again amplify its input by 1.6 and produce a

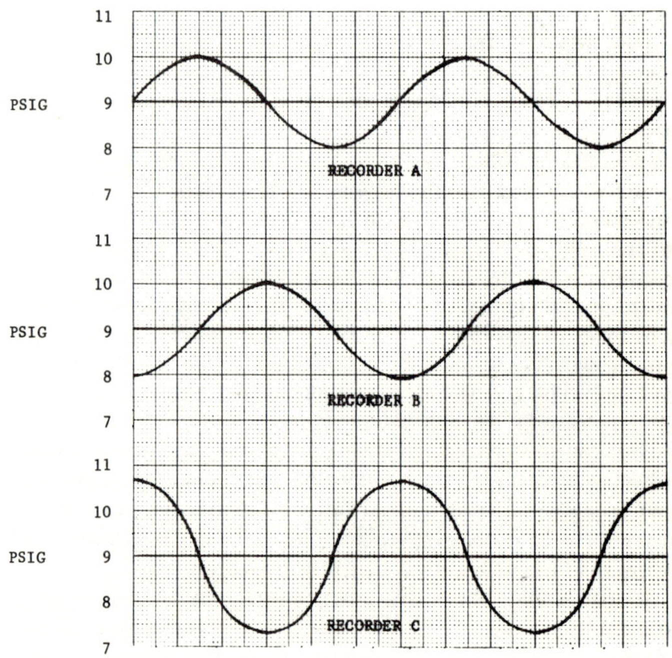

Figure 5-4. Chart records for 0.84 cpm.

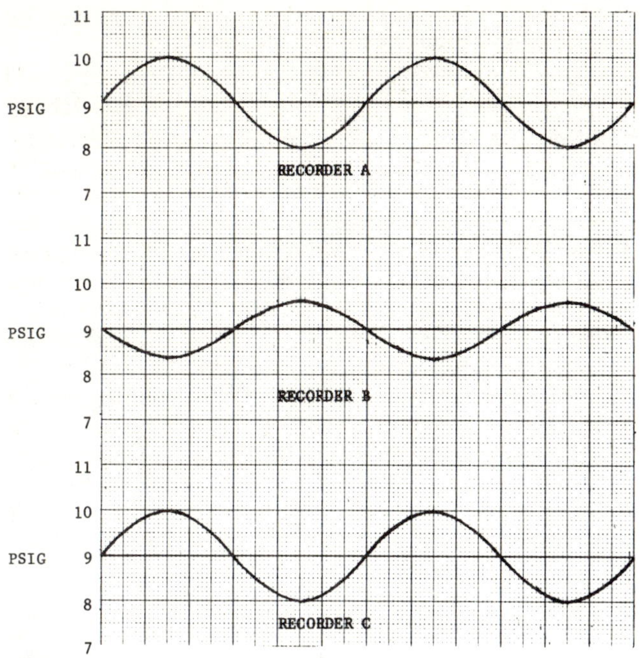

Figure 5-5. Chart records for 2.0 cpm.

phase reversal of 180°. The controller output amplitude is then 1.6 times 0.625, or 1.0. The result with the 180° phase reversal is as shown by the record of recorder C in Figure 5-5. The result is that the signal at recorder C is identical to the sine wave generator as shown on recorder A. If under these conditions valve A was instantly closed and valve B instantly opened, the system would continuously cycle. This condition of continuous cycling is of course the limiting condition between a stable, decaying system as shown in Figure 1-12c and an unstable sytem shown in Figure 1-12d. The problem now is to define "process" and controller parameters that cause this condition.

Under original test conditions having valve A open and valve B closed, the control loop is open. By defining all components of the control loop except the controller as the "process," it is now possible to define the open loop as the product of "process" and controller transfer functions. By similar analysis, it is easy to see that when valve A is closed and valve B is open, the control loop is closed. In the example of Figure 5-5 continuous cycling resulted when the loop was closed. To see why this happened and to establish limits, it is necessary to examine the open loop. The example of Figure 5-5 shows an open loop gain of 1 and a phase angle of −360° prior to reversing valves A and B and changing from an open to closed loop condition. Since every closed loop system must have the −180° of corrective action to work, this 180° lag is normally assumed to be present and is neglected. Stability limits are then stated as follows—when the open loop gain and phase angle equal 1 and −180°, respectively, the closed loop will cycle continuously.

Remembering that the open loop is the product of "process" and controller transfer functions provides the clue to how the controller gain of 1.6 was selected. Knowing that the proportional controller would not add any phase lag, the controller gain which will cause cycling may be found by referring to the process transfer function. Again, referring to the answer to Problem 7, Chapter 4, a phase lag of 180° is noted at 2.0 cpm. Since the controller contribution to phase lag is zero, the open loop phase angle in this case is equal to the "process" phase angle. Thus, if the product of "process" and controller gains is equal to 1 at this frequency and the loop is closed, the system will cycle continuously. This may be expressed as follows:

$$P_G \times G_U = 1, \tag{5-2}$$

where P_G = "process" gain or magnitude at −180°,
G_U = ultimate controller gain or proportional controller gain that will cause a closed loop system to cycle continuously.

The value of $P_G = 0.625$ may be read from the process frequency response plotted in Problem 7, Chapter 4. By using this value and Equation (5-2), the controller gain was selected.

$$G_U = 1/P_G = 1/0.625 = 1.6 = k_1 \tag{5-3}$$

Had the author selected any value other than $k_1 = 1.6$, the conditions of Figure 5-5 would not have existed. As will be shown in Section 5.3, a value of k_1 less than 1.6 would cause a decaying oscillation, while a k_1 greater than 1.6 would result in an increasing oscillation.

5.3. Ultimate Period and Controller Adjustment

When the open loop of Figure 5-5 is closed, the system cycles continuously. This statement, while true, only tells part of the story. The cycle as shown in Figure 5-5 will exist at 2 cpm or a period of 0.5 min per cycle. Like most physical systems, the closed loop control system exhibits a natural frequency, which in the present example is 2 cpm. The reciprocal of the natural frequency is termed the ultimate period and in this example is 0.5 min. The significance of ultimate period and the effect of controller gain may be demonstrated by using the spring-mass analogy. It should be noted that the spring-mass system is being used only as an analog. Position response records of a spring-mass system are similar to control reponses and, as such, prove helpful at this point.

If the spring-mass system of Figure 5-6 is disturbed by lifting the handle attached to the spring, the mass will respond in an oscillatory motion. A time response of this motion will appear as shown in Figure 5-7. The slow decay of this oscillation can be attributed to the damping of air currents, which produce a frictional resistance proportional to velocity. If this same test were conducted in a vacuum, the oscillation would be theoretically continuous. The frequency of this oscillation would be the natural frequency of the system. This frequency is of course some function of the spring force or spring constant and mass magnitude. In fact, this frequency can only be changed by changing either the spring constant or mass magnitude. The point to note here is that the same problem exists in a control system. Due to the characteristics of its components, each control loop has a

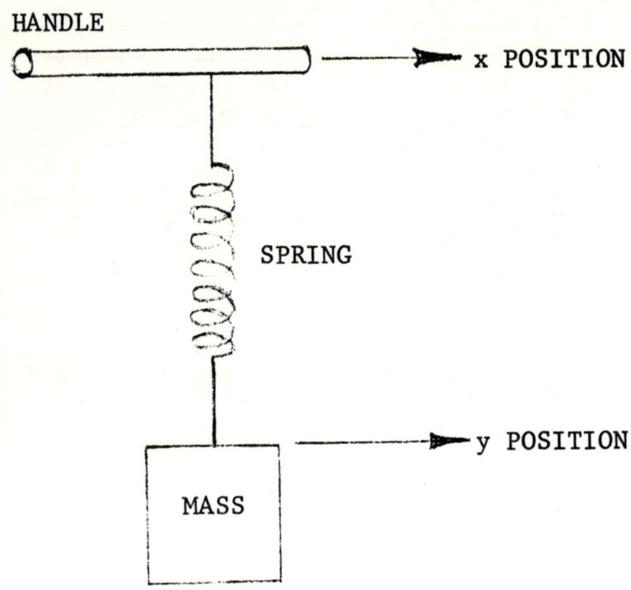

Figure 5-6. Spring-mass system.

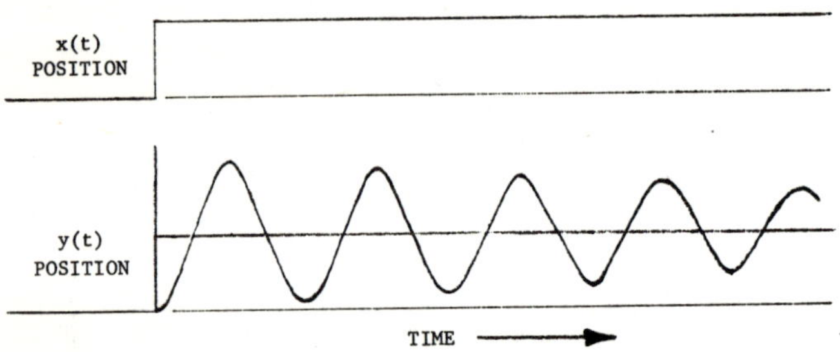

Figure 5-7. Time record of mass position in response to a step change in handle position.

natural frequency. To alter this frequency and change period in an effort to reduce waste product as noted in Figure 1-14, the characteristic of some component in the loop must be changed. This point will be considered in more detail in Chapter 12. The important concept at this time is that this natural frequency is equal to the frequency at which the "process" phase curve (obtained in the answer to Problem 7, Chapter 4) shows a phase lag of 180° and can be predicted if "process" transfer functions are known.

If a dashpot similar to a door closer is added to the spring-mass system of Figure 5-6, damping will be greatly increased over the simple air resistance previously noted. In fact, a very slow response of the mass position will result with no overpeak of the new position. In other words the door will not slam, but gradually close. The same result can be achieved in a control system as shown in Figure 1-12b. In the control system, however, damping is affected by changing controller gain. If, as noted previously, the controller gain in Figure 5-5 had been 1.5 instead of 1.6, a decaying exponential would have resulted similar to the response shown on Figure 5-7. A well-known rule of thumb to achieve good control or a quarter decay ratio as noted in Figure 1-13 is to adjust the gain of a proportional controller to 1/2 of the ultimate gain.[1] This is often referred to as the optimum gain:

$$G_O = G_U/2, \tag{5-4}$$

where G_O = optimum proportional controller gain.

This technique is known as the Ziegler-Nichols method for controller adjustment and will be used in the remaining portions of this text. As a rule of thumb, it will not always provide control with exactly quarter decay. It will in general, however, provide a good stable, control response that can be modified if necessary to provide a desired reponse.

As gain of the controller is reduced to achieve a decaying oscillation, it should also be noted that the frequency of the oscillation will decrease. This in turn increases the period and overall recovery time to set point or load disturbances. If controller gain is reduced to a small fraction of ultimate gain, the overdamped record of Figure 1-12b results and no overpeak is experienced. Although elimination of overpeak may look good to the beginner, the time-error or waste product record of this type of response can be equally as bad as the waste shown in Figure 1-12c, which shows an approximate quarter decay response. Because of this and the resulting offset to set point and load disturbances, which will be discussed in Chapters 8 and 9, the higher gain which gives quarter decay is the desired or so-called optimum adjustment. The increase in period is more difficult to define and this will also be discussed in Chapters 6 and 7. The basic idea to remember is that this period can never be less than the ultimate period.

By combining the "process" and controller it has been possible to establish stability limits and achieve one of the first control requirements, namely a technique of controller adjustment to achieve approximately quarter decay. A basic concept of system period has also been obtained, which shows the ultimate period as the lower limit. To establish the actual period and the control requirement of overpeak, it is now necessary to examine the closed loop in more detail in Chapter 6.

Chapter 5—Study Questions and Problems

(Answers appear on pages 156-157)

1. If controller output increases in response to an increase in the measured variable, this is a ________acting controller.

2. If an air-to-open control valve is used to control product flow in a pipe line, the controller action must be ________.

3. In a cooling process with an air-to-open valve, controller action must be ________.

4. In order to provide corrective action in the control loop, would there be an odd or even multiple of reverse-acting components?

5. The effect of corrective action is to produce a phase reversal of ________ degrees in the control loop.

6. The assumption or acceptance of a proportional controller implies that both gain and phase angle are "flat" or ________ within the frequency spectrum of the process.

7. The open loop is defined as the product of ________ and ________ transfer functions.

8. A closed loop system will cycle continuously if the open loop gain and phase angle are ________ and ________ degrees, respectively.

9. In defining stability limits the 180° phase lag of corrective action is ________.

10. In determining stability limits for a system the phase angle of a proportional controller is assumed to be ________. With this assumption the phase angle of the open loop is equal to the phase angle of the ________.

11. The ultimate controller gain or proportional controller gain that will cause cylcing is equal to one divided by the process gain or magnitude at ________ degrees.

12. If a "process" frequency response shows a gain or magnitude of 4 at a phase angle of −180°, what is the ultimate controller gain?

13. A system cycling continuously is said to be operating at its ________ frequency.

14. If a system has a natural frequency of 2.5 cpm, what is its ultimate period?

15. The natural frequency of a system controlled by a proportional controller can be determined from the ________ frequency response and is equal to the frequency at a phase lag of ________ degrees.

16. Reducing the ultimate controller gain by a factor of 2 should provide a closed loop control record having a ________ decay ratio.

17. As controller gain is reduced from its ultimate value, will the system period increase or decrease?

18. According to calculations, the ultimate controller gain for a particular system is 3.6. If the Ziegler-Nichols method for adjusting a proportional controller is used what will be the optimum gain value?

References

1. Caldwell, W.I., Coon, G.A., and Zoss, L.M., *Frequency Response for Process Control,* Chapter 2, New York: McGraw-Hill, 1959.
2. Ziegler, J.G., and Nichols, N.B., "Optimum Settings for Automatic Controllers," *Trans A.S.M.E.*, Vol. 64, 1942.

6 The Closed Loop Analog

In Chapter 5, a similarity in response was noted between a closed loop control system and simple spring-mass system. To achieve control objectives of period and overpeak, it is necessary to expand on this similarity and to examine both systems in detail. This chapter will consider the spring-mass system. Chapter 7 will deal with the closed loop control system, taking advantage of noted analogies to predict period and overpeak.

6.1. The Spring-Mass Equation

By application of Newton's law (force is proportional to mass times acceleration), it is possible to derive a second-order differential equation for the spring-mass system with damping. Using the system shown in Figure 6-1, the forces involved may be defined as follows:

- *Input force*—an input force is applied when the handle of the spring-mass system is moved. This force is a function of time, depending on how the handle is moved, and may be shown as $F(t)$. To establish some convention, assume that forces acting up are positive and down, negative. Thus, if the handle is quickly raised 1 inch, a positive force in the form of a step function is applied.
- *Spring force*—as the handle is moved, the spring is stretched. Actually the spring resists this motion and in so doing produces a downward or negative force. This spring force is proportional to position. The proportional constant is known as the spring constant and has units of pounds force per inch. This force may be expressed as $-ky$.
- *Damping force*—the dashpot resists any motion. When the motion is in an upward direction, this produces a negative force. This force, however, is proportional to the rate of change of position, or velocity. This effect can be easily verified by trying to rapidly close a door equipped with a plunger-type closer. The proportional constant in this case is the viscous damping factor $\bar{c}$, which has units of pounds force per inch per second. This force may be expressed as $-\bar{c}(dy/dt)$, where dy/dt is the rate of change of position with time, or velocity.

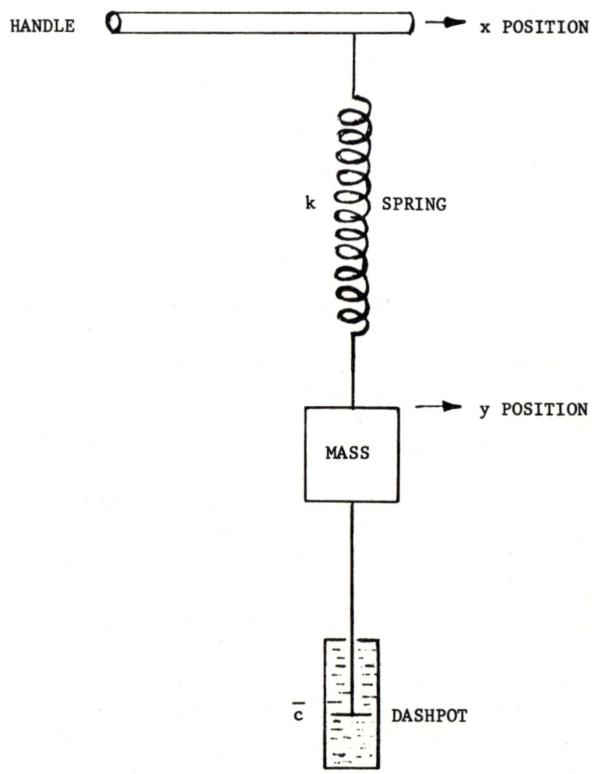

Figure 6-1. Spring-mass system with damping.

According to Newton's law, the product of mass times acceleration is equal to the sum of these forces, given the proper units. The mass is the object weight W divided by the acceleration constant[1] g_c, or $M = W/g_c$. Acceleration is the rate of change of velocity with time or dv/dt. However, since velocity is the rate of change of position with time, acceleration may be expressed as $[d\,(dy/dt)]/dt$ or d^2y/dt^2, which is the mathematical symbol for a second derivative.

If individual components are now combined according to Newton's law, the result is

$$(W/g_c)\,(d^2y/dt^2) = F\,(t) - ky - \bar{c}\,(dy/dt) \qquad (6\text{-}1)$$

To accomplish our control objectives, Equation (6-1) is rearranged as follows:

$$[F(t)]/k = x = [W/(g_c k)]\,(d^2y/dt^2) + (\bar{c}/k)\,(dy/dt) + y \qquad (6\text{-}2)$$

In Equation (6-2), the input force divided by the spring constant has been replaced by the equivalent handle position in Figure 6-1. In this form, a handle position change x becomes the input, and y the mass position change becomes the output. Using Laplace transform concepts of Chapter 3 and zero initial conditions, Equation (6-2) can be reduced to a transfer function relating output to input.

$$x = [W/(g_c k)]\,s^2y + (\bar{c}/k)\,sy + y \qquad (6\text{-}3)$$

Note that the acceleration or second-order term d^2 __$/dt^2$ is simply replaced by an s^2. Using the transfer function concept and factoring out y gives

$$y/x = 1/\{[W/(g_c k)]\,s^2 + (\bar{c}/k)\,s + 1\} \qquad (6\text{-}4)$$

Equation (6-4) is very similar in form to Equation (3-28). Here again, as time approaches infinity, s approaches zero and a steady-state solution results. Thus, if the handle or x position is raised 1 inch, and sufficient time is allowed for decay of the transient, the mass or y position will be raised 1 inch. The denominator or characteristic equation as in the case of Equation (3-28) describes the transient response and will be used to shed light on period and overpeak problems. The first step will be to examine roots of the characteristic equation.

Since the denominator or characteristic equation of Equation (6-4) is second-order, it will have two roots. Because this discussion is being conducted in the general form and specific values are not available for W, k, and $\bar{c}$, the only practical solution for the roots is the quadratic factor of Equation (3-20). By setting the characteristic equation equal to zero and comparing it with Equation (3-21), values of $a, b,$ and c may be obtained.

$$[W/(g_c\,k)]\,s^2 + (\bar{c}/k)\,s + 1 = 0 \qquad (6\text{-}5)$$

$$ax^2 + bx + c = 0 \qquad (3\text{-}21)$$

Then

$$a = W/(g_c k);\; b = \bar{c}/k;\; c = 1$$

Substituting these values in Equation (3-20) gives

$$s_{1,2} = \{-(\bar{c}/k) \pm ([\bar{c}^2/k^2] - [4W/(g_c k)])^{1/2}\}/[2W/(g_c k)] \qquad (6\text{-}6)$$

Combining terms reduces Equation (6-6) to

$$s_{1,2} = -[\bar{c}g_c/(2W)] \pm [g_c/(2W)]\,(\bar{c}^2 - [4Wk/g_c])^{1/2} \qquad (6\text{-}7)$$

The radical term in Equation (6-7) provides a means for explaining basic control objectives. There are three possible forms of the radical, each of which can be explained by mathematics and physical interpretation:

Case 1: Assume that $\bar{c}^2$ is greater than $(4Wk)/g_c$. Mathematically this results in two, real, unequal, and negative roots. A plot of these roots noted as s_1 and s_2 is shown in Figure 6-2. Physically for $\bar{c}^2$ to be greater than $(4Wk)/g_c$, a large, viscous damping must exist. The net result from a physical viewpoint is that the dashpot resists any mass or y-position response to a handle or x-position input change. As in the case of a plunger-type door closer, however, the piston slowly changes position, allowing the mass to respond to handle movement. In response to a step change in handle position, the mass movement is as shown in Figure 6-3. This is commonly known as an overdamped response and will occur whenever dominant roots of the characteristic equation are real, unequal, and negative.

Case 2: Assume that $\bar{c}^2$ equals $(4Wk)/g_c$. The mathematical result is that both roots are real, equal, and negative. Physically the viscous damping has been reduced to a very specific or

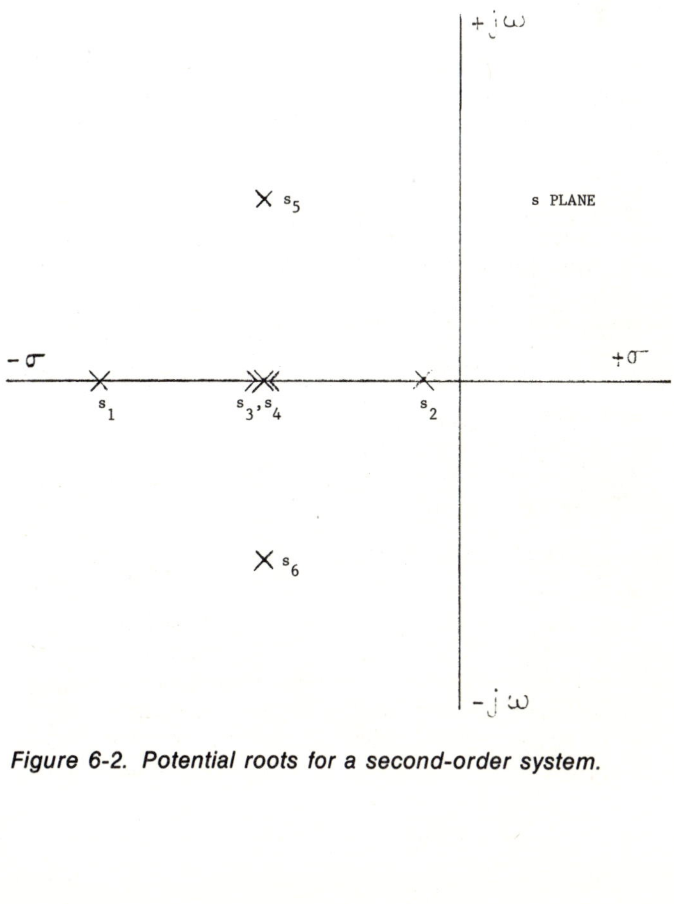

Figure 6-2. Potential roots for a second-order system.

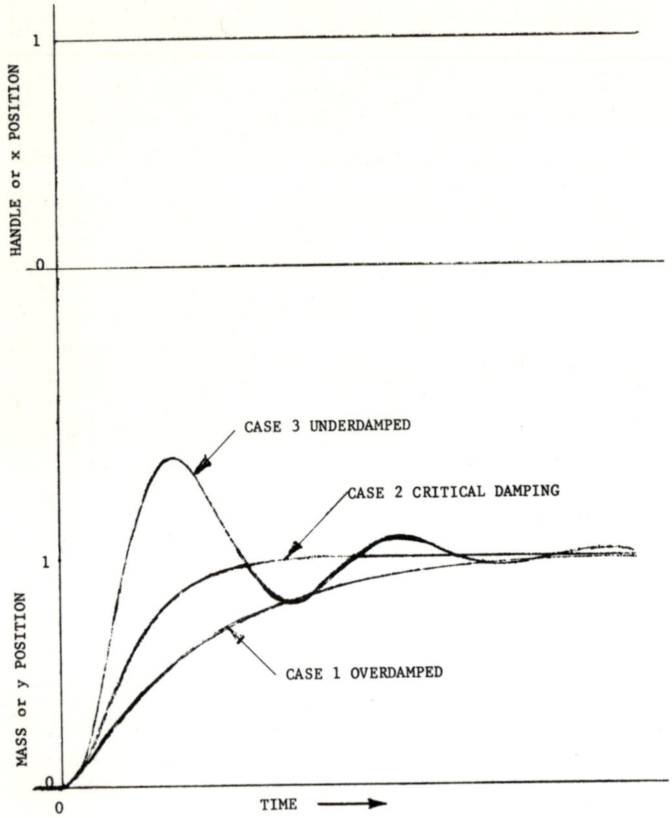

Figure 6-3. Typical spring-mass responses with damping.

"critical" value. The response of the mass to change in handle position is faster than in the overdamped case but is still limited to a change equal to the handle motion as shown in Figure 6-3. From a physical viewpoint a specific explanation of this condition is difficult if not impossible. It is however, easy to see that by experimentation with plungers and fluids it would be possible to get exactly this condition of minimum rise time and no overpeak. This would of course take some care and require critical conditions, as previously noted. As a result, this condition is commonly referred to as critical damping. The mathematical interpretation is noted in Figure 6-2 by roots s_3 and s_4. Since these roots are still on the real axis, they will not include oscillatory terms as previously noted in Figure 3-9 and, therefore, cannot produce an overpeak.

Case 3: Assume that $\bar{c}^2$ is less than $(4Wk)/g_c$. The mathematical result is a negative radical. Here again, application of complex numbers discussed in Chapter 3 solves the problem. To show technique, a simple exercise in solving square roots will be used. Assume that the square root of 256 is desired:

$$\sqrt{256} = \sqrt{(4)(64)} = \sqrt{4} \times \sqrt{64} = 2 \cdot 8 = 16 \qquad (6\text{-}8)$$

The result is trivial and is used only to show technique. If this same technique is now applied to Case 3 with the negative radical, the result is

$$\begin{aligned}(\bar{c}^2 - [4Wk/g_c])^{1/2} &\qquad (6\text{-}9)\\ &= [(-1)([4Wk/g_c] - \bar{c}^2)]^{1/2}\\ &= (-1)^{1/2}([4Wk/g_c] - \bar{c}^2)^{1/2}\\ &= j([4Wk/g_c] - \bar{c}^2)^{1/2}\end{aligned}$$

By substituting Equation (6-9) into (6-7), roots for Case 3 are

$$s_5, s_6 = -[\bar{c}g_c/(2W)] \pm j[g_c/(2W)]([4Wk/g_c] - \bar{c}^2)^{1/2} \qquad (6\text{-}10)$$

These roots form a complex pair as shown in Figure 6-2. Physically, viscous damping has been reduced to a very small value. When this happens, the mass will overpeak the initial position change and respond in an oscillatory manner. This type of response is also shown in Figure 6-3 and is known as an underdamped response.

The responses of Figure 6-3 are similar to the control responses of Figure 1-12b and c. The problem is to find the analogy between the spring-mass and closed loop control systems that will allow prediction of period and overpeak for control problems.

6.2. The General Equation, Frequency, and Time Solutions

To achieve control objectives of period and overpeak, it is convenient to express the denominator or characteristic part of Equation (6-4) into a more general form that can be related to closed loop control systems. This form is

$$\text{output/input} = y/x = 1/(T_n^2 s^2 + 2\zeta T_n s + 1) \qquad (6\text{-}11)$$

Comparing Equation (6-4) with Equation (6-11) demonstrates that

$$T_n^2 = W/(g_c k) \text{ or } T_n = (W/[g_c k])^{1/2} \qquad (6\text{-}12)$$

and

$$2\zeta T_n = \bar{c}/k \text{ or } \zeta = \bar{c}/(4Wk/g_c)^{1/2} \qquad (6\text{-}13)$$

At first glance it must appear as if Equation (6-11) was drawn from thin air, which of course it was not. In Case 3 of Section 6.1, for example, a special condition exists when viscous damping is zero. Using Equation (6-10) and setting $\bar{c}$ equal to zero gives

$$s_5, s_6 = \pm j(g_c k/W)^{1/2}, \qquad (6\text{-}14)$$

or two roots having only imaginary terms. From the mathematics and Figure 3-9, it may be concluded that a continuous cycle will result. It is also obvious from a physical viewpoint that with zero damping the system will oscillate continuously. The frequency of the oscillation is of course the natural frequency of the system as described in Chapter 5. If Equation (6-14) is now set equal to the general root expression of Equation (3-30), the result is

$$\pm \sigma \pm j\omega = \pm j\,(g_c k/W)^{1/2} \tag{6-15}$$

Substituting $2\,\pi f$ for ω and dropping σ, since the real portion of the spring-mass roots is zero, gives

$$j2\pi f = j(g_c k/W)^{1/2}, \tag{6-16}$$

which may be reduced to

$$f = 1/\{2\pi(W/[g_c k])^{1/2}\} \tag{6-17}$$

The frequency is the natural frequency of the system. For the general case of Equation (6-12) the radical is replaced by T_n. Thus, it is evident that T_n was selected to represent physical characteristics which influence the natural frequency of a system. In equation form this reduces to

$$f_n = 1/(2\pi T_n) \tag{6-18}$$

A similar situation results in the selection of ζ, which is defined as the damping ratio. In Case 2 of Section 6.1, it was assumed that $\bar{c}^2$ was equal to $(4\,Wk)/g_c$. This resulted in critical damping. To achieve this result the value of the damping factor must be

$$\bar{c} = [(4Wk)/g_c]^{1/2} \tag{6-19}$$

An examination of Equation (6-13) shows that ζ is in fact the ratio of actual damping to critical damping. Thus, ζ is often described in mathematical terms as

$$\zeta = \bar{c}_a/\bar{c}_c, \tag{6-20}$$

where $\bar{c}_a$ = actual damping factor,
$\bar{c}_c$ = critical damping factor.

If the system is overdamped as in Case 1, ζ will be greater than 1; if critical damping exists, ζ equals 1; and if underdamped, ζ is less than 1. Thus, a general equation has been defined which not only fits the spring-mass system, but may also be related to control systems in general. The next step is to obtain frequency and time solutions for Equation (6-11).

To obtain frequency solutions for Equation (6-11), the procedure of Chapter 3 will be used where $s = j2\pi f$. Making this substitution in Equation (6-11) gives

$$y/x = 1/[T_n^2\,(j2\pi f)^2 + 2\,\zeta T_n\,(j2\pi f) + 1], \tag{6-21}$$

which may be reduced to

$$y/x = 1/\{[1 - (2\pi f T_n)^2] + j4\pi f\,\zeta T_n\} \tag{6-22}$$

Selecting a particular value for T_n such as $1/(2\pi)$ (obviously, the author is lazy) and varying ζ results in a family of curves as shown in Figure 6-4.

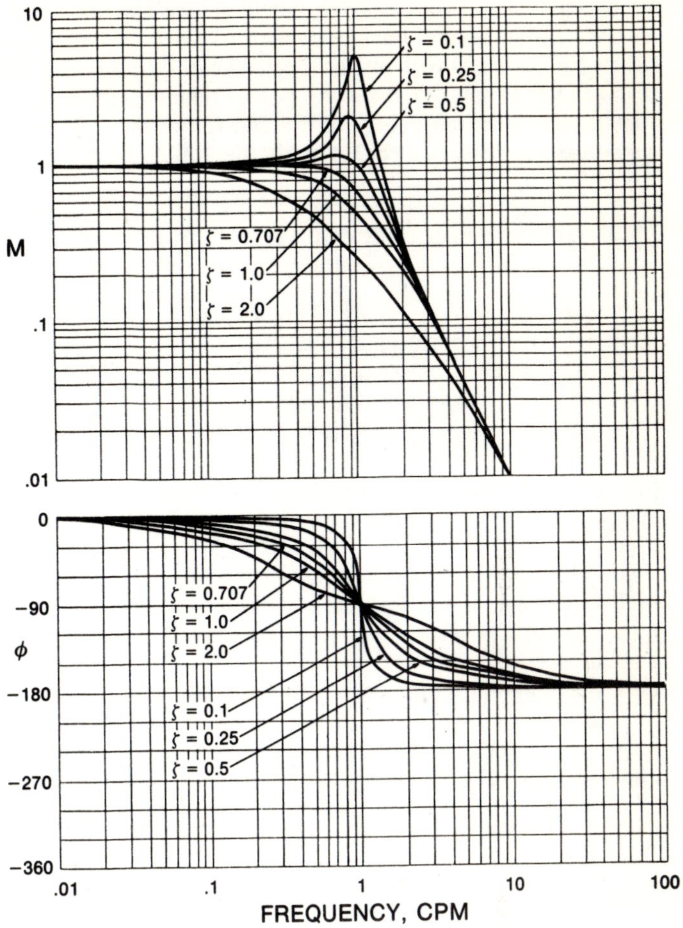

Figure 6-4. Frequency responses for a second-order system, $T_n = 1/(2\pi)$.

The time response curves to a step input are[2]

$$y = 1 - [1/(1 - \zeta^2)^{1/2}]\,e^{-\zeta t/T_n} \times \sin[(1 - \zeta^2)^{1/2}\,(t/T_n) + \psi], \tag{6-23}$$

where $\psi = \tan^{-1}[(1 - \zeta^2)^{1/2}/\zeta]$ and, for $0 < \zeta < 1$,

$$y = 1 - [1 + (t/T_n)]\,e^{-(t/T_n)}, \text{ for } \zeta = 1, \tag{6-24}$$

$$y = 1 - [T_a/(T_a - T_b)]\,e^{-t/T_a} + [T_b/(T_a - T_b)]\,e^{-t/T_b}, \tag{6-25}$$

and for $\zeta > 1$, $T_a > T_b$,

$$T_a = [\zeta + (\zeta^2 - 1)^{1/2}]\,T_n \tag{6-26}$$

$$T_b = [\zeta - (\zeta^2 - 1)^{1/2}]\,T_n \tag{6-27}$$

Equations (6-23), (6-24), and (6-25) are obtained using the inverse Laplace transform for special cases involving the damping ratio ζ. In Equation (6-25) where $\zeta > 1$, roots are factored as shown by Equations (6-26) and (6-27). In all cases these equations represent response to a unit step input as shown in

Figure 6-5. The procedure for obtaining the time-response equations using inverse Laplace transforms is detailed in almost all control texts and will not be repeated here. Emphasis in this text will be to show that there are specific relationships involving T_n and ζ that relate Figures 6-4 and 6-5 or the frequency response of Equation (6-22) with the time responses of Equations (6-23) through (6-25).

6.3. Time and Frequency Correspondence Based on T_n and ζ

Frequency and time curves of Figures 6-4 and 6-5 demonstrate intuitively a correspondence based on T_n and ζ. If this is true, then it would be possible to convert from frequency to time response using only values of T_n and ζ. Such a procedure would solve the remaining control objectives of period and overpeak. The reason for this chapter of course is that such a procedure does exist for a second-order system. For a pure, second-order system this conversion or correspondence is exact. However, as might be expected very few, real control systems are second-order. The saving salvation and only basis for devoting a chapter to this subject is that most actual systems may be approximated by a second-order system. The problem then is to determine the correspondence.

If the magnitude expression of Equation (6-22) is differentiated with respect to frequency and set equal to zero, it is possible to determine the frequency at which maximum magnitude occurs. The frequency for maximum magnitude is defined as the peak frequency, f_p. If the peak frequency is divided by the natural frequency f_n, the resulting ratio is

$$f_p/f_n = (1 - 2\zeta^2)^{1/2} \quad (6\text{-}28)$$

Likewise, the ratio of maximum or peak frequency magnitude to zero frequency or steady-state magnitude may be determined. This result is

$$\text{peak frequency magnitude/zero frequency magnitude} = 1/\left[2\zeta(1 - \zeta^2)^{1/2}\right] \quad (6\text{-}29)$$

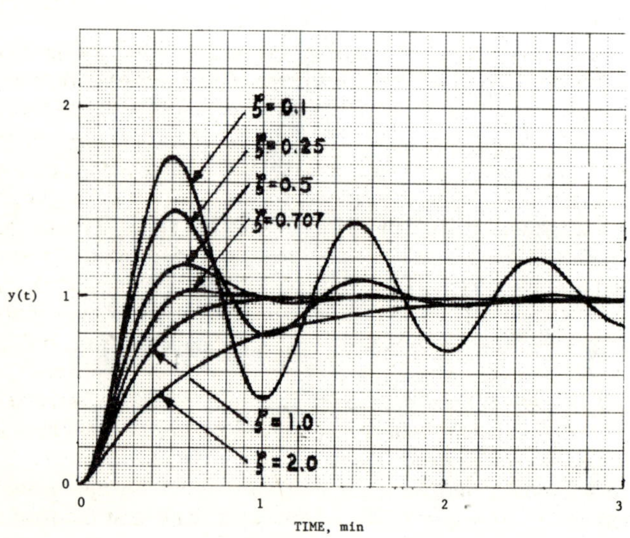

Figure 6-5. Time responses for a second-order system, $T_n = 1/(2\pi)$.

As can be seen Equations (6-28) and (6-29) are valid only when $\zeta < 1$. Since this is the only condition under which an oscillatory response exists with period and overpeak, this does not present any problem.

Using the same technique of differentiation to determine maximum values, it is possible to derive equations similar to Equations (6-28) and (6-29) for the time responses of Figure 6-5 and Equation (6-23). These are as follows:[3]

$$\text{Decay ratio} = \beta/\alpha = \exp - \left[2\pi\zeta/(1 - \zeta^2)^{1/2}\right] \quad (6\text{-}30)$$

$$\text{Time of maximum deviation} = X = 1/\left[2f_n(1 - \zeta^2)^{1/2}\right] \quad (6\text{-}31)$$

$$\text{Maximum deviation/Final value} = Y/Z = 1 + (\text{decay ratio})^{1/2} \quad (6\text{-}32)$$

$$f = (1 - \zeta^2)^{1/2} f_n \quad (6\text{-}33)$$

A plot of Equations (6-30) through (6-33) is shown in Figure 6-6 along with a time-response curve defining specific values. The problem now is to determine how Equations (6-28) through (6-33) and Figure 6-6 may be used to convert from frequency to time response, thus providing answers for overpeak and period.

As an example assume that the frequency response curve of $\zeta = 0.25$ shown in Figure 6-4 is available and that the time response is desired. There are three pieces of information that can be obtained from the frequency response. From Figure 6-4, the following values are obtained:

Zero frequency magnitude = 1
Peak frequency magnitude = 2.06
Peak frequency = 0.94 cpm

Using zero and peak frequency magnitudes and Equation (6-29), it is possible to solve for ζ. Because of the radical in Equation (6-29) this procedure again requires application of the quadratic formula. To avoid this complex procedure Equation (6-29) is usually approximated as follows:

$$\text{peak frequency magnitude/zero frequency magnitude} \cong 1/(2\zeta) \quad (6\text{-}29a)$$

The error in this approximation will be less than 10% for $\zeta \leq 0.435$. In the present example, Equation (6-29a) yields

$$\zeta = \text{zero frequency magnitude}/\left[2\,(\text{peak frequency magnitude})\right] = 1/\left[2\,(2.06)\right] = 0.243 \quad (6\text{-}34)$$

The difference between the known value of 0.25 and calculated value of 0.243 will usually prove to be negligible in actual system studies.

Using Equation (6-28), the calculated value of ζ, and the peak frequency value of 0.94 cpm read from Figure 6-4, it is now possible to calculate the natural frequency. It is obvious that for small values of ζ there is a negligible difference between peak and natural frequency values. Thus, it is possible to assume that $f_n = f_p$ for most plant studies.

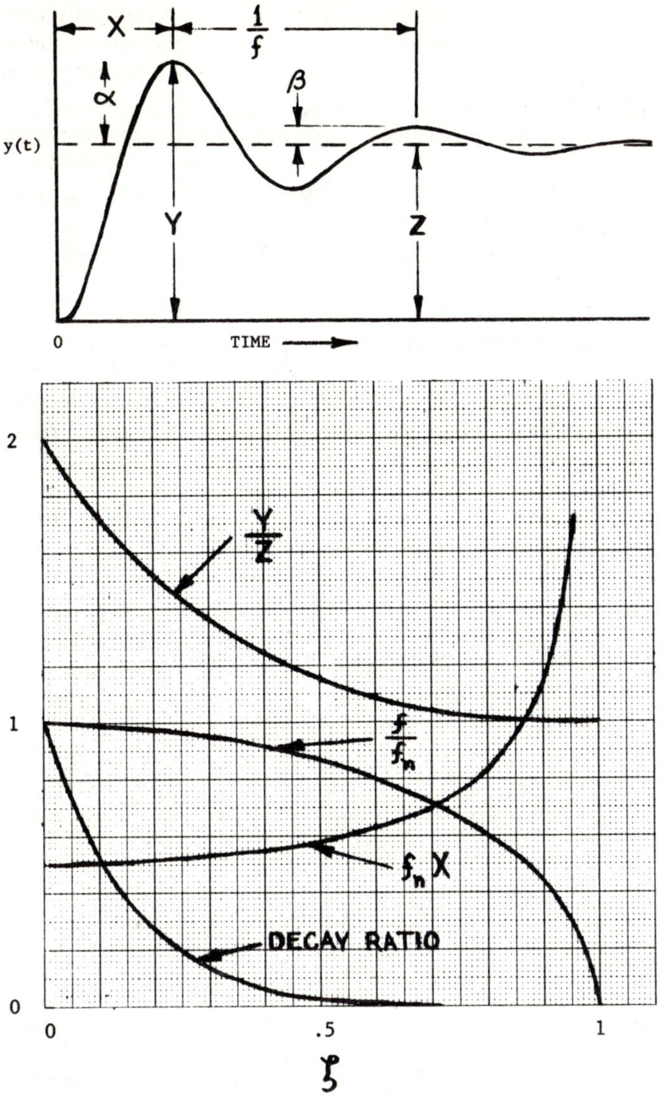

Figure 6-6. Time-response parameters as a function of ζ.

The final value Z may be determined from the zero frequency magnitude. The transfer function is the ratio of output-to-input magnitude. However, as frequency approaches zero, time approaches infinity as noted in Chapter 3. Thus, the final value or steady-state result will be equal to the zero frequency magnitude multiplied by the input change.

$$\text{Final value} = Z = (\text{Zero frequency magnitude}) \times (\text{Input step}) \tag{6-35}$$

In this example both the zero frequency magnitude and input step values are 1, thus the final value is 1. This will not always be the case as will be shown in Chapter 7.

Having obtained ζ, f_n, and Z from the frequency response, it is possible to predict the time response using Figure 6-6. To remove as much mystery as possible from Equations (6-30) through (6-33) and Figure 6-6, a physical interpretation of each curve will be attempted during its use. Using the ζ value of 0.24 and the Y/Z curve of Figure 6-6, the maximum deviation or overpeak may be found. According to Figure 6-6, $Y/Z = 1.45$ for a $\zeta = 0.24$. Since the final value Z equals 1, the time-response curve in Figure 6-5 should have a maximum deviation of 1.45. A check of Figure 6-5 for the curve of $\zeta = 0.25$ confirms this value. Thus, it is now possible to predict overpeak using the frequency response plot and Figure 6-6.

Assuming an input step of unity and remembering that ζ is a function of the damping, it is relatively easy to justify the Y/Z curve of Figure 6-6. With zero damping or $\zeta = 0$, it is an accepted fact that the mass position as shown in Figure 6-1 will not only oscillate continuously but will overpeak the input change. With zero damping no energy is lost, thus, the mass must overpeak by an amount equal to the input change. With unity input change and $\zeta = 0$, the mass position change must therefore be two units, or $Y/Z = 2$. As damping is added to the system, energy is lost in the form of heat, and the overpeak will be reduced. The limit of course is when $\zeta = 1$. In this case the system is critically damped and there is no overpeak. Under this condition, $Y/Z = 1$. Specific values between these limits are not possible using the physical approach; however, the general trend of the curve as defined by Equation (6-32) should now seem reasonable.

The decay ratio defined in Chapter 1 will now be examined. Using Figure 6-6 and a ζ of 0.24 results in a decay ratio of 0.2. By definition, the decay ratio equals β/α. From Figure 6-6 it can be seen that α equals the difference of maximum deviation and final value. In the example under study, $\alpha = 1.45 - 1.0 = 0.45$. The value of β is then $(0.45)(0.2) = 0.09$. To find the magnitude of the second peak, the value of β must be added to the final value. The second peak should then be $(1.00 + 0.09) = 1.09$, which again checks very closely with Figure 6-5.

The physical picture will aid in understanding the curve as developed from Equation (6-30). With zero damping a system cycles continuously. Under this condition each overpeak equals the previous value, and the ratio is 1. As damping is added to the system, the time response will decay. In fact, a point must be reached where a second overpeak no longer exists, and the decay ratio is zero. This physically realizable point can be shown to occur at $\zeta = 0.707$, using Equation (6-30).

Having now determined initial overpeak and decay, the next steps will be to show period and time of initial overpeak. From the curve of f/f_n in Figure 6-6, a value of 0.97 is obtained for a $\zeta = 0.24$. Here again, for small values of damping, the error is negligible, and usually for plant studies, f is assumed to be equal to f_n or f_p. In the present example this results in $f = 0.94$ cpm, or a time-response period of approximately 1.06 min. If Equations (6-28) and (6-33) had been strictly adhered to, the period would be 1.03 min. Since plant data can seldom be obtained to this accuracy, the difference in results is of academic interest only. From Figure 6-5 the first overpeak of the response for $\zeta = 0.25$ appears to occur at 0.5^+ min, while the second is between 1.5 and 1.6 min. The result is a period between 1.0 and 1.1 min. It is evident that assumptions regarding period are well within the ability to read data from charts.

The physical interpretation is again related to damping. With zero damping or $\zeta = 0$, the system will oscillate at the natural frequency, and thus $f/f_n = 1$. As damping is added, the system frequency decreases such that $f/f_n < 1$. As previously noted, a point is reached where the oscillatory response ceases to exist. Under this condition the response appears to have an infinite period. Since frequency is the reciprocal of period, $f = 0$ at this point. Thus, from a physical basis it is possible to see that f/f_n must go from 1 to 0 as ζ goes from 0 to 1.

The last curve in Figure 6-6 relates the time of maximum deviation or first overpeak to the natural frequency of the system. For the example calculation of $\zeta = 0.24$, the value of $f_n X$ from Figure 6-6 is 0.52. Using the earlier assumption that $f_p = f_n = 0.94$ cpm, and the fact that period is equal to the reciprocal of frequency, results in the following:

$$f_n X = f_p X = X/\text{period} = 0.52 \qquad (6\text{-}36)$$

Using the period previously calculated (1.06 min) results in a value of $X = (1.06)(0.52) = 0.55$ min. This checks as close as could be expected with the 0.5^+-min value of the first overpeak as read from Figure 6-5.

The physical rationale of the $f_n X$ curve of Figure 6-6 is again involved with damping. With zero damping ($\zeta = 0$) the system oscillates at the natural frequency. Thus, the first overpeak must occur at 1/2 the period, or $f_n X = 0.5$ as shown. As damping is added and increased, the oscillatory response disappears with the result that no overpeak exists. Another way of visual interpretation is to note that the peak occurs at an infinite time, or $f_n X = \infty$, as ζ approaches 1. This fact is confirmed by Equation (6-31), which has a denominator of zero when $\zeta = 1$.

The possibility of predicting time response from frequency response for the second-order spring-mass analog has been demonstrated. Using the final value, overpeak magnitudes, period, and time for the first overpeak, a typical time-response curve may be drawn. The initial response will inrease rapidly to the first overpeak in the form of a lazy S. After the first overpeak the response is an exponentially decaying sine wave as noted in Figure 6-5 and Equation (6-23). The overpeak values at period intervals define this decay and may also be used as a mirror image below the final value to determine the undershoots. The problem is now to relate this procedure to an actual control system as will be attempted in Chapter 7.

Chapter 6—Study Questions and Problems

(Answers appear on pages 157-158)

1. The objective in analyzing the spring-mass system is to obtain a mathematical and physical ________ of the closed loop control system.

2. Newton's law of motion may be used to describe the spring-mass system with damping and reduced into a ________ that relates mass position response to an input change in handle position.

3. The characteristic equation provides the answer to mass position during the ________ portion of the response.

4. If the roots of the spring-mass system are real, unequal, and negative, what type of response will result? Is this equivalent to having a system in which the damping factor is large or small?

5. For real, equal, and negative roots, will the mass position response show critical or noncritical damping?

6. In the case of complex roots will the mass position response be underdamped or overdamped? For this kind of response to occur, must the damping factor be small or large?

7. The value T_n is related to the natural frequency of a second-order system by the equation $f_n =$ ________.

8. The damping ratio ζ is defined on the basis of a second-order system as the ratio of ________ damping to ________ damping.

9. The general expression for a second-order system in transfer function form is: output/input = ________.

10. A typical frequency response curve for a spring-mass system with damping is shown below. Using this response, determine the damping ratio, natural frequency, and zero frequency magnitude.

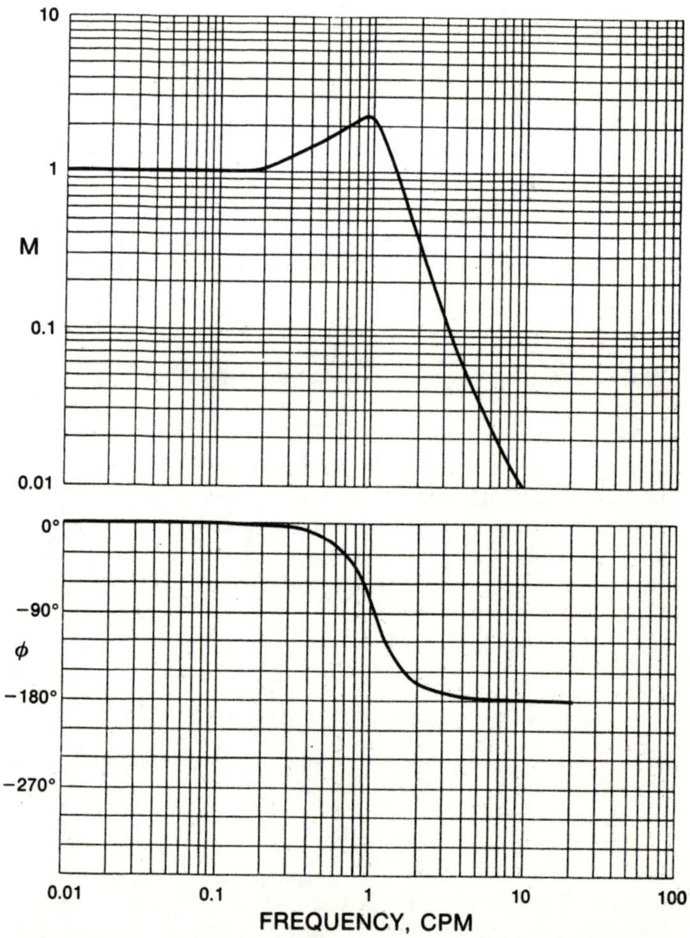

Figure 6-P10. Frequency response curve for a spring-mass system with damping.

11. Using the answers from Problem 10 and Figure 6-6, calculate the maximum deviation in response to a unit step input change.

12. From the previous problem answers and Figure 6-6, determine the decay ratio and magnitudes of the 1st, 2nd, and 3rd overpeaks.

13. Based on previous answers and Figure 6-6, determine the time response period.

14. From previous answers and Figure 6-6, determine the time of the first overpeak.

15. Using the answers from Problems 10 and 14, plot the time response to a unit step change for the system defined by the frequency response shown in the figure in Problem 10.

References

1. Solberg, H.L., Cromer, O.C., and Spalding, A.R., *Thermal Engineering,* Chapter 1, New York: John Wiley, 1960.
2. Coughanowr, D.R., and Koppel, L.B., *Process Systems Analysis and Control,* Chapters 7 and 8, New York: McGraw-Hill, 1965.
3. Caldwell, W.I., Coon, G.A., and Zoss, L.M., *Frequency Response for Process Control,* Chapter 10, New York: McGraw-Hill, 1959.

7 Closed Loop Dynamics

Having established basic stability criteria and a potential closed loop analog, it is now possible to examine actual closed loop dynamics.

7.1. The Closed Loop Equation

In Chapters 4 and 5, "process" and controller were combined (using a certain amount of witchcraft) to obtain a closed loop control system. The result was a system that cycled continuously at its natural frequency or ultimate period. The problem now is to develop a closed loop equation and relate this equation to the spring-mass analog of Chapter 6. A block diagram for the closed loop is shown in Figure 7-1. The simple heating system of Figure 4-1 is the example problem, and the reader is referred to Figures 4-2 and 5-1 for terminology. It is worth noting that in general terminology the controlled variable c is equal to Θ_2, the measured variable b is equal to P_3, and $P_4 = P_1$. Using these equalities it is possible to express "process" and controller Equations (4-8) and (5-1) as follows:

$$b / \Delta P_1 = \Delta P_3 / \Delta P_1 = g_2 g_3 g_4 h \qquad (7\text{-}1)$$

$$\Delta P_1 / e = \Delta P_4 / \Delta P_3 = g_1 \qquad (7\text{-}2)$$

Equations (7-1) and (7-2) may be combined to give the open loop transfer function:

$$b / e = (\Delta P_1 / e)(b / \Delta P_1) = g_1 g_2 g_3 g_4 h \qquad (7\text{-}3)$$

The open loop transfer function is by definition the product of all components in the loop. Equation (7-3) expresses in mathematical format the physical process of chasing signals around the loop that was undertaken in Chapter 5. It is also worthwhile noting that standard symbols b and e have been used to replace changes. This is to some extent a concession to existing ter-

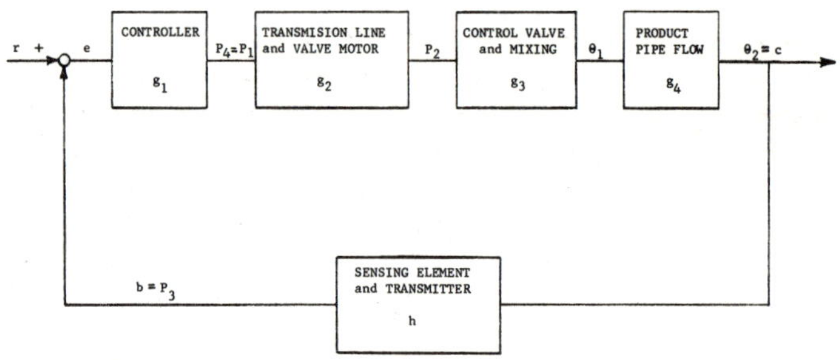

Figure 7-1. Closed loop block diagram for a simple heating system.

minology. It does, however, usually cause the reader some problems. As noted in Section 3.4, initial conditions or bias values are often neglected in control system analysis. This procedure was used, for example, in the original expression of Equation (5-1). As noted in Figure 7-1 and now expressed in Equation (7-2), change in controller output is some function g_1 of the error signal. Thus, to be totally correct, Equation (5-1) should have been expressed as follows:

$$P_{40} + \Delta P_4 = g_1 \left\{ (r_0 + \Delta r) - (P_{30} + \Delta P_3) \right\}, \qquad (7\text{-}4)$$

where added zero subscript implies initial condition. This, to say the least, is a cumbersome equation. If at time zero, initial conditions can be expressed as follows:

$$P_{40} = g_1 (r_0 - P_{30}), \qquad (7\text{-}5)$$

then controller response to change may be obtained by subtracting Equation (7-5) from Equation (7-4). The result is

$$\Delta P_4 = g_1 (\Delta r - \Delta P_3) \qquad (7\text{-}6)$$

If set point or reference is constant as assumed for Equation (5-1), then $\Delta r = 0$ and $\Delta P_4 = -g_1 \Delta P_3$. In this case the negative sign represents loop corrective action. The net effect is that the difference of actual values (such as r and P_3) is identical to the difference of initial or bias values plus changes such as ($r_0 + \Delta r$) and ($P_{30} + \Delta P_3$). For example, assume that initial values of r_0 and P_{30} are 9 psig. If P_3 changes to 11 psig, the difference is 2 psi. Using Equation (7-6), $\Delta r = 0$ and $\Delta P_3 = 2$ psi, producing an identical result. Thus deltas are normally dropped in writing equations, although the change in variable may be used in solving the equation. This is a fine point that does not present a problem from a theoretical viewpoint. The problem occurs the first time the reader tries to apply theory to an actual system.

One other important problem between theory and practice is worth noting at this time. In many tests the open loop equation will be defined as c/e. This is not an error, for in these cases authors have assumed that $h = 1$ and $b = c$. While unity feedback often occurs in servo systems, it is an improbable condition for a regulating or control system. The reason being that some measurement of the controlled variable must be made in order to accomplish regulation. Thus, for a control loop the open loop transfer function is b/e. This also points out the need for accurate and fast-responding sensing elements. The chart record or sensing element and transmitter output is *not* the controlled variable; it is, as defined, the measured variable. Unless the measured variable b is a reasonably good representation of the controlled variable c, good control is questionable. The chart record may be good, but what is actually happening to the controlled variable may be anyone's guess. A more detailed examination of this problem will be discussed in Chapter 10.

Having defined the open loop, the next step is to determine the closed loop response. In a regulator or control system this response is defined as b/r. From Figure 7-1 an equation for the summing point within the controller may be expressed as follows:

$$r - b = e \qquad (7\text{-}7)$$

Note that Δ values have been dropped to conform with standard textbook terminology. To obtain the closed loop response to a set point change, each term in Equation (7-7) is divided by r to give

$$1 - (b/r) = e/r \qquad (7\text{-}8)$$

Substituting Equation (7-3) for e in Equation (7-8) gives

$$1 - (b/r) = (b/r)(1/[g_1 g_2 g_3 g_4 h]) \qquad (7\text{-}9)$$

Factoring and rearranging terms gives

$$b/r = g_1 g_2 g_3 g_4 h/(1 + g_1 g_2 g_3 g_4 h) \qquad (7\text{-}10)$$

An inspection of Equation (7-10) shows that the closed loop transfer function is actually:

closed loop transfer function = open loop transfer function/(1 + open loop transfer function) (7-11)

The next step is to obtain the open loop transfer function and evaluate the closed loop transfer function.

7.2. The Closed Loop Frequency Response

The first step in obtaining the closed loop frequency response is to obtain the open loop response. As can be seen from Equation (7-3), the open loop response is the product of "process" and controller responses. The "process" response may be obtained from the figure obtained in the solution of Problem 7, Chapter 4. The controller as in Chapter 5 will be assumed to have only proportional action, thus $g_1 = k_1$. A value of $k_1 = 0.8$ will be used for controller gain. This value is calculated using Equation (5-4), and the ultimate gain of 1.6 established in Chapter 5. The result should be a response having approximately quarter decay, according to the Ziegler-Nichols technique for controller adjustment. The "process" and controller responses, and the open loop product are shown in Figure 7-2.

Development of the closed loop response presents a problem since vector addition must be accomplished. Equations (7-10) and (7-11) show that the denominator is equal to one plus the open loop transfer function. The Bode diagram as noted in Chapter 4 was specifically selected to handle vector multiplica-

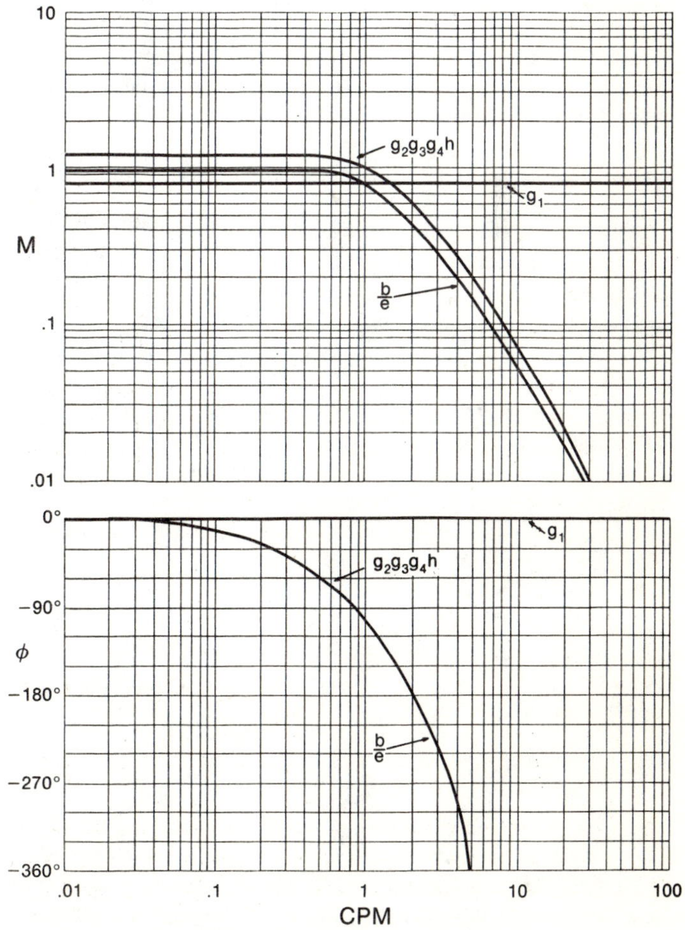

Figure 7-2. Open loop frequency response for a simple heating system.

tion and division, and no easy or practical means exists on this diagram for vector addition. One way to solve this problem is a graphical addition. Remembering that by definition a transfer function is the ratio of output to input, it is possible to define terms of Equation (7-11) as follows: Let the numerator or open loop transfer function equal the output; and the denominator or one plus the open loop transfer function equal the input. Output and input vectors may them be represented graphically. As an example assume a frequency of 0.5 cpm. From Figure 7-2 at 0.5 cpm, open loop magnitude or gain is approximately 0.94 and the phase angle is $-60°$. The resultant input vector as shown in Figure 7-3 has a magnitude of 1.7 and a phase angle of $-29°$. Since the output vector is the open loop, its value is 0.94 at a phase angle of $-60°$. To determine the closed loop value, output is divided by input. The process of vector division requires division of magnitudes and subtraction of phase angles. Using this procedure, closed loop values are

$$\text{closed loop gain} = 0.94/1.7 = 0.55 \quad (7\text{-}12)$$

$$\text{closed loop phase angle} = (-60°) - (-29°) = -31° \quad (7\text{-}13)$$

While this process will work, it is obviously too time-consuming to be practical. It was used here mainly to demonstrate a problem and potential solution. Closed loop values may also be calculated using complex number techniques. This process, however, can also be time-consuming unless a computer and program for Equation (7-11) are available. To overcome this problem, early researchers in feedback amplifier and control theory developed the Black-Nichols chart (or Nichols chart as it has come to be known).[1]

A Black-Nichols chart is shown as Figure 7-4.[2] This chart has x and y coordinates of open loop phase angle and open loop gain, respectively. The so-called $\mathcal{M}$ and α "circles" give closed loop gain and closed loop phase angle. The circle concept is a carry-over from earlier plots which produced families of circles for closed loop gain and phase angle on polar coordinates. To demonstrate the use of this chart, the open loop value at 0.5 cpm from Figure 7-2 is shown in Figure 7-4. The open loop value is located at a gain of 0.94 and phase angle of $-60°$. The closed loop gain value is between $\mathcal{M}$ circles of 0.5 and 0.6 and may be estimated at 0.55. The closed loop phase angle falls between α circles of $-30°$ and $-40°$ and may be estimated as approximately $-31°$. These results check with those of Equation (7-12) and (7-13). By repeating this process for the complete frequency spectrum, a closed loop response may be obtained.

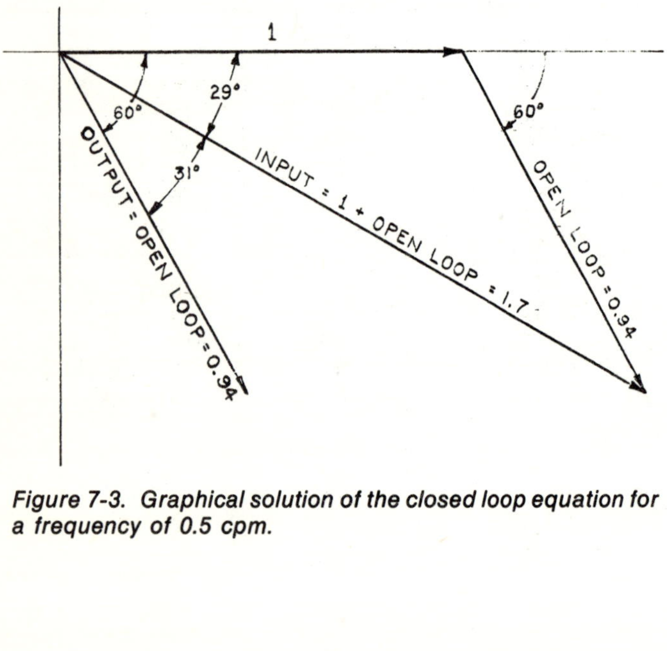

Figure 7-3. Graphical solution of the closed loop equation for a frequency of 0.5 cpm.

Application of the Black-Nichols chart will be demonstrated by converting the open loop response of Figure 7-2 to a closed loop response. To aid the reader in checking results, values taken from Figure 7-5 are itemized in Table 7-1. In comparing results the reader should not become overly disturbed by small variations in numbers. The procedure for developing "process" open loop and closed loop responses is graphical in nature, and plotting error is easily introduced. These errors, however, may be minimized by careful plotting, and in most practical applications the plots will be as valid as original data. Furthermore, objectives of a systems study are usually to overcome major problems, and this type of analysis is seldom hampered by small discrepancies. When values of open loop phase angle exceed $-180°$, the open and closed loop phase angle scales become mirror images. Points 13-15 show this effect both in Table 7-1 and on Figure 7-5. In the case of Point 14, for example, the open loop phase angle is $-240°$. Using the mirror image concept, this point is plotted at a scale reading on the Black-Nichols chart of $-120°$. This problem could be avoided by drawing the left half of the chart for phase angles between $-180°$ and $-360°$. To do this, however, would require either a fold-out chart, or including a magnifying glass with each book. The closed loop phase angle is read in the same manner. In the example of Point 14 the actual chart reading is $-102°$. Subtracting $-102°$ from $-360°$ gives $-258°$ as shown in Table 7-1.

Using data in Table 7-1, the closed loop frequency response may be plotted as shown in Figure 7-6. As the reader may have expected, the closed loop frequency response for the simple heating system is similar to the spring-mass frequency response curves drawn in Chapter 6. When this condition exists, techniques developed in Chapter 6 may be used to predict the time response for a control system.

7.3. The Closed Loop Time Response

Using Figure 7-6, values can be determined for zero frequency magnitude, peak frequency magnitude, and peak frequency. These values are

zero frequency magnitude = 0.48
peak frequency magnitude = 0.88
peak frequency = 1.65 cpm

The ζ value for the system may be evaluated using Equation (6-34):

$$0.88/0.48 \cong 1/(2\zeta) \text{ or} \quad (7\text{-}14)$$
$$\zeta = 0.48/[2(0.88)] \cong 0.27$$

The ζ value for quarter decay is equal to 0.215. With a ζ value of 0.27 the system will be more highly damped and have a decay ratio less than quarter decay. This is no particular problem and

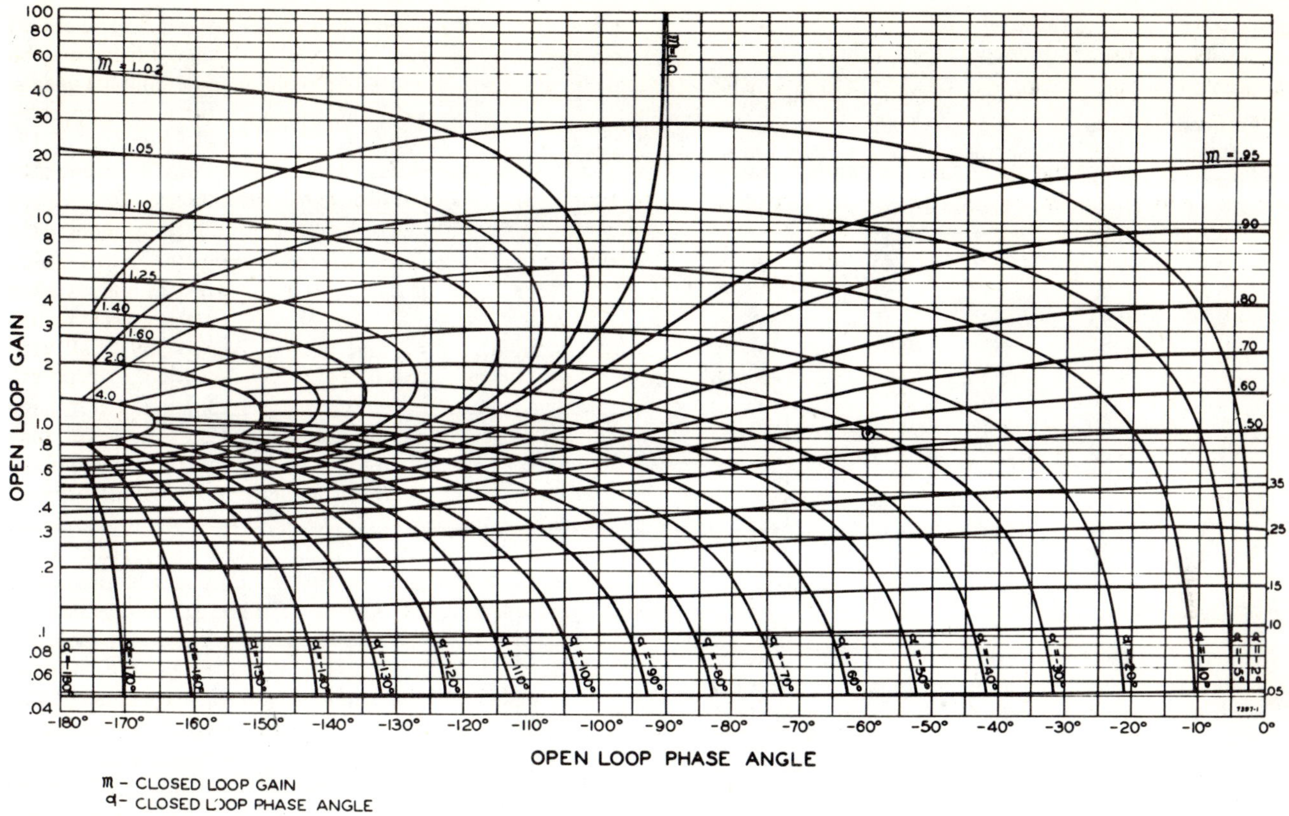

Figure 7-4. Black-Nichols chart.

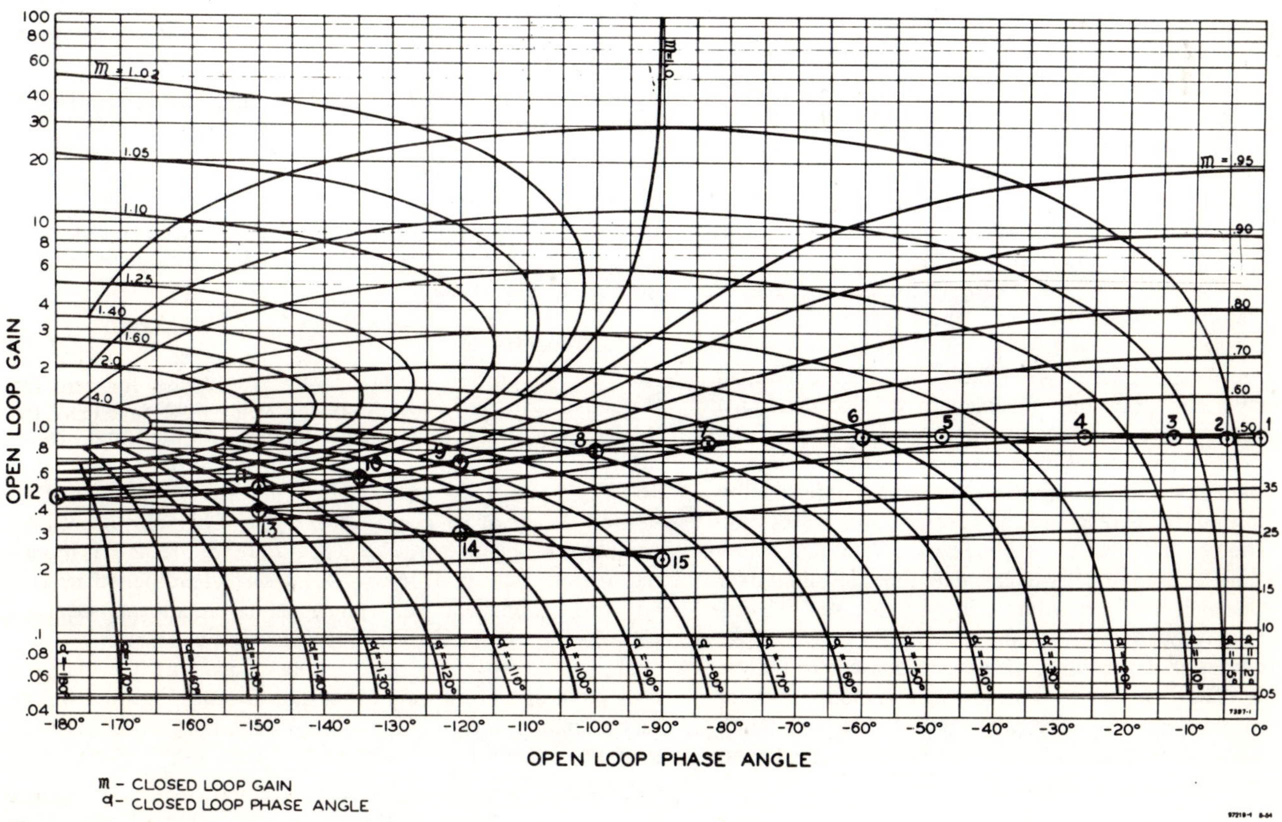

Figure 7-5. Open to closed loop conversation.

Table 7-1
Closed Loop Frequency Response Data

Point No.	Frequency (cpm)	Open Loop Gain	Open Loop Phase Angle	Closed Loop Gain	Closed Loop Phase Angle
1	0.01	0.96	0°	0.48	0°
2	0.05	0.96	−5°	0.48	−3°
3	0.10	0.96	−13°	0.49	−7°
4	0.20	0.96	−27°	0.50	−14°
5	0.40	0.96	−48°	0.53	−25°
6	0.50	0.94	−60°	0.55	−31°
7	0.80	0.87	−83°	0.61	−46°
8	1.0	0.80	−100°	0.68	−58°
9	1.2	0.70	−120°	0.79	−76°
10	1.4	0.60	−135°	0.82	−100°
11	1.65	0.52	−150°	0.88	−123°
12	2.0	0.46	−180°	0.80	−180°
13	2.3	0.40	−210°	0.59	−228°
14	3.1	0.31	−240°	0.35	−258°
15	3.4	0.23	−270°	0.23	−284°

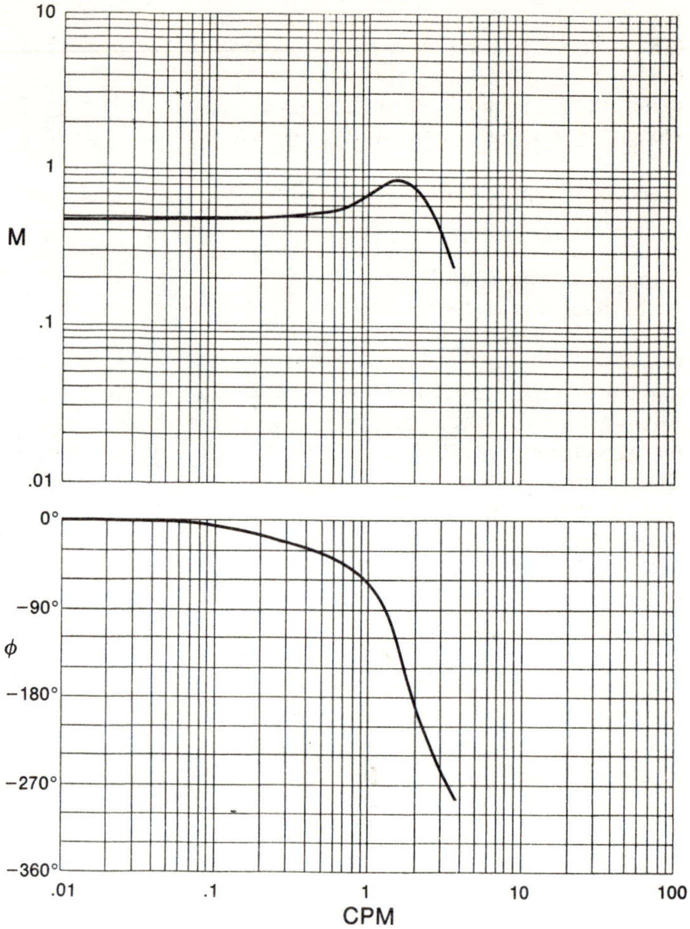

Figure 7-6. Closed loop frequency response for a simple heating system.

merely demonstrates the "ball park" approach of the Ziegler-Nichols method for controller adjustment. In most applications controller settings will be adjusted in the field using some trial-and-error procedure. No one absolute technique is available for all systems. The Ziegler-Nichols technique will normally result in a reasonable value that may be "trimmed" to produce a satisfactory result.

It should also be remembered that although the closed loop frequency response of Figure 7-6 looks like the second-order responses of Chapter 6, it results from a more complex system and will not, therefore, show the degree of correlation that resulted in Chapter 6. One of the more obvious discrepancies is the relation of peak and natural frequencies. According to Equation (6-28), for example, peak frequency is equal to $f_n \sqrt{(1 - \zeta^2)}$. From earlier work in Chapter 5, the natural frequency of this system was found to be 2 cpm. This value can be checked by noting the frequency of the open loop response at −180° in Figure 7-2. Using the 0.27 value of ζ determined in Equation (7-14), a natural frequency of 2 cpm, and Equation (6-28), the calculated peak frequency is 1.93 cpm. This does not agree with 1.65 cpm read from Figure 7-6 or Table 7-1. This is the main discrepancy of the spring-mass analog with real systems. In real systems having more than second-order transfer functions, the resonant peak of the closed loop frequency response shifts to lower frequencies more rapidly than the second-order analog. The period in the time response, however, also increases more rapidly. The result is that peak frequency of the closed loop response may be used directly to calculate time for initial overpeak and period. This procedure will be shown as the time response is developed.

The first step in determining the time response is to find the final value. The final value or steady-state response is determined from the zero frequency gain or magnitude. The fundamental concept is that as frequency approaches zero, time approaches infinity. Thus, at infinite time,

$$b/r = 0.48 \tag{7-15}$$

If a reference or set point change of 10°F is made (for example from 160° to 170°F), the final value of the measured variable may be calculated as follows:

$$b = 0.48\,(10°F) = 4.8°F \tag{7-16}$$

For actual systems the zero frequency magnitude is not necessarily 1.0 as with the spring-mass analog, and the input change need not be unity. Adding the final value of 4.8°F to the bias of 160°F gives a value of 164.8°F. The discrepancy between a desired change of 10°F and a response of 4.8°F may or may not come as a shock to the reader. The difference of 5.2°F is termed the "offset or steady-state error" and is a real problem that exists with proportional control. A physical interpretation of offset and a means for eliminating it will be discussed in Chapter 9.

Having established ζ, peak frequency, and the final value, it is now possible to find the time response using Figure 6-6. The ratio of maximum deviation to final value for $\zeta = 0.27$ is equal to approximately 1.41, reading from Figure 6-6. The maximum deviation is then

$$\text{maximum deviation} = 1.41\,(4.8°) = 6.8°F \tag{7-17}$$

Assuming the 160°F bias, the maximum deviation is 160° + 6.8° = 166.8°F. The first overpeak exceeds the final value by 6.8° −4.8° = 2°F. The decay ratio for $\zeta = 0.27$ is approximately 0.17 as shown in Figure 6-6. Second and third overpeaks along with absolute values may then be calculated as follows:

$$\text{second overpeak} = (0.17)(2°) = 0.34°\text{F} \quad (7\text{-}18)$$
$$\text{absolute value} = 160° + 4.8° + 0.34° = 165.14°\text{F}$$

$$\text{third overpeak} = (0.17)(0.34°) = 0.06°\text{F} \quad (7\text{-}19)$$
$$\text{absolute value} = 160° + 4.8° + 0.06° = 164.86°\text{F}$$

Equations (7-18) and (7-19) clearly show increased damping, as previously noted. If a control chart range of 100°F with limits between 100° and 200°F had been selected for this system, both the second and third overpeaks would be less than 1% of range. Thus, accuracy indicated in Equations (7-8) and (7-19) is used to show method, rather than any belief that these numbers are real. It does show, however, that the system is very stable.

The value for $f_n X$ (as read from Figure 6-6 for $\zeta = 0.27$) is 0.52. Assuming that $f = f_n = f_p = 1.65$ cpm, period and first overpeak may be calculated as follows:

$$\text{period} = 1/f_p = 1/1.65 \cong 0.61 \text{ min} \quad (7\text{-}20)$$

$$X = 0.52/f_n = (0.52)(\text{period}) = (0.52)(0.61) \quad (7\text{-}21)$$
$$\cong 0.32 \text{ min}$$

Times for overpeaks and undershoots may then be calculated as follows:

$$\text{first overpeak} = X = 0.32 \text{ min} \quad (7\text{-}22)$$
$$\text{first undershoot} = 0.32 + 0.5\,(0.61) \cong 0.63 \text{ min}$$
$$\text{second overpeak} = 0.32 + 1.0\,(0.61) \cong 0.93 \text{ min}$$
$$\text{second undershoot} = 0.32 + 1.5\,(0.61) \cong 1.24 \text{ min}$$
$$\text{third overpeak} = 0.32 + 2.0\,(0.61) \cong 1.54 \text{ min}$$

A time response may now be constructed using data found in Equation (7-16) through (7-22). The procedure is the same as explained in Chapter 6, and the result is shown in Figure 7-7.

As noted earlier, the degree of agreement between predicted and actual system responses will not be as good as was experienced in Chapter 6. In Chapter 6 the second-order analog was used with a second-order system, and results were understandably good. In using the second-order analog to predict results for real systems, exact correspondence does not exist. Results are with few exceptions, however, adequate for engineering purposes. A comparison of predicted and actual records for a real system is shown in Figure 7-8. For most process control applications, discrepancies noted in Figure 7-8 would be negligible. Objectives of a study, however, normally determine required accuracy and will be discussed in more detail in Chapters 11 and 12.

7.4. A Practical Analysis of the Closed Loop

Having now established basic concepts of a closed loop or feedback control system, a short review is justified to see what

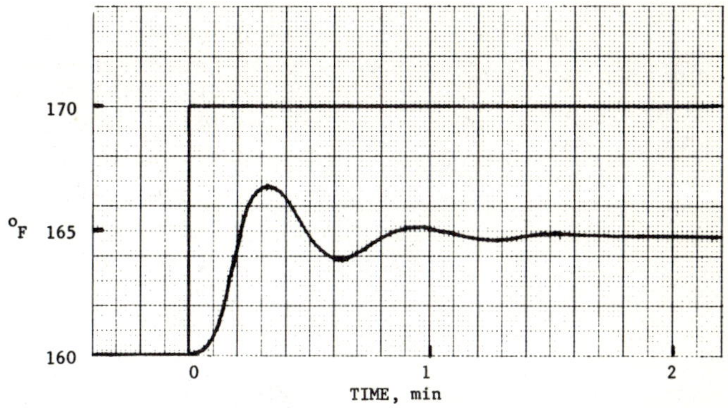

Figure 7-7. Predicted time response for a simple heating system.

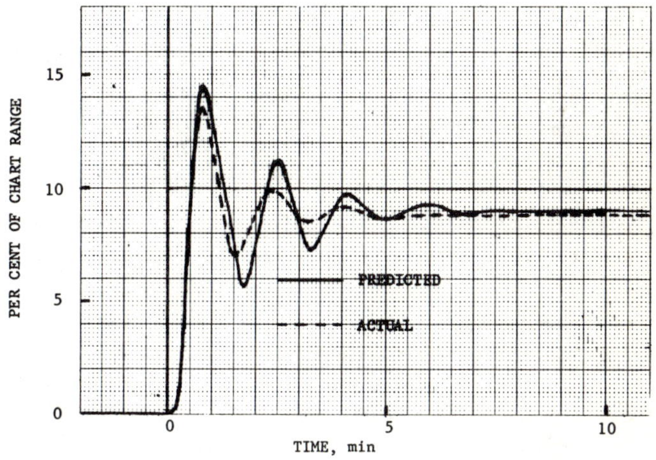

Figure 7-8. Predicted and actual time responses for a closed loop control system.

has been accomplished and why certain steps were necessary. Much of what will be discussed has been noted earlier in the text; now it will have some meaning.

The time response as developed in Chapter 7 was the original objective. Assuming that original component transfer function equations are available, one of two paths are open to the reader. The classical route, which is taken in many control texts, is to substitute individual transfer functions into Equation (7-10) and solve for the closed loop transfer function. The input reference or set point disturbance is then defined (usually as a step input) and Laplace transfomed to give an equation for the measured variable b in the "s-domain." Using inverse Laplace transform techniques a time-response equation is obtained. This procedure was used in Chapter 6, without detailed explanation, to obtain Equations (6-23) through (6-25) from Equation (6-22). For the second-order system of Chapter 6, this is a relatively simple mathematical exercise. As the order of the system increases, complexity of the inverse transformation also increases. The original equation in the s-domain usually requires a partial fraction expansion and considerable mathematical manipulation to obtain the time-response equation. When dead time lag is present, as it is in most process control systems, the procedure becomes even more complex. The result is the reader usually falls asleep or gives up as may be the present case.

The alternative route is the one that has been used in this text. Individual component transfer functions have been plotted and combined using graphical techniques to obtain the "process" transfer function. Combining controller and "process" plots yields the open loop transfer function. Through use of the Black-Nichols chart, the open loop response is converted to a closed loop response. A time response may then be approximated from the closed loop using the second-order analog. The procedure, as noted, does not give an exact solution. It does, however, provide a means of obtaining an approximate solution that can be understood and used *when needed*. Very few systems will require a detailed analysis such as demonstrated up to this point. The individual steps, however, were necessary as a teaching tool to explain basic concepts and to help in developing thinking patterns that may be applied in troubleshooting and designing control systems.

One item that requires further analysis is the problem of offset. In both text and problem examples, a steady-state error or offset occurred. Physically this may be explained by looking more closely at the proportional controller. If in Equation (7-4), $(P_{40} + \Delta P_4) = P_4 = P_1, g_1 = k_1, r_0 + \Delta r = r$, and $P_{30} + \Delta P_3 = P_3 = b$, then the equation becomes

$$P_1 = k_1 (r - b), \tag{7-23}$$

and since $r - b = e$, controller output is equal to the proportional gain times the error signal. While this equation may be used for study of control systems in a steady-state analysis, it is incorrect. For example assume that in the simple heating system the set point and measured variable are both equal to 160°F. Under this condition the error is zero and controller output P_1 must be zero. With zero controller output the steam valve is closed, and steam flow is zero. The only way that a temperature of 160° can be measured at the sensing element with no steam flow is to have the product enter at 160°F. What actually happens is that every proportional controller must have some zero adjustment. When the controller is installed, the zero is adjusted to provide the required controller output to bring the measured value equal to the reference or set point. In equation form this would look as follows:

$$P_1 = k_1 (r - b) + C, \tag{7-24}$$

where C = proportional controller zero constant. In a dynamic analysis this value may be neglected since it is a constant and does not change. It does help, however, to know it is there and to see how an original controller output is obtained.

An adjustment of C will now be assumed such that when $r = b = 160$°F, $P_1 = 9$ psig. A controller output of 9 psig will open the steam valve to 50 percent. If the reference or set point is now changed to 170°, the controller output will increase in an effort to supply more steam to the product. If and/or when the product temperature gets to 170°F, controller output as shown by Equation (7-24) must return to 9 psig since $r - b$ is again zero. The explanation is that a given product flow cannot be heated an additional 10°F without increased steam flow. If the system is stable, the result is as shown in Figure 7-7. Controller output is increased by an amount equal to $k_1 (r - b)$, where $r - b$ is equal to the steady-state error or offset. Since the product of $k_1 (r - b)$, or $k_1 e$ will be constant in the steady state, the error or offset may be reduced by increasing k_1. Increasing the controller gain in the simple heating system or in any system, however, will reduce damping and cause more overpeaking. The only solution is the addition of reset control action that will be discussed in Chapter 9.

The same conclusion concerning offset may be reached using Equation (7-10). According to Equation (7-10), offset will be eliminated if in the steady state the ratio of b/r equals 1. To achieve this result, zero frequency open loop gain must be very large. If, for example, zero frequency open loop gain equals 1000, then $b/r = 1000/1001 \cong 1$. Unfortunately, in both the simple heating system and Problem 10, zero frequency open gain is too small to produce a unity ratio of b/r in the steady state. Equation (7-10) is not specific in requiring high controller gain. Any combination of component open loop gains that will yield a large value for the zero frequency gain is acceptable for a set point change as shown by Equation (7-10). To show why controller gain should be high requires an analysis of load changes. This subject will be treated in the next chapter.

Chapter 7—Study Questions and Problems

(Answers appear on pages 158-159)

1. In a given control loop, the open loop transfer function is the product of all component ________.

2. In controller terminology variable changes are often used, although symbols indicate actual values. This procedure is valid since the difference of actual values is the same as the difference of initial or bias values plus ________.

3. Is the recorded value in a control system the controlled variable?

4. Express the closed loop transfer function in equation form as a function of the open loop transfer function.

5. Write open and closed loop transfer functions for the closed loop block diagram below.

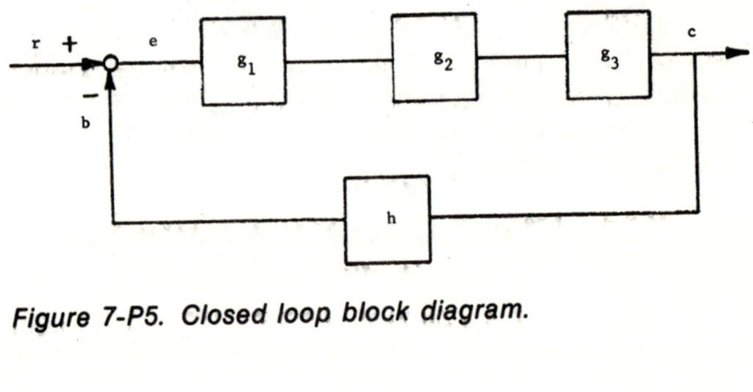

Figure 7-P5. Closed loop block diagram.

6. Assume that for the figure in Problem 5 the process transfer function is $g_2 g_3 h$ and that individual components have the following values: $g_2 = 5/(1 + 0.4s)$; $g_3 = 3/(1 + 1.2s)$; $h = 0.12/(1 + 0.1s)$. Using these values, plot the process frequency response.

7. Using the process frequency response plotted in the solution to Problem 6 and assuming a proportional controller, calculate the controller setting for approximately quarter decay. G_O = ________.

8. The open loop is the product of ________ and ________ transfer functions. Combine the answers of Problems 6 and 7 to obtain the open loop frequency response.

9. Using a Black-Nichols chart convert the open loop frequency response plotted in the solution to Problem 8 to a closed loop frequency response and plot the results.

10. Using the closed loop response obtained as a solution to Problem 9, construct a closed loop time response based on a reference or a set point change of 10°F and starting at a bias of 140°F.

11. The solution to most control problems will usually require that a ________ of the closed loop be obtained.

12. A time response for a closed loop control system may be obtained using ________ or ________ techniques.

13. Does the second-order analog technique used to obtain a time response from a closed loop frequency response provide an exact or approximate solution?

14. A set point change in a control loop using a proportional controller will normally produce an ________ of the measured variable.

15. According to the closed loop equation, offset following a set point change will be zero for all practical purposes if the zero frequency open loop gain is very much greater than ________.

References

1. James, H.M., Nichols, N.B., and Phillips, R.S., *Theory of Servomechanisms,* New York: McGraw-Hill, 1947, and Dover, 1965
2. Black-Nichols Chart, Form 97219-1, available from Taylor Instrument Co., 95 Ames St., Rochester, NY 14601.

8 Load Changes in the Closed Loop System

In servomechanism studies the main problem is determination of controlled variable response to a change in reference. A typical example is numerical control of a milling machine. The reference is continuously changing, and the milling cutter must duplicate this change to produce a satisfactory product. In this type of system few, if any, external or load disturbances affect the system. A regulator or process control system, on the other hand, will normally have a fixed reference or set point. This does not mean that the reference or set point cannot be changed. Reference or set point changes do occur, but are usually of an intermittent nature and remain constant for relatively long periods of time. In a regulator or process control system, load variations usually present the primary problem. Electrical power generation is a typical example of a regulator or process control system. A reference or set point of 120 v and 60 Hz is undoubtedly fixed. The problem is to maintain this set point in spite of continuously changing load demands. Technique for load change analysis is the objective of this chapter.

8.1. Load Change Transfer Functions

Load changes, as discussed in Chapter 1, may enter the closed loop at almost any location. For the simple heating system of Chapter 4, 5, and 7, load changes result from steam supply pressure, line pressure, product flow rate, and product inlet temperature variations. Utility disturbances to control components can also cause problems. Control component manufacturers, however, have taken steps to minimize this type of problem and, with exception of utility failure, may normally be discounted. For a general analysis of load changes, assume a system as shown in Figure 8-1. Potential load changes are shown as U_3 and U_4. Although the block diagram is identical to the simple heating system example, it is a fundamental configuration that could represent almost any closed loop system. Plus and minus signs on U_3 and U_4 determine load effect on the controlled variable and are determined from load equations as noted in Chapter 1. If, for example, this diagram were to be used for the simple heating system, U_3 could be taken as a steam supply pressure change and U_4 as a product flow rate change. An increase in steam supply pressure causes increased steam flow, which will result in a product temperature or controlled variable increase. Thus, U_3 is shown as being positive. Increasing U_4 or product flow rate, however, causes a decrease in the controlled variable and is assigned as negative. In the analysis that follows, the reader should not confine his thinking to any isolated example such as the simple heating system. The intent is to show both positive and negative load effects for a general system. The simple heating system has been used only to provide a practical example of how signs are determined.

Load change transfer functions are developed in the same manner as the closed loop transfer function in Chapter 7. In obtaining b/r in Chapter 7, no loads were indicated. Loads

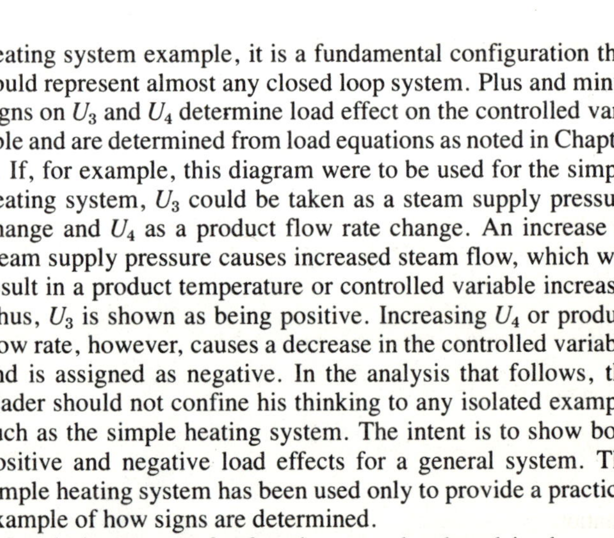

Figure 8-1. Closed loop control system with load changes.

existed but were assumed constant. To obtain the transfer function for b/U_3, it is now necessary to assume r and U_4 are constant. If no change occurs in r and U_4, they may be assumed to equal zero, as previously discussed in Section 7.1. The transfer function may then be derived as follows:

$$a + U_3 = d \tag{8-1}$$

Values of a and d from Figure 8-1 are

$$a = g_1 g_2 e \tag{8-2}$$

$$b = g_3 g_4 h d \tag{8-3}$$

Since $r = 0$, $-b = e$, and Equation (8-2) is equivalent to

$$a = -g_1 g_2 b \tag{8-4}$$

Substituting Equations (8-3) and (8-4) into Equation (8-1) yields

$$-g_1 g_2 b + U_3 = b/(g_3 g_4 h) \tag{8-5}$$

Combining terms and dividing by U_3 gives

$$1 = (b/U_3)\{g_1 g_2 + [1/(g_3 g_4 h)]\} \tag{8-6}$$

Obtaining a common denominator gives

$$b/U_3 = g_3 g_4 h/(1 + g_1 g_2 g_3 g_4 h) \tag{8-7}$$

The format of Equation (8-7) is similar to the closed loop equation of Chapter 7 and again involves the problem of vector addition and division. In this instance, however, no common open loop term exists as it did in Chapter 7. To overcome this problem, Equation (8-7) is multiplied by $g_1 g_2/g_1 g_2$ to give

$$b/U_3 = (1/g_1 g_2)\{g_1 g_2 g_3 g_4 h/(1 + g_1 g_2 g_3 g_4 h)\}, \tag{8-8}$$

which may then be reduced to

$$b/U_3 = [1/(g_1 g_2)]\,[b/r] \tag{8-9}$$

Equation (8-9) shows that response to a load change at U_3 may be determined by dividing the closed loop reponse by $g_1 g_2$. Since division is the reciprocal of multiplication, this process may be accomplished on a Bode plot. In vector multiplication, magnitudes are multiplied and angles added. For vector division, magnitudes are divided and angles subtracted. On the logarithmic magnitude scale of the Bode plot, division is accomplished by subtracting distances rather than adding, as was the procedure in multiplication. To demonstrate this procedure the simple heating system will again be used.

The closed loop response b/r from Figure 7-6, controller response g_1 from Figure 7-2, and transmission line and valve motor response g_2 from the figure in the solution to Problem 4, Chapter 4, are replotted in Figure 8-2. To divide b/r by $g_1 g_2$, positive magnitude and phase angle values of g_1 and g_2 are assumed negative, and negative values are assumed positive. On the magnitude or gain plot of Figure 8-2, neither g_1 nor g_2 has

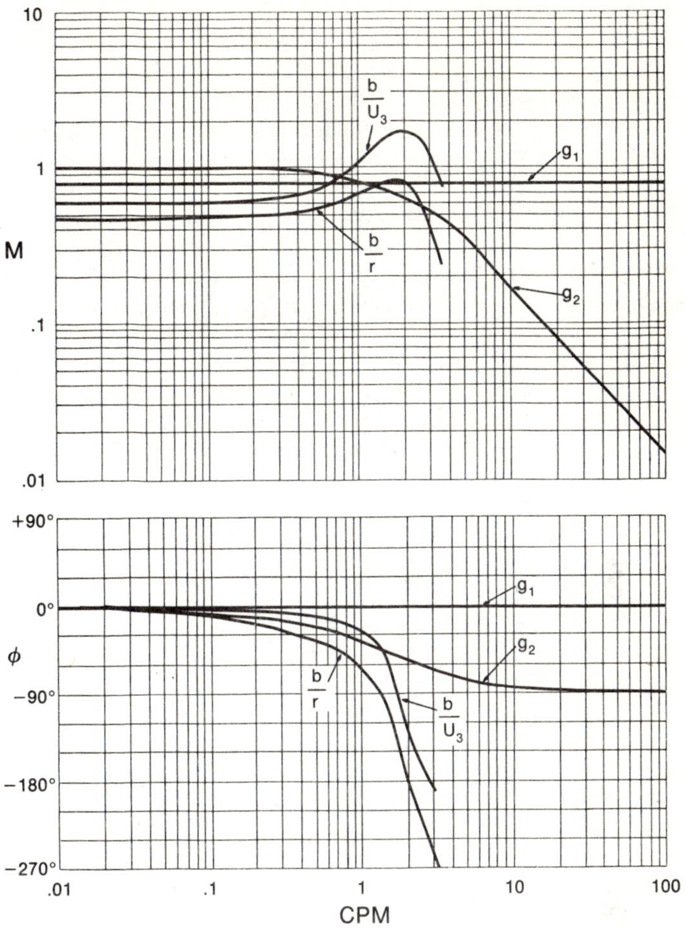

Figure 8-2. Frequency response plot for load change at U_3.

values greater than 1. Thus, all values of g_1 and g_2 being less than 1 must be graphically added to b/r. For example, at 0.01 cpm the values of g_1 and g_2 are 0.8 and 1, respectively. The value of b/r is 0.48. Dividing 0.48 by (0.8) (1) gives 0.6. Using a magnitude or gain of 1 as the reference, the same result is obtained by adding logarithmic values on Figure 8-2. Since the value of g_2 is 1, it has no effect. Adding the distance that g_1 lies below the unity reference to b/r gives 0.6 for b/U_3. At frequency values greater than 0.3 cpm, the effect of g_2 must also be considered. To check technique, the reader should consult Problems 1 and 2 at the end of this chapter.

8.2. Analysis of Load Change Transfer Functions

An analysis of the load change transfer functions b/U_3 and b/U_4 and their response plots (see Figure 8-2 and the solution to Problem 2 of this chapter) yields several interesting conclusions. The effect of sign convention, as previously noted in Section 8.1, is obvious in the transfer function equations. Equation (8-8) shows a positive result, indicating an increase in the measured variable when load U_3 is increased. In the case of b/U_4 the result is negative, indicating a decrease in b for an increase in U_4. The sign convention, however, has no effect on the frequency response plots. Thus, the effect on output must either be determined from transfer function equations or by a physical examination of the load change effect. This is a minor point that may cause problems in early attempts on real systems.

One major conclusion involves format of load change transfer functions. Although two examples would naturally not be considered a sufficient sample on which to draw general conclusions, it is a fact that all load change transfer functions may be modified to include the closed loop transfer function. As in the case of both b/U_3 and b/U_4, however, an additional term will also be included. To determine what this addtional term will be requires introduction of "forward" and "feedback" elements. The forward element is the product of transfer functions between the load change entry point and output measurement. Using this definition and Figure 8-1, the forward element for U_3 is g_3g_4h. The feedback element is the product of transfer functions between the output and load change. Referring again to Figure 8-1 and load U_3, the feedback element is g_1g_2. For load U_4 the forward element is g_4h, and the feedback element is $g_1g_2g_3$. An examination of Equation (8-8) for b/U_3 and the solution to Problem 1, Chapter 8, for b/U_4 shows that in each case the term by which the closed loop transfer function is multiplied is equal to 1/feedback element. Thus, a general transfer function equation for load changes may be expressed as follows:

$$b/U = (1/\text{feedback element})\ \{\text{closed loop}\} \tag{8-10}$$

This format for load change transfer functions becomes very helpful in determining time responses as will be shown in the next section.

Transfer functions plots for b/U_3 and b/U_4 (shown in Figure 8-2 and the solution to Problem 2 of this chapter, respectively) reveal quite clearly a limitation of the spring-mass analog in predicting time-response records. As noted in Chapter 7, a real process is very seldom a pure, second-order system. Because of this, predicted responses will normally deviate to some extent from actual responses as shown in Figure 7-8. For set point or reference changes this deviation is usually not critical, since the closed loop response is very similar to a second-order response. Dividing the closed loop by the feedback element to obtain a load change response, however, normally creates a distorted curve. This effect is clearly shown in the solution to Problem 2 of this chapter. The effect on b/U_3 is marginal since g_1 is a constant, and the small time constant of g_2 produces a minor effect as can be seen in Figure 8-2. The combined influence of both g_2 and g_3 (with its larger time constant), however, can be clearly seen in the b/U_4 response in the solution to Problem 2 of this chapter. This does not mean that these curves are in error, but only that the spring-mass or second-order analog can no longer be used to predict the time response. Thus, some additional technique will be required to predict load change results. A method is available and will be discussed in Chapter 12.

8.3. Influence of Transfer Function Time Constants on Load Change Time Responses

Although load change time responses can not normally be predicted using the spring-mass or second-order analog, some interesting conclusions may be drawn from transfer function equations. As an example assume the system shown in Problem 3, of this chapter, the transfer equations in the solution to Problem 3, and the following values for each of the transfer functions:

$$g_1 = k_1 = 9 \tag{8-11}$$

$$g_2 = k_2/(1 + T_2s) = 1/(1 + 0.1s) \tag{8-12}$$

$$g_3 = k_3/(1 + T_3s) = 1/(1 + 0.4s) \tag{8-13}$$

$$g_4 = k_4/(1 + T_4s) = 2/(1 + 1.25s) \tag{8-14}$$

$$h = k_h = 1 \tag{8-15}$$

In assuming these values, proportional controller gain has been selected to give approximately quarter decay. The sensing element and transmitter transfer function h has been assigned the value of unity, thus $b = c$. The problem of the sensing element and transmitter will be examined in Chapter 10.

If load changes of 40% of range are now introduced in sequence at each load junction, the experimental time responses are as shown in Figure 8-3. Steady-state offset may be calculated for each load change equation by using only steady-state gain terms. This is the same procedure that was used in Chapters 6

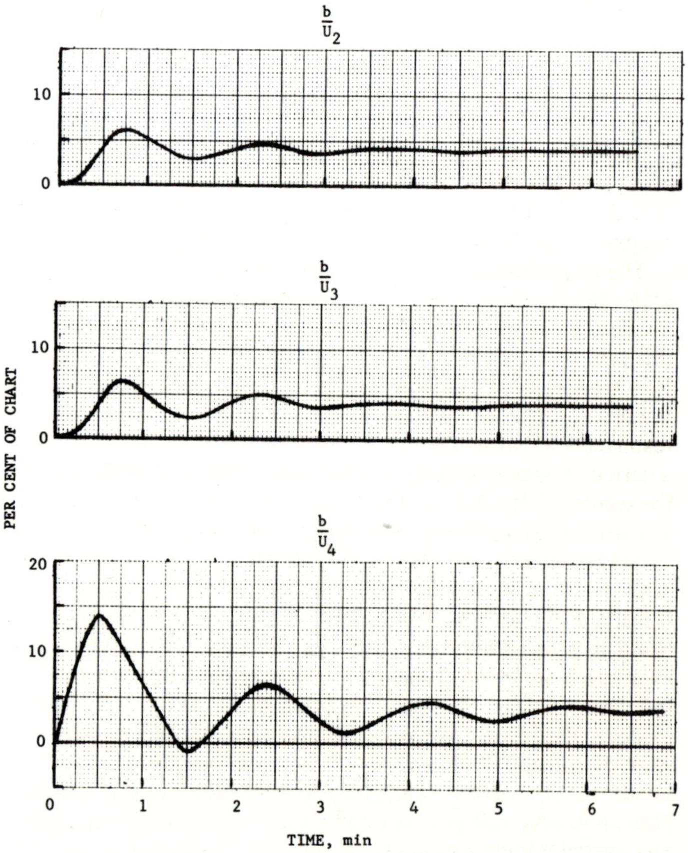

Figure 8-3. Load-change time-response records.

and 7 and stems from an earlier concept in Chapter 3 that when time approaches ∞, s approaches zero. In the steady state, time is assumed infinite, and $s = 0$. Thus, only gain terms are required to solve for the final value. Steady-state values are then

$$b/r = k_1k_2k_3k_4k_h/(1 + k_1k_2k_3k_4k_h) \quad (8\text{-}16)$$
$$= (9)(1)(1)(2)(1)/[1 + (9)(1)(1)(2)(1)] = 18/19$$

For a 40% change in U_2 the final value of b is

$$b = (1/g_1)(b/r)U_2 = (1/9)(18/19)40 = 4.21\% \quad (8\text{-}17)$$

Results for b/U_3 and b/U_4 will have the same values since the gain terms of both g_2 and g_3 equal 1. This condition was selected on purpose since only time-constant effects are being analyzed at this time. Gain effects will be examined in the next section of this chapter.

Time-response records in Figure 8-3 show an offset or final value of just over 4%, which agrees very closely with the calculated value of 4.21%. The error is well within measurement accuracy of experimental equipment. A more interesting observation at this time, however, is the transient portion of the curve. For b/U_2 the closed loop function b/r is divided by only g_1. Since g_1 is a constant, shape of the closed loop frequency response is unaltered. Position on the Bode diagram is of course altered, but shape remains constant. Because of this, the spring-mass or second-order analog may be used to predict the time response. The same damping ratio would apply with overpeaks, decay, and times, following procedures used in Chapters 6 and 7. The one exception to this rule is the offset value. For a load change the controlled variable should return to the original set point. Thus, for a load change the final value and offset are equal. The result would be a symmetrical response as shown in Figure 8-3. In the case of b/U_3, where b/r is divided by g_1g_2, the result is very similar. The time constant of g_2 is only 0.1 min and will do little to alter the basic shape of b/r. This fact can be checked from the response of b/U_3 as shown in Figure 8-3.

The response of b/U_4, however, clearly shows the distortion effect noted in Section 8.2. To obtain b/U_4, the closed loop b/r is divided by $g_1g_2g_3$. The time-constant value of g_3 is 0.4 min and in this case would cause a shift of b/r and considerable distortion. It is worth noting that actual size of the time constant is not important. What is important is the ratio of corner frequency to resonant peak of the closed loop. From the time response of b/U_2 a period of approximately 1.5 min may be estimated. Based on the second-order analog, the resonant peak frequency is equal to 1/1.5 or 0.67 cpm. Corner frequencies for g_2 and g_3 are $f_{c2} = 1/(2\pi T_2) = 1/(2\pi\, 0.1) = 1.6$ cpm and $f_{c3} = 1/(2\pi T_3) = 1/(2\pi\, 0.4) = 0.4$ cpm, respectively. Thus, in the case of b/U_3 the effect of g_2 at 1.6 cpm is well beyond the 0.67 cpm of the resonant peak and causes a negligible influence. For b/U_4, however, the 0.4-cpm location of g_3 will have a definite influence on the resonant peak. The result as shown in Figure 8-3 is clearly evident. The initial overpeak is greater, time for the initial overpeak is reduced, and the first full period is increased over what would be expected in a normal, second-order approximation. Note, however, once the initial distortion is overcome, the time record settles into a fundamental response based on the dynamic characteristics of the system. This shows that, like all solutions to differential equations, the solution for b/U_4 has both a steady-state and transient solution. The transient portion for b/U_4, however, has multiple terms which account for the initial distortion followed by the characteristic decay to steady state.

To emphasize time-constant effects T_2 and T_4 will now be interchanged such that

$$g_2 = k_2/(1 + T_4s) = 1/(1 + 1.25s) \quad (8\text{-}18)$$

and

$$g_4 = k_4/(1 + T_2s) = 2/(1 + 0.1s) \quad (8\text{-}19)$$

All other transfer functions will remain as previously defined. The offset or final value will be as calculated in Equation (8-16) since gain or magnitude values have not been changed. The effects of load changes at U'_2, U'_3 and U'_4, where the prime sign is used to indicate the revised system, are shown in Figure 8-4. The results, to say the least, are somewhat startling.

The time response for b/U'_2 is the same as for b/U_2. Neither g_2 or g_3 are involved in the feedback element. The load change sees the same transfer function in both cases, in spite of the time-constant changes and this result can be easily observed from

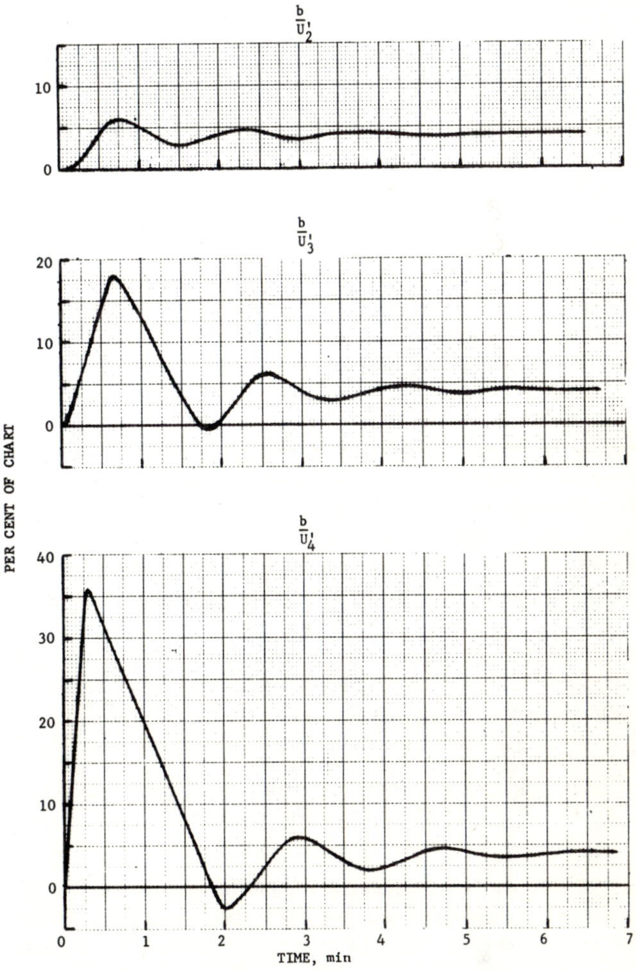

Figure 8-4. Load-change time-response records.

either a mathematical or physical interpretation. The influence of the large time constant in the g_2 component is obvious in the b/U'_3 time response. With the 1.25-min time constant in the g_2 transfer function, the corner frequency occurs at approximately 0.13 cpm. This is well below the resonant peak frequency of 0.67 cpm, and distortion of the frequency response is evident. The effect is even more pronounced in the case of b/U'_4. Here, both g_2 and g_3 are in the feedback element. The combined effect of the two, large time constants of 1.25 and 0.4 min in distorting the frequency response curve is readily shown on the time response.

A second and probably more fruitful method of examining this problem is to follow the load change around the loop. A load change at U_2 is actually filtered by the forward element $g_2g_3g_4h$ and produces a very slow and small change in the measurement b. A change in b, however, is instantly detected and acted upon by the controller, providing corrective action to compensate for the change in U_2. The identical procedure exists for U'_2. The result is a nullification of the load change and a limited disturbance of b. A similar situation occurs when U_3 is the load change. Controller corrective action to counteract the effect of U_3 is delayed by the g_2 component. However, since T_2 is small, this effect is of negligible consequence and good control is maintained. When the load is at U_4, controller corrective action is delayed by both g_2 and g_3. Since T_3 is a larger time constant, corrective action is delayed for a longer time and U_4 has greater influence on b. This causes a larger initial overpeak and results in poor control as shown in Figure 8-3.

When T_2 and T_4 are interchanged, the effect of time constants on load changes is even more clearly shown. For a load at U'_3 the forward element g_3g_4h provides only a limited filtering or delaying action. The largest time constant T_4, however, is now associated with g_2 and is in the feedback element. The result is that controller corrective action is now filtered or delayed, which allows U'_3 time to cause a major disturbance in b. When the load enters at U'_4, the worst possible set of conditions exist. The small time constant of 0.1 min associated with g_4 provides a negligible filtering of U'_4, while the two, large time constants of g_2 and g_3 in the feedback element literally hold back controller corrective action. As a result, the major portion of the load change is transmitted to the controlled and measured variable as shown in Figure 8-4.

The conclusion is relatively obvious—for good control, large time constants should be downstream of load changes or in the forward element; and small time constants should be upstream of load changes or in the feedback element. When these conditions cannot be obtained because of process considerations, special control techniques, which will be discussed in Chapter 13, must be applied.

8.4. Influence of Transfer Function Gain on Load Change Time Responses

The effect of transfer function gain terms on load change time responses may be seen in both the steady-state and transient portions of the response. In Chapter 7, it was noted that closed loop zero frequency gain or b/r should equal 1 to eliminate offset. To achieve this goal, open loop gain had to be very much greater than 1, with no conditions as to the location of this gain in the loop. The closed loop equations in the solution to Problem 3 of this chapter, which are given here for convenience, tell a different story:

$$b/U_2 = (1/g_1)(b/r) \tag{8-20}$$

$$b/U_3 = \left[1/(g_1g_2)\right](b/r) \tag{8-21}$$

$$b/U_4 = \left[1/(g_1g_2g_3)\right](b/r) \tag{8-22}$$

To eliminate offset following a load change, the measured variable b must return to its original value. Stated another way, the change in b in response to a load must be zero. If an ideal design is assumed for which $b/r = 1$, then the only solution that will give $b/U_n = 0$ is a feedback element zero frequency gain of infinity. The practical solution is not infinity, but a number very much greater than 1. An examination of Equations (8-20) through (8-22) reveals that the only foolproof way to do this is to make controller gain large. The reason being that g_1 (the controller transfer function) is the only component that is in the feedback element for all load changes. Since there are no restrictions concerning gain location for a set point change, making the controller gain as large as possible will minimize offset for both set point and load changes. The allowable limit on gain is still determined by stability considerations as noted in Chapter 5.

Effect of gain on the transient portion of the response may be observed by following the load change around the control loop. The time response b/U'_4 as shown in Figure 8-4 is a good example of what should not be done. For U'_4 the downstream or forward element is g_4h. Since h has been assumed equal to 1 in this discussion, g_4 is the only component in the forward element that has an effect on the load change. As previously shown, when the time constant of g_4 is 0.1 min, very little filtering or delaying load action occurs. This, however, is not the only problem. As shown in Equation (8-19), $k_4 = 2$. Thus, the g_4 component not only allows rapid transmission of the load change but actually amplifies the change by a factor of 2. A more desirable condition would be to attenuate the load change by having $k_4 \leq 1$. If k_4 was equal to 1, the controller gain could be increased to 18. As shown in Equation (8-16), the result for reference or set point changes would not change. For load changes, however, the final value or offset would be reduced by a factor of 2 as shown by Equation (8-17). More important than offset would be the increased gain in the feedback element. Controller corrective action in the feedback would produce twice the compensating action and give much better control.

The conclusion is that for good control, gain in the forward element should be small to attenuate load changes, while gain in the feedback element should be large to amplify corrective action and reduce offset. The relative influence of both transfer function gain and time-constant values are shown in Figure 8-5. From Figure 8-5 it is obvious that for good control, large time constants and small gains should be in the forward element or downstream of load changes. Likewise, small time constants and large gains should be in the feedback element or upstream of load changes. If these criteria are met, both initial overpeak and offset are reduced as shown in Figure 8-5.

Unfortunately, theoretical ideals are not always possible to obtain in practice. Physical process demands, mechanical limi-

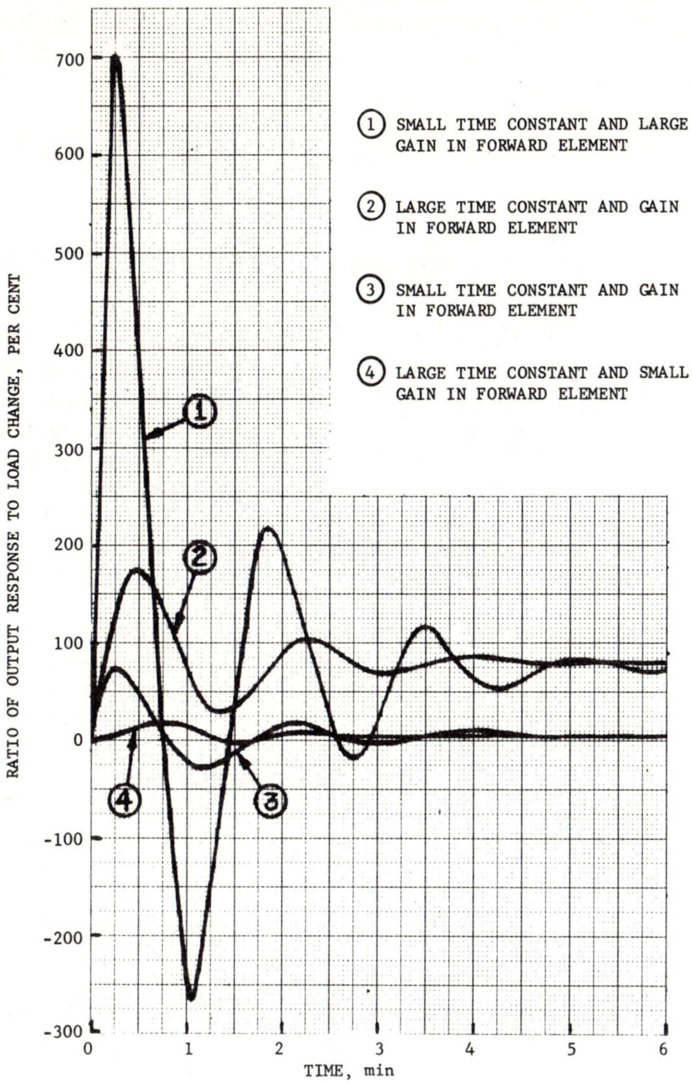

Figure 8-5. Relative influence of transfer function gain and time-constant values on load-change time responses.

tations, and structural requirements quite frequently do not coincide with good control practice. As a result many modifications have been developed in both controller transfer functions and general application techniques. Such modifications represent the practical side of control technology and will now be discussed in Chapters 9 through 14.

Chapter 8—Study Questions and Problems

(Answers appear on page 160)

1. Using the notation in Figure 8-1, develop the transfer function for b/U_4.

2. Note $b/U_4 = -(1/g_3)(b/U_3)$. Using the response shown in Figure 8-2 for b/U_3 and assuming that $g_3 = 5/(1 + 0.4s)$, plot the frequency response for b/U_4.

3. Develop transfer functions for b/r, b/U_2, b/U_3, and b/U_4 based on the block diagram for a closed loop control system with load changes as shown below.

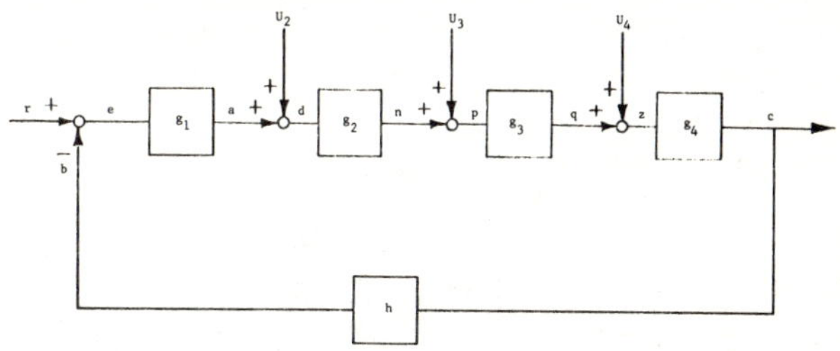

Figure 8-P3. Block diagram for closed loop control system with load changes.

4. The final value in response to a load change in a control loop using only proportional control is also the ________.

5. The final value or offset for a load change may be determined by substituting the ________ or ________ values in the load change transfer function.

6. To achieve good control following load changes, large time constants should be in the ________ element, and small time constants should be in the ________ element.

7. If the corner frequency of the transfer function in the feedback element is less than the ________ frequency of the closed loop frequency response, the conventional second-order shape of the closed loop frequency response will be altered.

8. If the load change frequency response is distorted, may the second-order of spring-mass analog be used to predict the time response?

9. The physical effect of a large time constant in the feedback element is to filter or delay controller corrective action. Does this cause an increase or decrease in initial overpeaking? Increase or decrease in time for the initial overpeak? Increase or decrease in the first full period of the time response?

10. A large time constant in the forward element allows time for controller corrective action to compensate for the load change. What effect does this large time constant in the forward element have on the load change?

11. To minimize offset following a load change in a control loop using only proportional controller action, should gain in the feedback element be small or large?

12. The one loop component that will always be in the feedback element for all load changes is the ________.

13. To obtain good control following a load change, should the forward element gain be small or large in comparison to the feedback element gain?

9 Controllers and Controller Transfer Functions

From the preceding chapters on theory, it is evident that a "process" with one big time constant located just upstream of the controlled variable would be easy to control. If the component also has low gain, the "process" would exhibit ideal conditions. Contrary to most rules, such ideal conditions do occasionally prevail in process control. Probably, the most common example is a home heating system. The house transfer function has small gain and a large time constant, due primarily to its massive thermal capacity. From a physical viewpoint, it takes a long time and many units of thermal energy to change the house temperature. If smaller time constants associated with the furnace and sensing element are combined with the house transfer function, a "process" frequency response (Figure 9-1) results. If the Zigler-Nichols method for calculating controller settings is used (Chapter 5), it is easy to see that a high-gain or on-off controller may be used. These calculations are shown in Figure 9-1. Although the on-off controller causes a cyclic time response, temperature amplitudes are small, and satisfactory control is achieved. This is an ideal system that satisfies theory, compensates for ambient load changes, is relatively inexpensive, and seldom requires maintenance.

When "process" conditions are not ideal and controller gain must be reduced to obtain stability, problems of offset and overpeak are created. To overcome these problems special controller functions have been introduced and added to the controller. The study of these controller functions, their purpose, and means of adjustment are the objectives of this chapter.

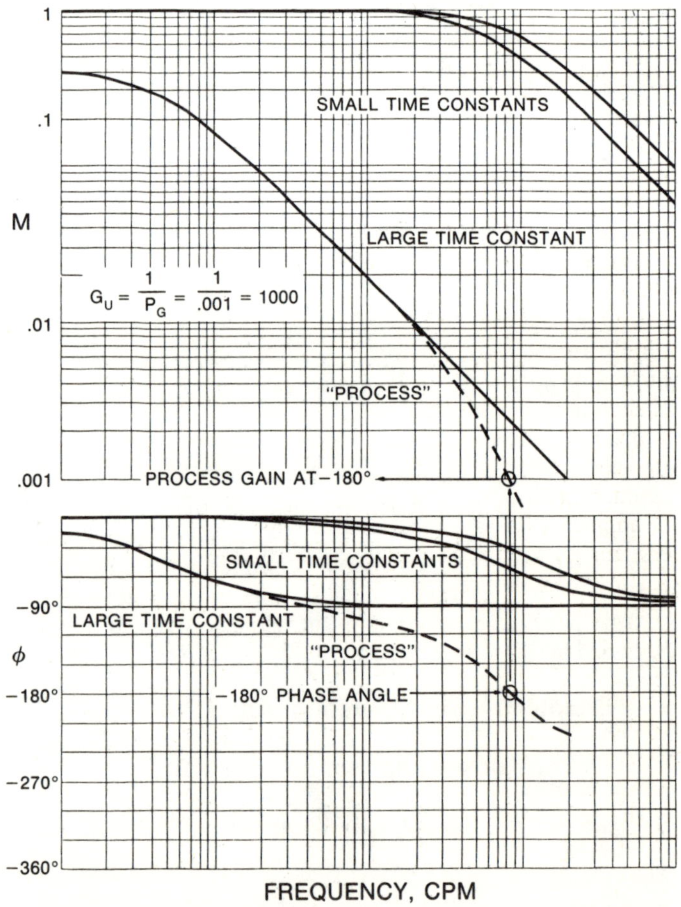

Figure 9-1. Ideal on-off control system (values are relative and not for a specific system).

9.1. The Proportional Controller

In order to explain specialized controller functions, it is necessary to examine the proportional controller in more detail. Due to the variety of commercial controllers that are available, it

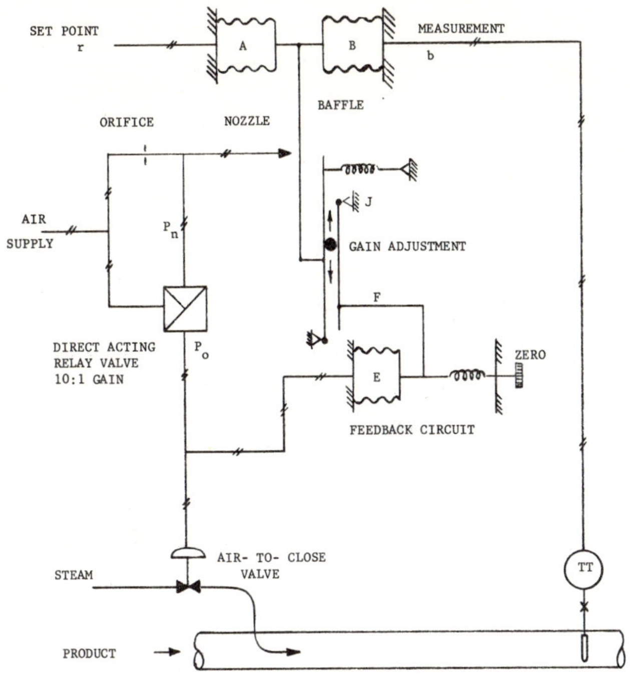

Figure 9-2. Proportional controller.

would be a major task to explain each mechanism in detail. For this reason, a pseudo controller mechanism will be used in which component objectives will be emphasized. To see how a particular controller accomplishes each of these objectives, the reader should refer to specific instruction forms or company training literature.

The pseudo controller, along with a simple heating system, is shown in Figure 9-2. Each controller must have an error-detecting stage that determines the difference between the measured variable and reference or set point. This function is performed as shown in Figure 9-2 by the A and B bellows. Any change in measurement or set point causes a baffle movement which is sensed at the nozzle. The combination of nozzle, baffle, and orifice is the heart of any pneumatic controller. Its counterpart in other controller systems is a high-gain transducer or amplifier. By proper selection of nozzle and orifice diameters, for example, sensitivity or gain may be as high as 8000 psi/in.[1] With this high sensitivity it is only necessary to have a change of 0.0015 in. at the baffle to change nozzle pressure P_n from 3 to 15 psi. Thus, it is easy to see how small measurement changes at bellows B could cause an on-off response in P_n.

Although a high-gain or on-off controller could be made by connecting P_n directly to the control valve, this is seldom done. Most high-gain transducers or amplifiers are signal rather than power devices, and as such perform poorly under direct loading. In the nozzle-baffle-orifice combination, for example, air flow to inflate the control valve motor would be limited by the orifice. The combination of orifice resistance and valve motor capacity would produce a big time constant. This result, as noted in Chapter 8, would be contrary to good design. To overcome this problem a power amplifier is usually included in the control circuit.

In pneumatic controllers the power amplifier is usually referred to as a relay valve. The relay valve in Figure 9-2 is a direct-acting amplifier with a gain of 10. Thus, a change of 2 psi in P_n causes a change of 20 psi in output pressure P_o. The relay valve has very little flow resistance in comparison to the orifice and can quickly fill the control valve motor under most circumstances. There are many variations in relay valves or power amplifiers and also in their characteristics.[2] The reader is again referred to manufacturer's instruction forms for specific details.

With a gain of 10 in the relay valve, a baffle movement of only 0.00015 in. will cause P_o to vary from 3 to 15 psig. To obtain proportional control, it is fairly obvious that some technique must be found to reduce this high gain. This is achieved through a feedback circuit as shown in Figure 9-2. Since controller output is a function of baffle motion, reducing overall baffle motion should reduce output and provide reduced controller gain. To show how this works assume the gain adjustment ball is raised to pivot point J, and that a measurement change at bellows B causes a baffle movement of 0.00015 in. at the nozzle. This causes a 12-psi change from 3 to 15 psig in P_o as previously noted. Although P_o is fedback to bellows E and drives lever F, there is no effect on the baffle since the gain adjustment ball is at pivot point J. Controller gain or sensitivity is

$$\text{Case I: } P_o/b = 12 \text{ psi}/0.00015 \text{ in.} = 80{,}000 \text{ psi/in.} \quad (9\text{-}1)$$

If the gain adjustment ball is lowered, any motion of lever F will reduce overall baffle movement at the nozzle. For example, assume the ball is positioned such that a 12-psi change in bellows E causes a baffle movement of 0.010 in. at the nozzle. Since this motion is in opposition to the input, an input signal of 0.01015 in. of baffle movement must be generated by bellows B to produce a resultant baffle movement of 0.00015 in. For this condition, controller gain or sensitivity is

$$\text{Case II: } P_o/b = 12 \text{ psi}/0.01015 \text{ in.} = 1{,}182 \text{ psi/in.} \quad (9\text{-}2)$$

The result is that controller gain or sensitivity has been reduced by a factor of 68. It should be noted that nozzle-baffle sensitivity or gain has not been changed. The feedback effect has reduced overall baffle motion such that a greater change is required at bellows B to cause the same output at P_o.

The zero spring in the feedback circuit was discussed in Chapter 7, and particularly in Equation (7-24). Changing the zero does not affect gain. It does, however, cause a change in P_o. Thus, a way is provided to initially adjust P_o to a value that will cause measurement b to equal the reference or set point r for a given load.

In Chapter 5 where system stability was explained, direct and reverse controller action was considered. The controller in Figure 9-2 is direct-acting in that an increase in measurement causes an increase in output. Action of this controller may be reversed by changing inputs to bellows A and B. With the measurement sensed at bellows A, an increase in measurement causes a decrease in P_o. All other components act as previously described. Most commercially available controllers may be reversed by similar techniques which are equally as simple.

While a pneumatic circuit was used to explain proportional control, the reader should recognize that the same results can be obtained using electrical and hydraulic components.[1] The basic components, such as error detector, high-gain transducer or amplifier, power amplifier and feedback circuit, however, will normally exist in all controller forms.

There are of course other specialized forms of proportional controllers such as a time-proportioning controller. In this type of controller the output is either on or off, rather than throttled. Proportional effect is achieved by regulating the ratio of on-to-off time. This type of controller has some advantages in electrical heating applications but is normally not used to regulate valves. For these special types of proportional controllers, the reader is again referred to manufacturer's literature.

9.2. Proportional Controller Units

In the early stages of controller development manufacturers working independently labeled proportional controller adjustments with a variety of units and ranges. These varied from simple increase and decrease arrows to some generalized concepts that are presently in use. The problem of defining an adjustment unit stems from the variety of temperature, pressure, flow, and other sensing elements that may be used with a common controller. In general the adjustment unit depends on controller input and output ranges and not on the actual variable that causes this input. Keeping in mind that a transfer function is the output-to-input ratio, this arrangement is not too surprising. The three most common units, namely gain, proportional band, and sensitivity, will be described. For other units the reader will have to depend on manufacturer's literature.

As noted in Chapter 1 Figures 1-5, and 1-6 the measured variable may be directly connected to a controller or relayed through a transmitter. Although the resultant effect is the same in either case, the different techniques are helpful in explaining units. The two possible techniques are shown in Figure 9-3. Many early controllers, as well as today's models, had their proportional control action calibrated in terms of proportional band or throttling range. By definition, proportional band is the percent of input range required to cause a 100% change in output. To see how this works assume the direct connected system of Figure 9-3a. For a controller dial setting of 100% proportional band, the measured variable would have to change 100°F (from 100 to 200°) to create an output change of 12 psi (from 3 to 15 psig). If the controller dial was changed to a proportional band (PB) of 50%, a measurement change of 50°F would cause the same 12-psi change at controller output. With a controller dial setting of 200%, a 100°F change in measured variable would cause a 6-psi change at controller output.

A similar procedure was used for controllers calibrated in terms of sensitivity. In this case, however, chart width was included in the analysis. Assume again the direct connected system of Figure 9-3a, with a 4-in. chart width. A 100°F change in the measured variable for 100 to 200°F, would drive the recorder pen full range of 4 in. To obtain a change of 12 psi at the output, a sensitivity S of 3 psi/in. is required. Thus, sensitivity was defined as *the output change in psi per inch of pen movement.* Using a 4-in. chart and a 12-psi output range, a sensitivity dial setting of 3 psi/in. produces the same effect as a 100% PB. By changing the dial setting to 6 psi/in., an output change of 12 psi

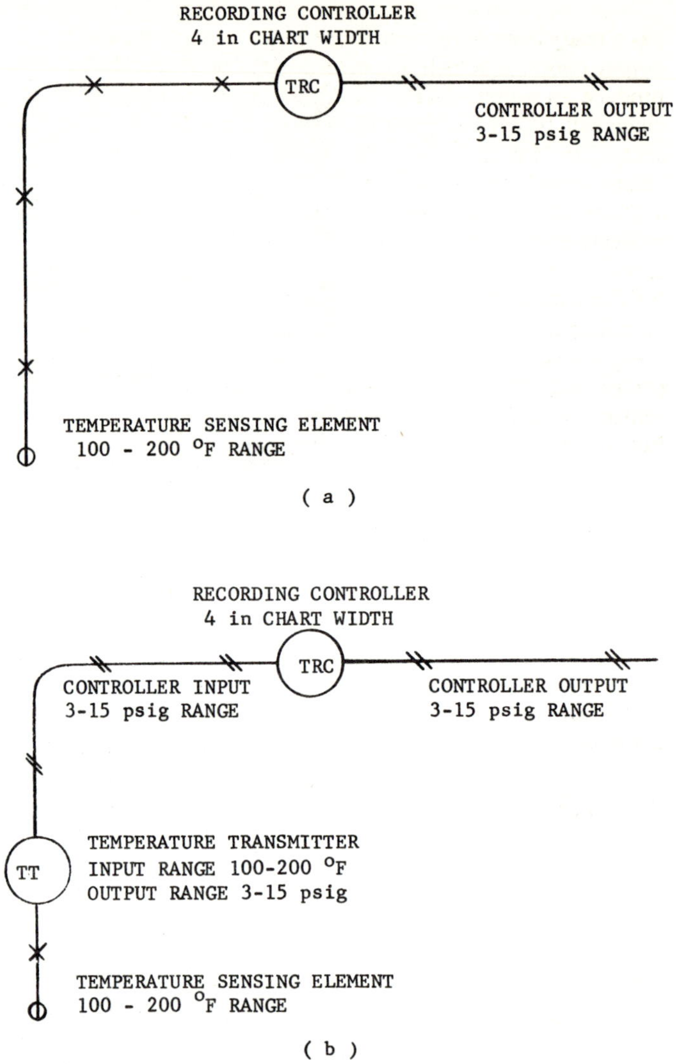

Figure 9-3. Measured variable inputs: (a) direct connected, (b) transmitter interface.

can be obtained for only 2 in. of chart movement or a measured variable change of 50°F. This same result was obtained for a 50% PB setting. As suspected, there is relationship between proportional band and sensitivity. Unfortunately for the control user, the two units are inversely related as shown by Equation (9-3):

$$S\,(\text{psi/in.}) = (100\%)\,(\text{Controller Output Range, psi}) / \left[(\text{PB}\%)\,(\text{Recorder Chart Width, in.})\right] \qquad (9\text{-}3)$$

Using Equation (9-3), a 4-in. chart width, and a controller output range of 12 psi, equivalent sensitivity for a 200% PB is 1.5 psi/in. In the case of an indicating controller, chart width is replaced by indicator scale width or height.

With the advent of transmitters some manufacturers changed to the gain unit for the proportional dial. While this is very convenient, it does not preclude use of sensitivity or proportional band. This may be shown by reference to Figure 9-3b. It has already been shown that for the 4-in. chart width and 12-psi output range, a sensitivity of 3 psi/in. or PB of 100% will produce an output change of 12 psi for a measured variable change of 100°F. Using the transmitter, however, the controller sees a 3 to 15 psig or 12-psi change when the measured variable changes from 100 to 200°F or 100°F. Since gain G is defined as *a dimensionless ratio of output to input,* controller gain dial must be set at 1 to produce an equivalent output change of 12 psi. It is easy to see that when the measured variable changes only 50°F or 6 psi at the transmitter output (controller input), controller gain must be 2 to produce a 12-psi output change. As previously noted, this same result was obtained for a sensitivity of 6 psi/in. and a PB of 50%. Thus, gain is directly proportional to sensitivity and inversely proportional to proportional band. Equivalence is established in the following equations:

$$G = (S\text{, psi/in.}) \times (\text{Recorder Chart Width, in.})/(\text{Controller Input Range, psi}) \quad (9\text{-}4)$$

$$G = [(100\%)/(PB\%)] \times [(\text{Controller Output Range, psi})/(\text{Controller Input Range, psi})]$$

For standard 12-psi input-output ranges and a 4-in. chart, Equations (9-4) and (9-5) show that gain is equal to 1/3 of a given sensitivity setting and 100 divided by proportional band.

The basic technique in establishing Equations (9-3) through (9-5) is the block diagram. Each controller, for example, may be described as shown in Figure 9-4a as having input and output scale factors, and a multiplier. Using a direct connected controller as shown in Figure 9-3a and the proportional band unit, scale factors are

$$\text{Input Scale Factor} = 100\% \text{ of input}/[100°\text{F (input range)}]$$

$$(\text{output/input}) \text{ multiplier} = (100\%/PB\%)(\text{output/input})$$

$$\text{Output Scale Factor} = 12 \text{ psi (output range)}/100\% \text{ of output}$$

Scale factors are determined by controller input and output ranges; the multiplier by defining the proportional controller unit. In the case of proportional band, for example, it is the percent of input (PB%) required to cause an output change of 100%. Thus, the multiplier is 100% (output)/PB% (input). The actual controller output change for given input signal, using units of Figure 9-4b, may then be calculated as shown in Equation (9-6):

$$\text{output, psi} = (100\%/100°\text{F})(100\%/PB\%)(12 \text{ psi}/100\%)(\text{input, °F}) \quad (9\text{-}6)$$

This same concept may be used for all controllers. It should be noted, however, that in some cases factors may be dimensionless and equal to unity. A controller using sensitivity as the proportional unit is an example of this condition. Assuming a direct connected sensing element, the result is as shown in Figure 9-4c. In this case the output scale factor is actually a part of the multiplier and could be omitted. It is included in this example only to show general procedure.

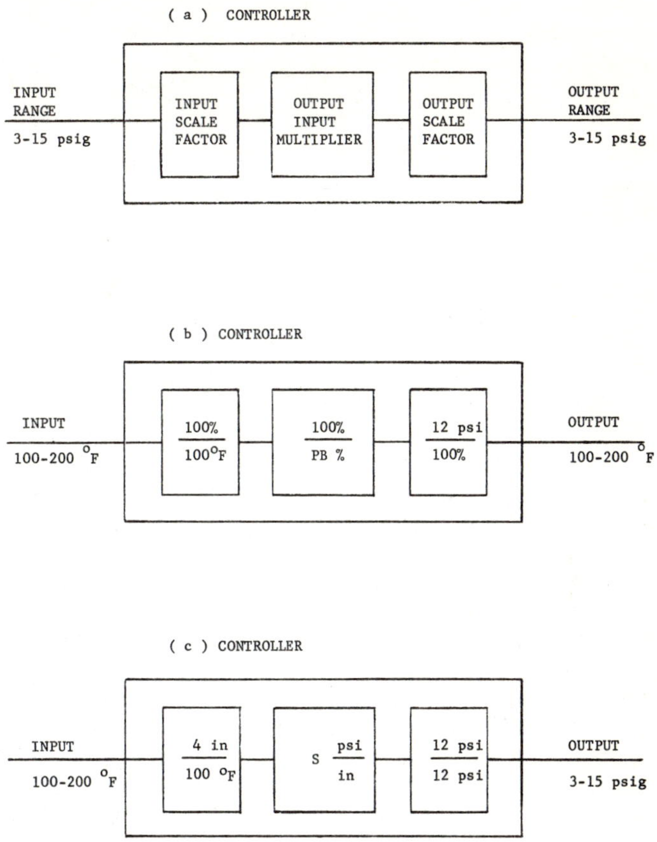

Figure 9-4. Controller unit description.

9.3. Reset or Integral Control, Time Response

The basic problem of a proportional controller is offset, which occurs following set point or load changes. As noted in Chapters 7 and 8, offset can be eliminated or reduced to a negligible value by using high gain. In most systems, however, this is an unsatisfactory solution since high gain will cause an unstable process. What is needed is a controller that will initially respond with low gain to achieve stable control, and then change to high gain to overcome offset. A technique for accomplishing this type of response is shown in Figure 9-5.

As the reader will note, the only difference between Figures 9-5 and 9-2 is that the zero adjustment spring of the proportional controller has been replaced by bellows H and value I. If value I is closed at some fixed pressure level, it is evident that bellows H will react in the same manner as the zero spring of Figure 9-2. Thus, with valve I closed, the controller of Figure 9-5 is nothing more than a proportional controller. If valve I is completely open, the forces of bellows E and H balance, and feedback is removed from the circuit. Lever F under this condition remains fixed, and, independent of the gain adjustment, the controller reverts to an on-off or high-gain controller. Having established potential of both low and high gain in one controller, the next step is to note how the transition takes place.

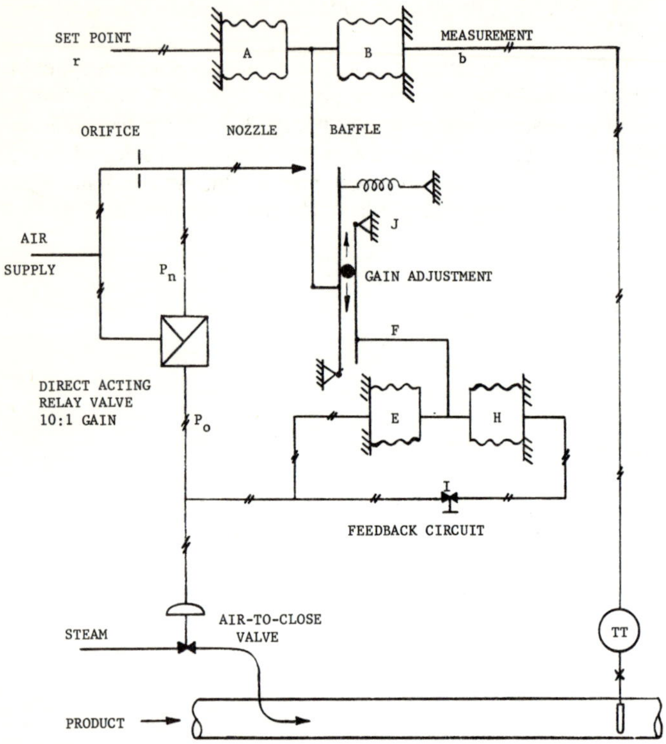

Figure 9-5. Proportional plus reset controller.

Assume equilibrium process conditions with a controller output pressure of 9 psig, and valve I opened approximately to its midposition. If product flow is increased, product temperature will decrease, and the baffle will move away from the nozzle. Initial response is a decrease in output pressure P_o to the control valve motor and to bellows E. The pressure feedback to bellows *E* limits the output pressure change by repositioning the baffle through the gain adjustment mechanism. To be more specific, the controller provides proportional response. Due to the change in output pressure, a pressure differential now exists across valve I. This causes air to flow from bellows H through valve I and has the same result as resetting the zero spring. This moves the baffle away from the nozzle and produces an additional decrease in output pressure. The effect is to increase controller gain. This procedure of course takes time, which allows added steam flow to increase product temperature and reposition the baffle through bellows B. The pattern of events is hard to follow due to continuous changes and can be seen much better by examining the controller output to a fixed input change. Important points to note are (1) initial proportional response, (2) gain increase as a function of flow to or from bellows H, and (3) resetting effect of bellows H. It should be evident at this point why this type of control response was called automatic reset and has evolved to reset control. If it is not, the picture may be improved by looking at a controller response to a step input change.

Assume that the measurement signal to bellows B is replaced by a constant pressure source of 9 psig, which is equal to the set point pressure in bellows A. Output pressure could have any value, but for convenience assume this to be 9 psig also. If the proportional gain adjustment is set at 2 and pressure to bellows B is instantly increased from 9 to 10 psig and held at 10 psig, the output will instantly change from 9 to 11 psig as shown in Figure 9-6. This change results from proportional control action, which causes a 2-psi change in output for an error of 1 psi multiplied by a gain of 2. Bellows H is still at 9 psig, and with the output pressure at 11 psig, a 2-psi drop exists across valve I. Thus, air will flow to bellows H in an attempt to equalize the pressures. As pressure increases in bellows H, lever F is moved to the left, causing the baffle to further reduce nozzle flow. This produces an increase in P_o which causes more flow to bellows H, yielding the result shown in Figure 9-6. The effect of course is like being a little pregnant and continues until output reaches the supply pressure limit. The rate of rise is dependent on the setting of valve I and, as previously noted, may vary from proportional response only when valve I is closed to an on-off response when valve I is completely open. Figure 9-6 shows the concept of low initial gain to achieve a stable record and finally high gain to overcome offset. The output pressure record also shows why this type of controller response is known as integral control. If the time-error product is summed from time zero, the total (since error is constant) would appear as a ramp when plotted against time. This is exactly the response of P_o from 11 psig to the air supply limit. Since summation is equivalent to integration, this controller response is also referred to as integral control.

Having described the controller physical response, it is now necessary to look at the mathematical description. The equation for an ideal proportional plus reset controller[2] is

$$\Delta P_o = Ke + (K/T_r)\int e\,dt, \tag{9-7}$$

where ΔP_o = output pressure change, psi,
K = controller proportional gain,
e = error signal, $r - b = e$,
T_r = reset time, min,
t = time, min.

After examining Figure 9-6 it should not be much of a surprise to see a combination of proportional plus integral terms in Equa-

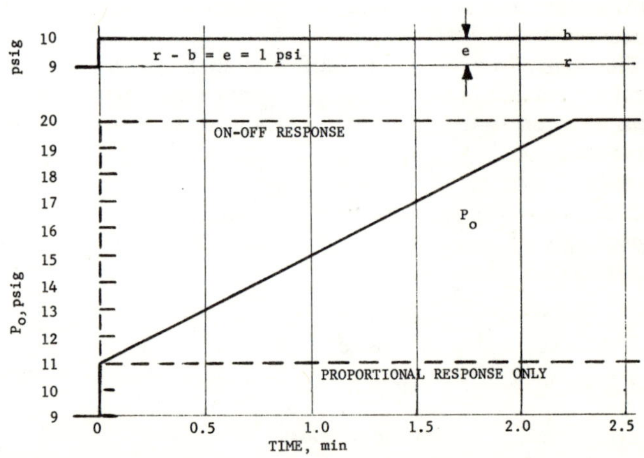

Figure 9-6. Proportional plus reset controller output response to an input step change.

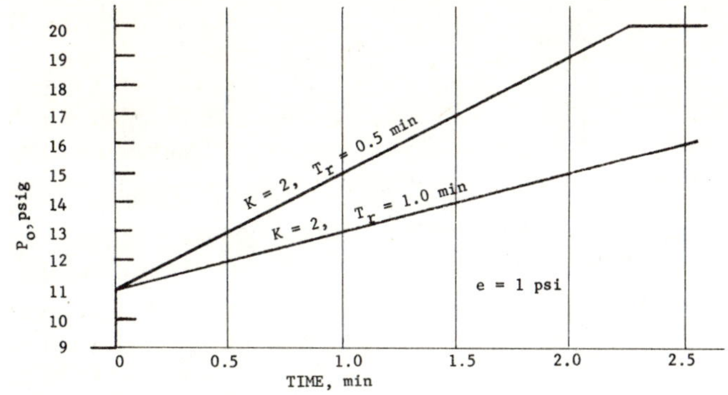

Figure 9-7. Proportional plus reset controller output responses to an input step change for different values of T_r.

tion (9-7). The only new term is reset time T_r. To see what this means from both a mathematical and physical viewpoint, assume $K = 2$, $e = 1$ psi (as previously defined for Figure 9-6), and $T_r = 0.5$ min. If these values are substituted in Equation (9-7) and integration is performed between limits of 0 and t, the result is

$$\Delta P_o = 2\,(1 \text{ psi}) + \left[2/(0.5 \text{ min})\right] \int_0^t (1 \text{ psi})\,(dt \text{ min}) \tag{9-8}$$

Since error is constant at 1 psi, the result is

$$\Delta P_o = (2 + 4t) \text{ psi} \tag{9-9}$$

A plot of Equation (9-9) is shown in Figure 9-7. The reader will note that initial proportional response is 2 psi and occurs in theory instantaneously. This of course happens only in theory. In an actual controller some small rise time is required; however, this can normally be neglected. Also note that at 0.5 min controller output has changed an additional 2 psi from 11 to 13 psig. Stated in another way, reset or integral action of the controller has taken 0.5 min to repeat or produce an output change of 2 psi, which is equivalent to the 2-psi change of proportional action. Thus, T_r or reset time is the time required for reset action to repeat the change caused by proportional action. Some controller dials are labeled min/repeat; however, in most cases the repeat term is deleted.

This time effect is directly related to the setting of valve I, as previously noted. For example, assume valve I was changed from its midpoint to a more closed position. Due to increased valve resistance, a different reset time would result. The effect is identical to a first-order time constant discussed in Chapters 2 and 3, where valve I provides resistance and bellows H capacitance. If, for example, T_r is assumed equal to 1 min, Equation (9-9) changes to

$$\Delta P_o = (2 + 2t) \text{ psi} \tag{9-10}$$

and is as shown in Figure 9-7. Thus, as shown in Figure 9-7, reset time is a measure of how fast gain is changed. Since system stability is a function of controller gain, there must be some relation between process characteristics and reset time. This is true and will be examined in the next section.

9.4. Reset or Integral Control, Frequency Response

The transfer function for a proportional plus reset controller in the frequency domain may be obtained by transforming Equation (9-7) using techniques of Chapter 3. In the transformation of Equation (9-7), it is necessary to transform an integral term, which was not discussed in Chapter 3. One way to handle an integral term is to remove it from the equation by differentiating with respect to time. The result is

$$d(\Delta P_o)/dt = K(de/dt) + (K/T_r)\,e \tag{9-11}$$

Equation (9-11) may now be expressed in operational form by substituting s for d/dt:

$$s(\Delta P_o) = Kse + (K/T_r)\,e \tag{9-12}$$

This may be written as

$$\Delta P_o = Ke + (K/T_r)\,(e/s) \tag{9-13}$$

The result in Equation (9-13) could have been obtained directly from Equation (9-7) by substituting $1/s$ for $\int$ _____ dt. While no attempt will be made at a mathematical proof of this substitution, the reader should recognize that integration is the reciprocal of differentiation. As such, the substitution of $1/s$ (or the reciprocal of s) is a logical substitution for the integral term. For a more rigorous discussion, check a text dealing with Laplace transforms.

The controller transfer function of output-to-input change ratio may now be obtained from Equation (9-13) as follows:

$$\Delta P_o/e = K\left\{1 + \left[1/(T_r s)\right]\right\} \tag{9-14}$$

To obtain the frequency response, let $s = j2\pi f$:

$$\Delta P_o/e = K\left\{1 + \left[1/(j2\pi f T_r)\right]\right\} \tag{9-15}$$

Equation (9-15) may be rewritten in a more conventional form by removing j from the denominator of the bracketed term. This may be done by multiplying numerator and denominator by j:

$$\Delta P_o/e = K\left\{1 + (j/j)\left[1/(j2\pi f T_r)\right]\right\} \tag{9-16}$$

Since $j = \sqrt{-1}$, $j^2 = -1$, and Equation (9-16) may be reduced as follows:

$$\Delta P_o/e = K\left\{1 - j\left[1/(2\pi f T_r)\right]\right\} \tag{9-17}$$

Equation (9-17) is now in a vector form of $a - jb$, which was discussed in Chapter 3. To obtain the controller frequency response, magnitude or gain and phase angle of Equation (9-17) must be determined as a function of frequency. These are

$$M = K\left[1 + 1/(2\pi f T_r)^2\right]^{1/2} \tag{9-18}$$

and

$$\Phi = -\tan^{-1}\left[1/(2\pi f T_r)\right] \tag{9-19}$$

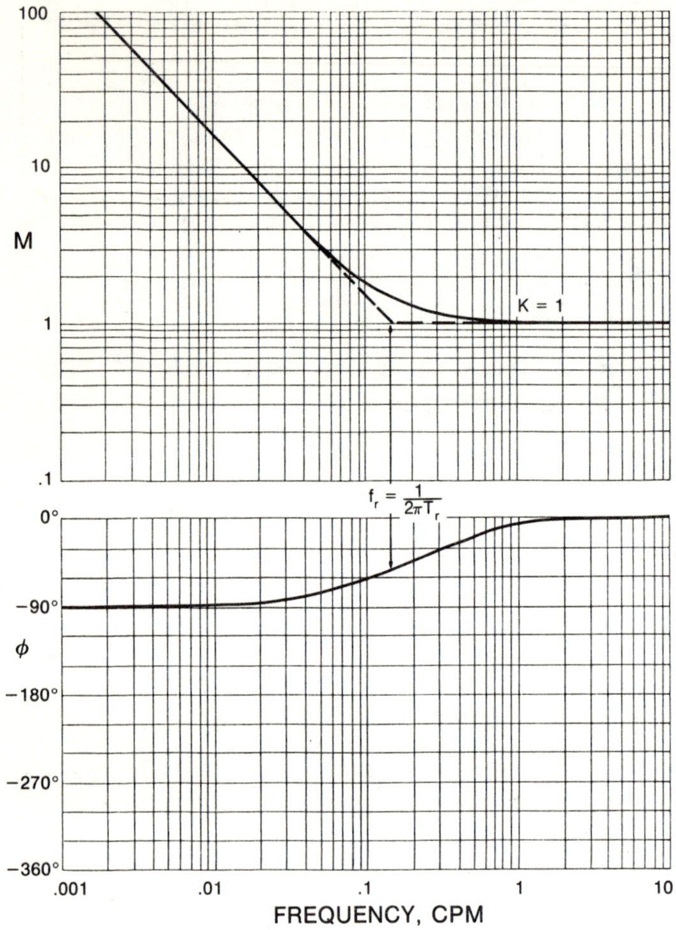

Figure 9-8. Proportional plus reset controller frequency response for K = *1 and* T_r = *1 min.*

If the reader has questions concerning Equation (9-18) and (9-19), Chapter 3 should be reviewed.

A Bode or frequency response plot of Equations (9-18) and (9-19) for $K = 1$ and $T_r = 1$ min is shown in Figure 9-8. Without too much effort the reader should recognize this as being very similar to the first-order lag template developed in Chapter 3. Resultant curves are in fact identical, with the exception that phase angle is reversed and magnitude is reversed and upside down. This should not come as too much of a surprise, since the proportional controller was converted to proportional plus reset by installing a needle valve (resistance) and bellows (capacitance) in the feedback path (see Figure 9-5). To show this relationship in more detail assume that f approaches infinity in Equations (9-18) and (9-19). When this happens magnitude or gain equals K and phase angle equals 0°. In other words, at high frequencies or zero time, the controller reacts as a porportional controller as shown in Figure 9-6. The reader should remember that as f or s approaches infinity, time goes to zero.

Now assume that f approaches zero. From Equations (9-18) and (9-19) it is easy to see that magnitude or gain approaches infinity, and phase angle approaches a lag of 90°. This is equivalent to the high gain that develops after a given time period, as shown in Figure (9-6). If frequency is valued at $1/(2\pi T_r)$, values of magnitude or gain and phase angle are $1.414K$ and −45°, respectively. Since 1.414 is the reciprocal of 0.707, it is easy to see this is the corner frequency condition as described for a first-order lag. Thus, a proportional plus reset controller may be defined by proportional gain and reset time on a Bode plot in the same manner that magnitudes or gain and time constant were used to define exponential lags. In fact, the same template may be used. The reset corner is equal to $1/(2\pi T_r)$ and high-frequency gain is equal to K.

The controller response shown in Figure 9-8 for Equations (9-18) and (9-19) is for a theoretical controller. In a real controller zero frequency gain will not be infinite but will be limited by the high-gain amplifier section of the controller. Thus, an actual controller will appear as shown in Figure 9-9. In a pneumatic controller this zero frequency gain is usually in the range of 300 to 500 and may for all practical purposes be assumed infinite. The same conclusion applies to electronic and other types of controllers, which may and do have higher zero frequency gains. Thus, as the reader will soon see, a theoretical controller may be used in process analysis without introducing any appreciable error. It is more important for the reader to note that (a) changing controller gain will raise or lower gain on a Bode

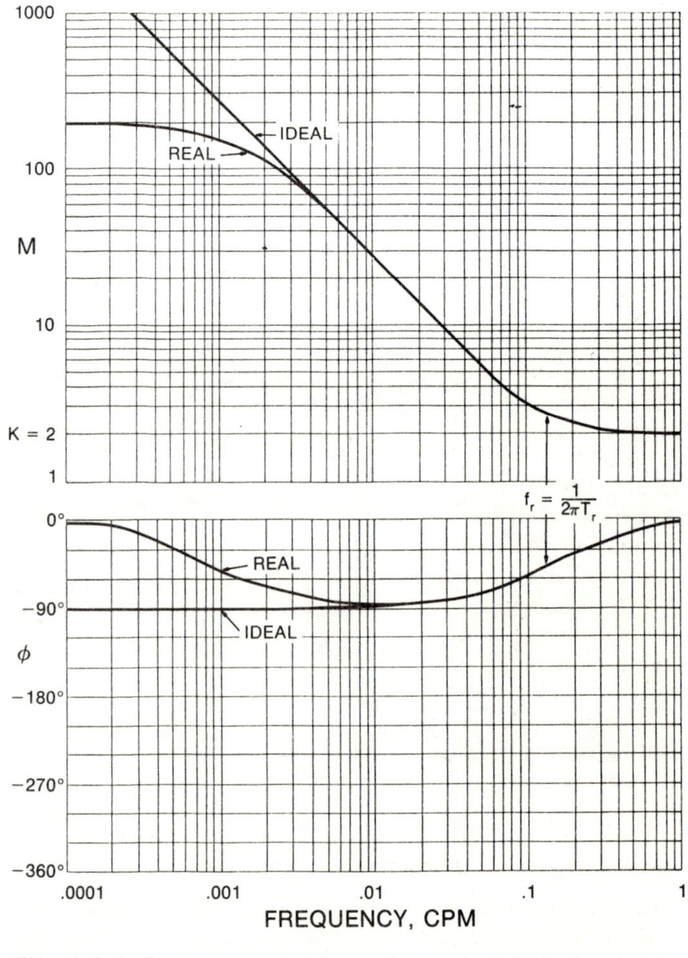

Figure 9-9. Frequency responses for real and ideal proportional plus reset controllers.

diagram but have no effect on phase; whereas (b) changing reset time will move both gain and phase curves to higher or lower frequencies. These concepts will become important in analyzing controller gain and reset time settings for good control.

9.5. Adjusting the Proportional plus Reset Controller

The problem in adjusting a proportional plus reset controller is to find a combination of proportional gain and reset time that will provide initial stability and overcome offset as noted in Section 8.3. Numerous methods have been proposed in the literature to accomplish this task.[4,5] Probably best known is the Ziegler-Nichols method, which was also used in Chapter 5. For a proportional plus reset controller, this method recommends the following settings for optimum control:

$$G_O = 0.45 G_U \tag{9-20}$$

and

$$T_r = P_U/1.2, \tag{9-21}$$

where P_U = ultimate period, min per cycle.

Ultimate period is defined in Section 5.3 as the reciprocal of natural frequency, or period, when the closed loop cycles with only proportional control.

Equations (9-20) and (9-21) result from a combination of theory and practice and are intended to give "ball park" answers. As a rule some additional field adjustment will usually be required, but this is not uncommon with any technique. To see how this method works and why particular numbers were selected assume the simple heating process shown in the solution to Problem 7, Chapter 4. From results of Chapter 7 and particularly Figure 7-7, it should be evident to the reader that this "process" is a prime candidate for reset control.

The first step is to determine G_U and P_U. From the figure in the solution to Problem 7, Chapter 4, "process" gain at a phase angle of −180° is 0.625. Using the same procedure as in Chapter 5 for proportional control, $G_U = 1/0.625 = 1.6$ (see Equation (5-3)). The ultimate period is defined as the cyclic period or reciprocal of natural frequency. Again using the figure in the solution to Problem 7, Chapter 4, the natural frequency or "process" frequency at −180° is 2.0 cpm. As shown in Figure 5-5, if the proportional controller is set for ultimate gain, the closed loop will cycle at 2 cpm or an ultimate period of 0.5 min. These same results may be determined experimentally in the field by locking out reset and increasing proportional gain until the system cycles continuously. Reset may be locked out by closing valve I in Figure 9-5. With valve I closed bellows H acts similar to a spring and the PI controller reacts as a proportional controller.

Having established values of G_U and P_U, it is now possible to calculate G_O and T_r. According to Equations (9-20) and (9-21),

$$G_O = 0.45\, G_U = 0.45\,(1.6) = 0.72, \tag{9-22}$$

$$T_r = P_U/1.2 = 0.5/1.2 \cong 0.42 \text{ min}, \tag{9-23}$$

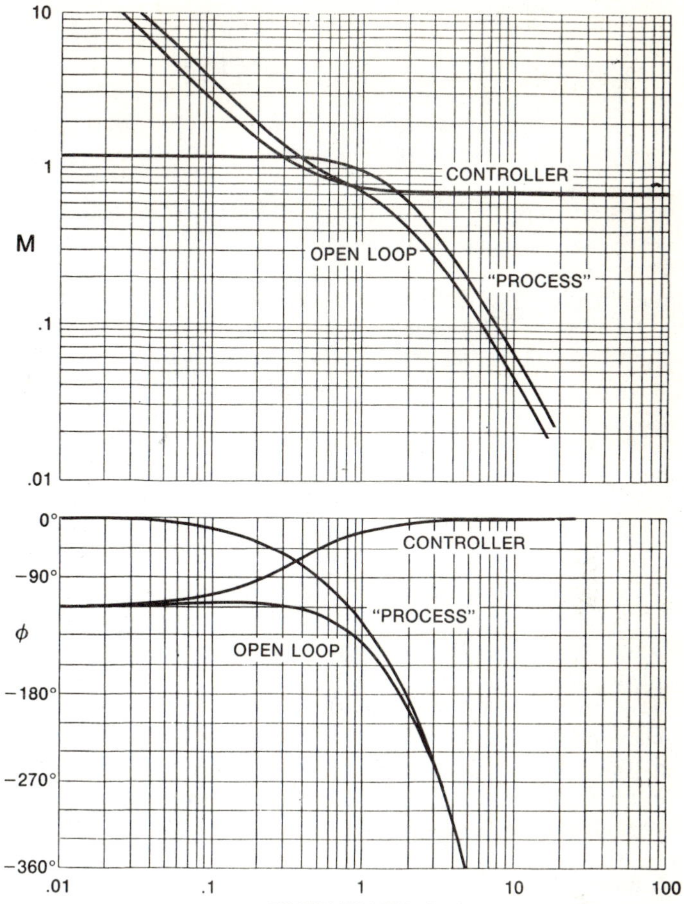

Figure 9-10. Process, PI controller, and open loop frequency responses.

which gives a controller proportional gain $K = G_O = 0.72$, and $T_r = 0.42$ min. Using techniques established in Section 8.4, a Bode plot for the controller may now be established:

$$\text{reset corner} = 1/(2\pi T_r) = 1/[2\pi(0.42)] = 0.38 \text{ cpm} \tag{9-24}$$

The resultant plot and original "process" response from the figure in the solution to Problem 7, Chapter 4, are shown in Figure 9-10. By combining "process" and controller plots, the open loop frequency response can be obtained and is also shown in Figure 9-10. Using the open loop frequency response, it is possible to explain the reset controller in more detail and also to achieve an understanding of numbers in Ziegler-Nichols' Equations (9-20) and (9-21).

It is obvious from Equation 9-8 that to get a high zero frequency gain it was necessary to accept phase lag in the controller. From a purely logical consideration this would appear to be a step in the wrong direction. The result shown in Figure 9-10, however, shows that this phase lag can be tolerated at low frequencies without causing any problems. The problem occurs if reset is too fast. A short reset time T_r, as shown in the figure solution to Problem 13 of this chapter, moves the reset corner to a higher frequency and introduces phase lag at the $-180°$ frequency. Actually, as shown in Figure 9-10, the controller adds approximately 10° of lag at this frequency, namely, 2 cpm. This amount of lag is considered acceptable and is the basis for selecting 1.2 in Equation (9-21). The particular result for a given "process" of course depends on the relation of "process" phase curve and ultimate period and explains why only ball park answers are expected with this technique. The 10° of additional phase lag from the controller, when added to the "process" curve, lowers the so-called phase crossover frequency or frequency at $-180°$. As shown in Figure 9-10, crossover frequency is lowered from 2.0 to about 1.7 cpm. At this lower frequency "process" gain is greater than at 2.0 cpm. Thus, to keep approximately the same closed loop stability, it is necessary to reduce proportional gain from 0.5 to 0.45 G_U. As any experienced instrument engineer will note, accuracies implied by these numbers are questionable. The important concept is that, with reset control, proportional gain must be lowered to maintain stability similar to what would be obtained using only a proportional controller.

Reducing proportional gain and phase crossover frequency has two detrimental effects. With a lower proportional gain there is usually more overpeaking, and with a lower crossover frequency the closed loop period is longer. If the reader doubts these results, a review of Chapters 6 and 7 is recommended. These results are accepted, however, to obtain the desired objective of eliminating offset. A comparison of closed loop time-response records for a process set point change, using P and PI controllers, is shown in Figure 9-11. The record for PI control clearly shows increased overpeaking, a longer period, and elimination of offset. Results for a load change in the same control loop are shown in Figure 9-12. The records of Figures 9-11 and 9-12 show that increases in period and overpeak are a small price to pay for elimination of offset. It should be emphasized that elimination of offset is the only reason for using reset action.

As with proportional control, manufacturers did not standardize on units for reset action. Mathematics of Equation

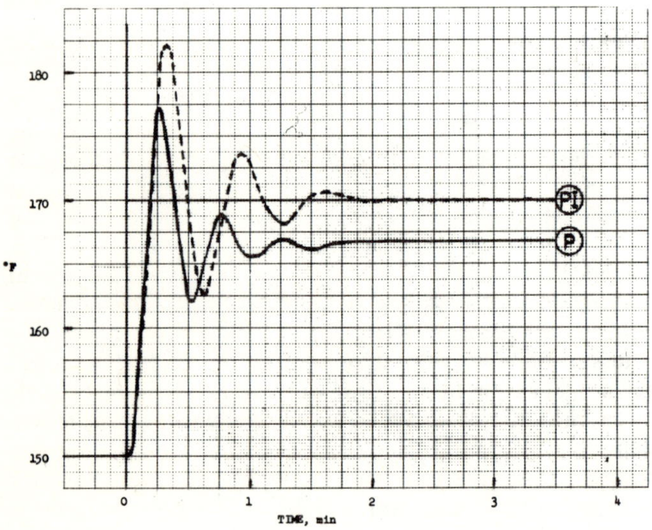

Figure 9-11. Time responses for set point changes using P and PI controllers.

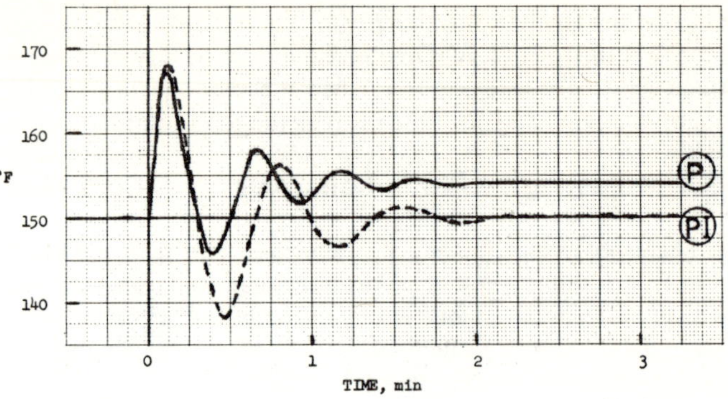

Figure 9-12. Time responses for load changes using P and PI controllers.

(9-7), however, clearly limit the reset "constant" to some function of time. The alternate controller expression is

$$\Delta P_o = Ke + KI \int e \, dt, \tag{9-25}$$

where I = reset rate, min.$^{-1}$

An inspection of Equations (9-7) and (9-25) quickly shows that reset time and reset rate are reciprocals. Stated another way, T_r is the time required for reset to repeat the change caused by proportional action, and reset rate I is the number of times the proportional action change is repeated per minute. From a functional viewpoint, $T_r = 1/I$. Since settings are reciprocals, it is important to note how a particular controller is calibrated. In comparison to the variety of proportional action terms, the problem with T_r or I is relatively simple.

9.6. Rate or Derivative Control, Time Response

Two original control objectives discussed in Chapter 1 were reduction of period and initial overpeak. Although reset action eliminates offset, it does nothing to help problems of period and initial overpeak. In fact, as shown in Figure 9-12, period and initial overpeak are increased when reset action is used. The objective of rate or derivative control is to reduce initial overpeak and period. To reduce initial overpeak and period the controller must provide fast corrective action following a set point or load change. This basic concept is similar to the discussion on load changes in Chapter 8. Good control resulted when the controller could provide corrective action at the point of disturbance and null out the disturbance effect. One way to get fast corrective action is to use high gain. As previously noted, this would cause a stability problem. The solution is to apply high gain to reduce initial overpeak and then reduce gain to maintain stability. Although this appears to be in direct conflict with reset action, the rationale is sound and will be explained in detail. Necessary controller modifications to achieve this type of response are shown in Figure 9-13.

The basic controller circuit shown in Figure 9-13 is identical to the proportional controller of Figure 9-2 with one exception. In Figure 9-13 valve D has been added in the feedback line to bellows E. If valve D is closed at some fixed pressure level,

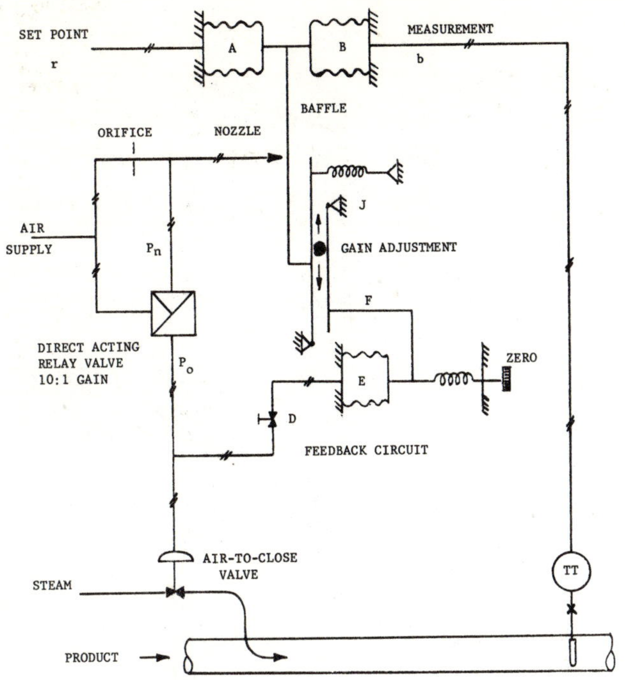

Figure 9-13. Proportional plus rate controller.

proportional feedback is eliminated and the controller acts as a high-gain on-off controller. With valve D open, proportional control results. Thus, by varying the opening of valve D, controller gain can be varied from a high-gain on-off response to reduced gain of proportional action.

One problem in analyzing rate or derivative control is examining initial response. Assume valve D to be somewhere near its midposition with the controller in equilibrium and an output pressure of 9 psig. As before, an increased product flow rate causes a decrease in product temperature and moves the baffle away from the nozzle. The initial response is a decrease in output signal P_o to the control valve motor. With a simple proportional controller, the pressure in bellows E also decreases and limits change in P_o by repositioning the baffle through the gain adjustment mechanism. With valve D partially closed, however, a pressure decrease in bellows E is delayed. This in turn delays proportional feedback action and allows a greater change in output pressure. This is what provides high initial gain. If the original change is sufficiently large or in the form of a step input, it is possible to drive output pressure P_o to saturation conditions, namely zero or 20 psig depending on direction of the input change. For this reason it is common practice to analyze rate response using a ramp-type input. Initial response in output pressure will still occur very quickly and be greater than what would be obtained using only proportional control. The change in P_o, however, causes a pressure drop across valve D and air now flows from bellows E. This initiates feedback action and a return to proportional control. The result is that change in controller output pressure is a function of rate of change of input or error signal and positioning of valve D. To examine rate or derivative response in detail it is necessary to examine the controller equation.

The equation for an ideal proportional plus rate controller is

$$\Delta P_o = Ke + KT_d(de/dt), \tag{9-26}$$

where ΔP_o = output pressure change, psi,
K = controller proportional gain,
e = error signal, $r - b = e$,
T_d = rate time, min,
t = time, min.

Equation (9-26) has a standard proportional response term and the additional rate term. As might be expected, the rate term includes rate of error change de/dt and rate time T_d, which is a function of valve opening. To examine this equation in detail, assume that the error signal is changing as a ramp function, as previously noted. Mathematically, this is expressed as $e = At$. Assuming values of $K = 2$, $A = 2$ psi/min, and $T_d = 0.5$ min, Equation (9-26) may be reduced as follows:

$$\Delta P_o = 2(2t) + 2(0.5)\left[d(2t)/dt\right] \tag{9-27}$$

A plot of error versus time in Figure 9-14 shows that slope of the error function of de/dt is equal to 2. Thus, Equation (9-27) may be reduced to

$$\Delta P_o = 4t + 2 \tag{9-28}$$

A plot of P_o assuming an initial equilibrium pressure of 9 psig, is shown in Figure 9-14. In order to explain rate time in more detail, response of a simple proportional controller is also shown

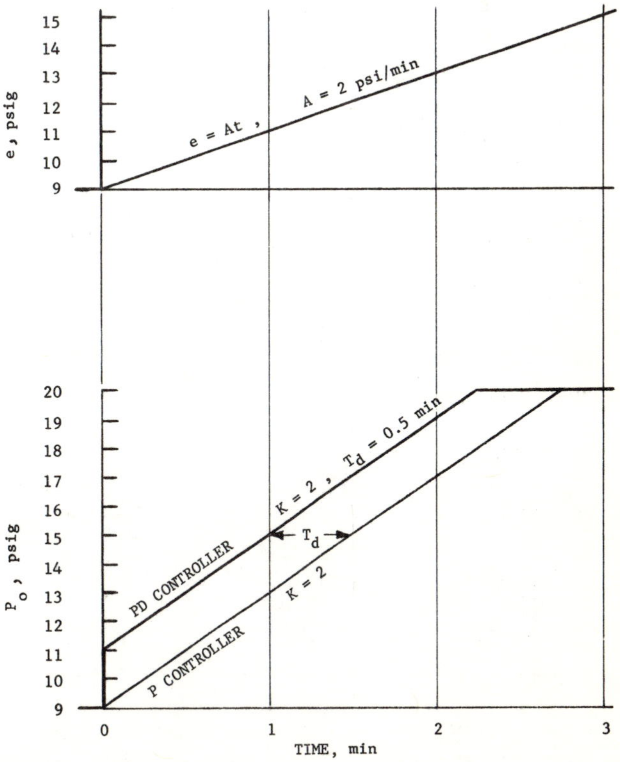

Figure 9-14. Proportional plus rate controller output response to a ramp input.

in Figure 9-14. The effect of rate response is to produce an initial change proportional to the rate of change of error signal de/dt. Magnitude of this initial change depends on de/dt and proportional constants K and T_d. This is why a step input cannot be used to study rate action. For a step input, de/dt equals infinity, and controller output would saturate at zero or supply pressure. Following this initial change the controller behaves as a proportional controller as seen in Equation (9-28) and in Figure 9-14. This is true only if the error continues at the same rate. Rate or derivative time T_d as shown in Figure 9-14 is defined as the time by which output of a proportional plus derivative controller leads the output of a proportional controller. As originally assumed, $T_d = 0.5$ min. By providing an initial gain increase and then responding with basic proportional action, the PD controller is able to reduce initial overpeaking and period and still maintain stability.

If proportional gain is reduced from $K = 2$ to $K = 1$, the result is as shown in Figure 9-15. General shape of the PD curve remains fixed and still leads proportional only response by T_d. Initial gain response is reduced, however, and also slope of the proportional response. If T_d is changed from original value of 0.5 to 0.25 min, results are as shown in Figure 9-16. In this case, only magnitude of the initial response is reduced. From Figures 9-15 and 9-16 it is easy to see that adjusting proportional gain has much the same influence as if only a proportional controller were used. Adjustment of T_d, on the other hand, determines initial amplitude or lead and must again be determined from process dynamics. This adjustment or setting of rate time can be critical, and this is especially true if rate action is improperly

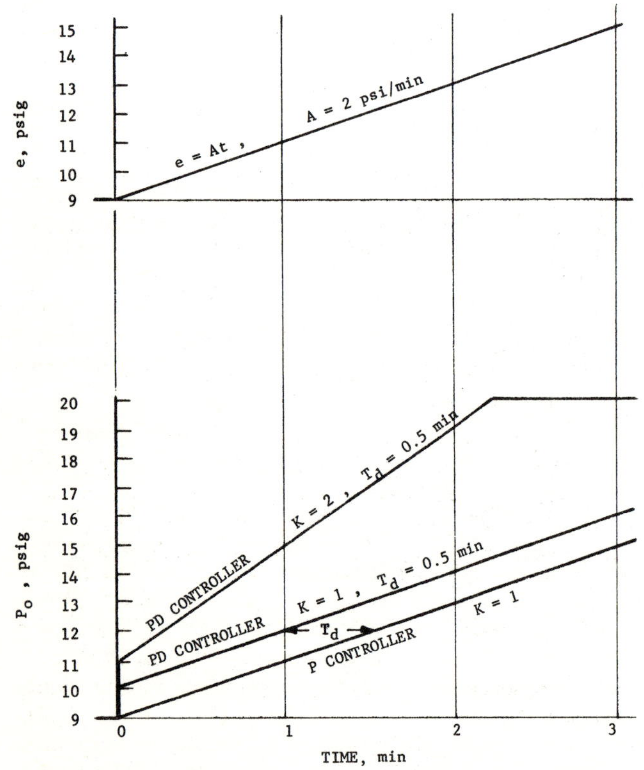

Figure 9-15. Proportional plus rate controller output responses to a ramp input for different values of K.

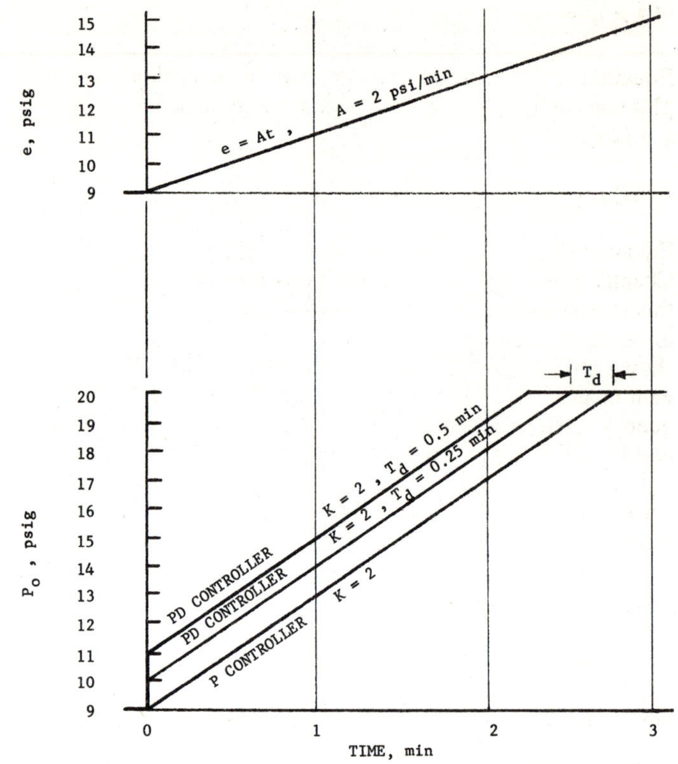

Figure 9-16. Proportional plus rate controller output responses to a ramp input for different values of T_d.

employed. This, however, is not the fault of the controller and will be discussed in more detail in the next section.

In Figures 9-14 through 9-16 input error has been changed at a constant rate. Controller outputs shown are the response to this type of signal or open loop results. In a closed loop application, the error signal would change due to controller action on the "process." One important concept to note is that rate action only functions when error is changing as shown by Equation (9-26). For this reason rate action must be used in conjunction with a proportional controller, since without proportional action the rate controller would be satisfied with any constant error signal. This also means that offset can exist with a proportional plus rate action controller. For this reason it is very common to have a PID (proportional, reset, rate) controller. Rate action functions to reduce initial overpeaking and period, and reset to eliminate offset. A clear interpretation of these functions is important to proper controller application and will be discussed under three term controllers.

9.7. Rate or Derivative Control, Frequency Response

The transfer function for a proportional plus rate action controller in the frequency domain may be obtained by transforming Equation (9-26). The result is

$$\Delta P_o = Ke + KT_d se \tag{9-29}$$

Here again d_____/dt has been replaced by s as detailed in Chapter 3. Output-to-input ratio may now be found as follows:

$$\Delta P_o / e = K(1 + T_d s) \tag{9-30}$$

Equation (9-30) is the transfer function for an ideal proportional plus rate controller. In order to obtain the frequency response, let $s = j2\pi f$:

$$\Delta P_o / e = K(1 + j2\pi f T_d) \tag{9-31}$$

Equation (9-13) is in the vector form of $a + jb$, as discussed in Chapter 3. Rate action provides a very interesting result and for this reason merits a detailed review of Equation (9-31). Based on a vector diagram, Equation (9-31) may be expressed as shown in Figure 9-17. As can be easily seen from Figure 9-17, gain or magnitude M and phase angle Φ are functions of frequency f. This assumes that K and T_d are fixed. Influence of K and T_d will be observed shortly. The most interesting observation, however, is that phase angle is always positive. In fact, Φ varies from 0 to 90° as frequency changes from 0 to infinity, respectively. In other words, output ΔP_o leads input e. This confirms time-response plots of Section 9.6. Values for M and Φ using Figure 9-17 are

$$M = K\left[1 + (2\pi f T_d)^2\right]^{1/2} \tag{9-32}$$

and

$$\Phi = \tan^{-1}(2\pi f T_d) \tag{9-33}$$

A Bode or frequency response plot for $K = 2$ and $T_d = 0.5$ min is shown in Figure 9-18. As for reset, the reader should recognize this plot as a simple variation of the first-order lag. This should not come as any surprise, since the rate circuit is established by placing a resistance (valve D) in series with a capacitance (bellows E) as shown in Figure 9-13. Thus, it is again possible to use the first-order or exponential lag template to sketch a PD controller. In this case the rate corner or corner frequency is equal to $1/(2\pi T_d)$. This result is shown in Figure 9-18.

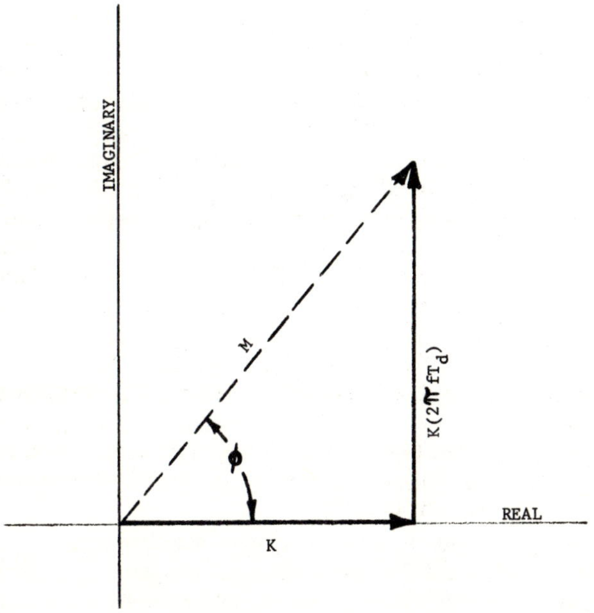

Figure 9-17. Vector diagram of PD controller transfer function.

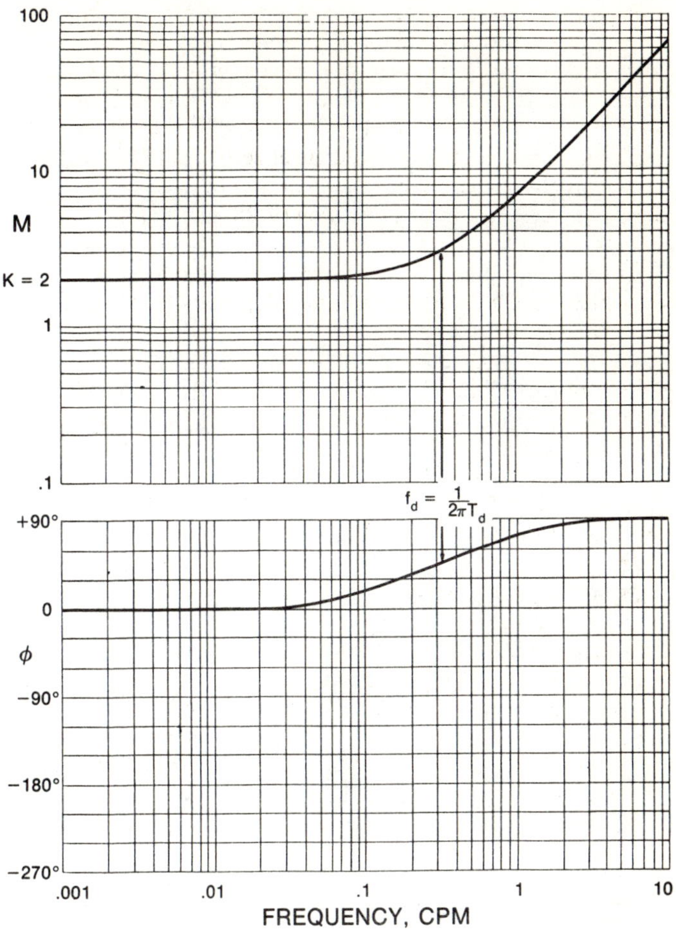

Figure 9-18. Proportional plus rate controller frequency response for $K = 2$ and $T_d = 0.5$ min.

Effects of K and T_d on the frequency response of a PD controller are very similar to what was experienced with a proportional plus reset controller. As can be seen from Equations (9-32) and (9-33), changes in controller proportional gain affects only gain or magnitude. Changes in rate time, however, influence both magnitude and phase. These effects will again become important in the discussion on controller settings.

One problem that exists with an ideal PD controller is noise. Many industrial processes such as flow control loops can be very noisy. Noise may result from vapors entrained in liquid flow or from pulses developed from pumps and compressors. While noise in general is objectionable, it is particularly detrimental if a rate controller is used. Noise by definition is an extraneous signal at frequencies in excess of fundamental process signal frequencies. As can be seen from Figure 9-18, rate controller action provides high gain at high frequencies. This amplification of noise in a controller with rate response can cause a variety of problems ranging from loss of control to excessive valve wear. For this reason rate control is not applied to processes with

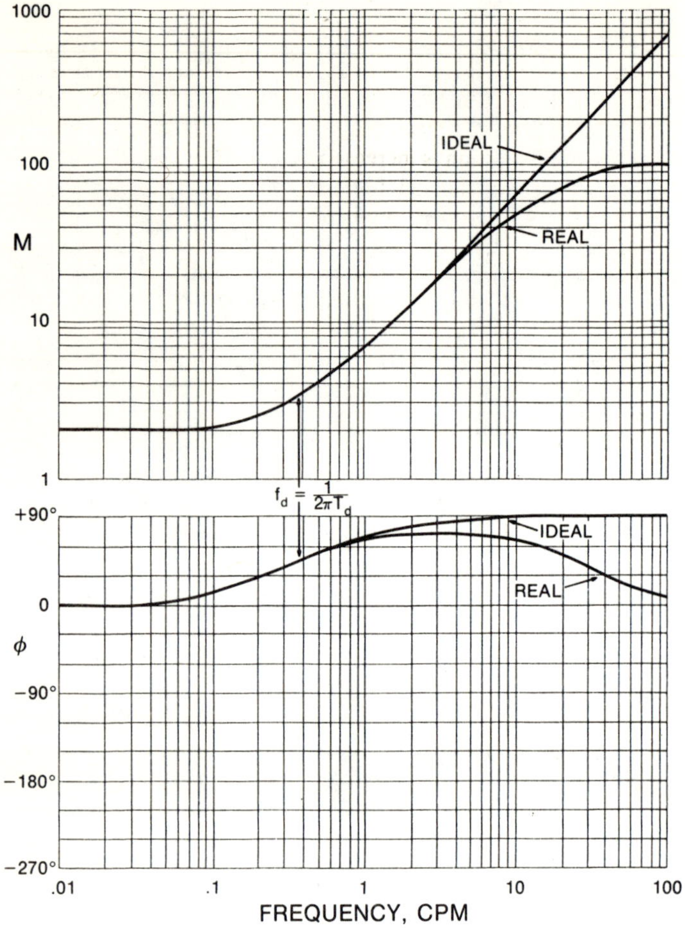

Figure 9-19. Frequency responses for real and ideal proportional plus rate controllers.

known noise problems. Even under the most ideal conditions theoretical or ideal rate response is not used. Most industrial controllers have a built in lag circuit which reduces high-frequency gain. A real or practical controller response is shown in Figure 9-19. This is a practical compromise which reduces the noise problem while maintaining desirable advantages of rate response.

Advantages of reducing initial overpeaking and period will become more evident in the following discussion on controller adjustments. Since the noise problem occurs at frequencies beyond normal process signals, it can usually be neglected in the controller adjustment procedure. Thus, as was the case for reset, a simple, ideal rate response transfer function may be used.

9.8. Adjusting the Proportional plus Rate Controller

As with proportional plus reset control, the Ziegler-Nichols method will be used to establish controller settings. The problem is to find a combination of gain and rate time that will minimize overpeaking, reduce period, and maintain stable control. For proportional plus rate control, the following initial adjustments are recommended:

$$G_O = 0.6G_U \tag{9-34}$$

and

$$T_d = P_U/8 \tag{9-35}$$

Values for G_U and P_U were found in Section 9.5 to be 1.6 and 0.5 min, respectively. As noted in Section 9.5 these values were determined from the "process" frequency response shown in the figure in the solution to Problem 7, Chapter 4. The reader may check that figure and Section 9.5 if a review of these calculations is required.

Having established G_U and P_U, it is possible to calculate G_O and T_d from Equations (9-34) and (9-35):

$$G_O = 0.6\,G_U = 0.6(1.6) = 0.96 \tag{9-36}$$

$$T_d = P_U/8 = 0.5/8 = 0.06 \text{ min} \tag{9-37}$$

The rate corner may now be calculated in order to establish a Bode plot for the controller:

$$\text{rate corner} = 1/\left[2\pi T_d\right] = 1/\left[2\pi(0.06)\right] = 2.67 \text{ cpm} \tag{9-38}$$

The resultant controller frequency response and original "process" response from the figure solution to Problem 7, Chapter 4, are shown in Figure 9-20. By combining controller and "process" plots, the open loop frequency response for the system is obtained. The open loop response is also shown in Figure 9-20. With this plot it is now possible to explain rate controller action in more detail and also achieve an understanding of empirical constants in Equations (9-34) and (9-35).

An examination of Figure 9-20 shows that the open loop phase angle curve has been shifted to higher frequencies. This is a result of phase lead from the rate controller. In the application of reset control as shown in Figure 9-10, the controller added phase lag and caused a reduction in crossover frequency (frequency at −180°). This, as the reader should remember, caused an increase in period. By increasing crossover frequency as shown in Figure 9-20, period is reduced. It is worth noting that, at 2 cpm in Figure 9-20, the controller provides approximately 35° phase lead. One technique for adjusting a rate controller suggests 30° phase lead at the "process" crossover frequency. To apply such a method for field tuning is of course out of the question. Reciprocal of "process" crossover frequency is ultimate period, and a major clue to the Ziegler-Nichols technique of Equation (9-37). Using $P_U/8$ to determine T_d will normally result in a rate setting that produces approximately 30° phase lead at this freuency. This technique assumes process uniformity that does not exist. In general, however, a stable setting will be determined which can be tuned to obtain desired degree of stability.

The increase in crossover frequency also has a gain effect. "Process" gain at the original crossover frequency of 2 cpm is 0.6, as shown in Figure 9-20. At the new crossover frequency of approximately 2.5 cpm, "process" gain is about 0.48. Thus, it should be possible to increase controller gain and maintain

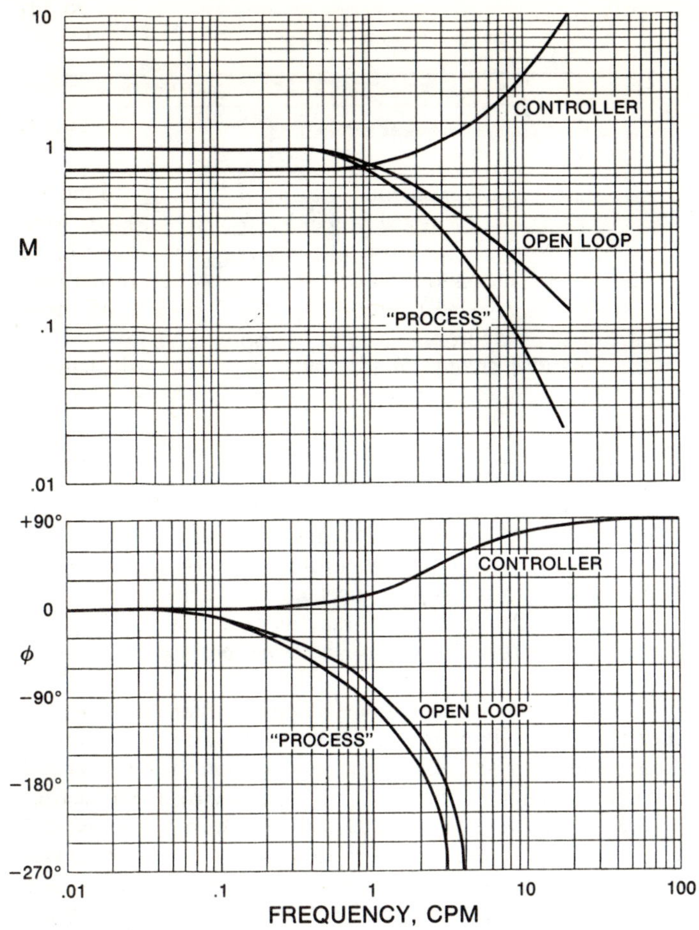

Figure 9-20. Process, PD controller, and open loop frequency responses.

equivalent stability. This is the justification for the 0.6 value in Equation (9-36). The reader will remember this value as 0.5 for proportional and 0.45 for proportional plus reset control. This technique again assumes a uniformity of process responses, which will not always be true. In fact, the results of Figure 9-20 are not the most desirable for rate control and will be discussed in more detail.

As shown in Figure 7-2 for proportional control and in Figure 9-10 for PI control, an open loop gain of 0.5 is expected when the open loop phase angle is −180°. This is commonly referred to as a gain margin of 0.5 and will provide a closed loop time response having approximately quarter decay. As shown in Figure 9-20, open loop gain is 0.7 when open loop phase angle is −180°. Although a stable response will result since gain is less than 1, it will be considerably more cyclic and generally unacceptable. To obtain the desired quarter decay it would be necessary to reduce controller gain. The fact that so-called optimum or ideal settings are not always obtained is in no way a condemnation of the method. The Ziegler-Nichols technique is very useful in obtaining approximate controller settings, and field tuning should be expected as emphasized by this example.

The increase in crossover period and higher controller gain which can normally be used with a properly applied PD controller will also reduce initial overpeaking. Theory is similar to that applied for analysis of load changes in Chapter 8. As noted in Chapter 8, overpeaking was a function of how much and how fast corrective action could be applied to a disturbance. Since rate control provides faster response and increased controller gain, corrective action is increased both in time and magnitude. The result is a reduction in initial overpeaking. Key words in this discussion are, "a properly applied PD controller." Actually, the "process" of Figure 9-20 is a poor candidate for rate action, as will now be shown.

The "process" transfer function of Figure 9-20 was given in Equation (4-9) and originally plotted in Problem 7, Chapter 4.

$$\Delta P_3/\Delta P_1 = (1/[1+0.1s])(10)(e^{-0.15s})(0.12/[1+0.05s]) \qquad (4\text{-}9)$$

In this process the dominant component is a dead time lag of 0.15 min. As shown in Figure 9-20, this produces a very steep phase slope in the vicinity of the process crossover frequency. As previously noted, rate response is adjusted to produce approximately 30° phase lead at this frequency. Adding this small phase lead to the "process" curve is similar to filling a bathtub with an eyedropper. Some improvement is made, but effort may not be worth the result. Thus, the first conclusion is that rate action is of minor benefit if the "process" is dominated by dead time lag.

In a more positive approach assume a "process" transfer as follows:

$$\Delta P_3/\Delta P_1 = (2/[1+1.2s])(1/[1+0.4s])(e^{-0.1s}) \qquad (9\text{-}39)$$

The "process" frequency response is shown in Figure 9-21. From Figure 9-21 process gain and frequency at −180° are approximately 0.13 and 0.85 cpm, respectively. Ultimate gain and period may then be calculated as follows:

$$G_U = 1/P_G = 1/0.13 = 7.7 \qquad (9\text{-}40)$$

$$P_U = 1/0.85 = 1.2 \text{ min} \qquad (9\text{-}41)$$

Controller settings may now be calculated from Equations (9-34) and (9-35).

$$G_O = 0.6 G_U = 0.6(7.7) = 4.6 \qquad (9\text{-}42)$$

$$T_d = P_U/8 = 1.2/8 = 0.15 \text{ min} \qquad (9\text{-}43)$$

and

$$\text{rate corner} = 1/[2\pi T_d] = 1/[2\pi(0.15)] = 1.1 \text{ cpm} \qquad (9\text{-}44)$$

Frequency response plots of controller and open loop are also shown in Figure 9-21.

Results shown in Figure 9-21 clearly indicate what can happen when rate control is properly applied. With proportional control only, this system has a crossover frequency of 0.85 cpm

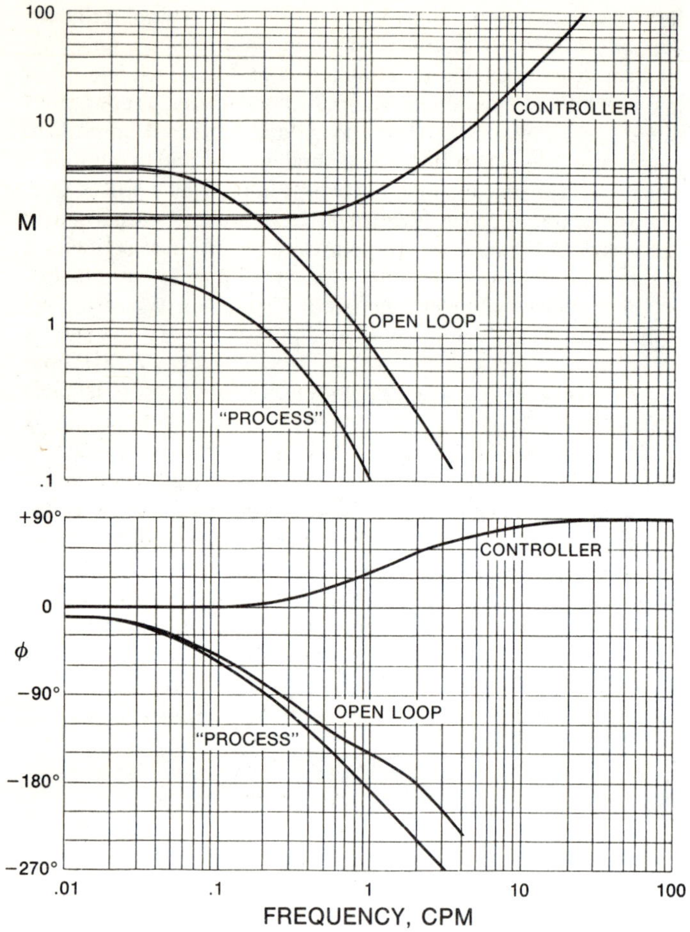

Figure 9-21. The PD controller applied to "process" of Equation (9-39).

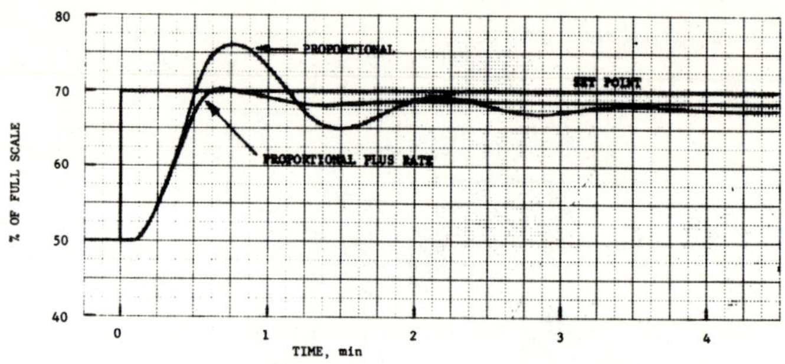

Figure 9-22. Proportional and proportional plus rate control responses to a change in set point.

or an ultimate period of 1.2 min. As shown in Figure 9-21, the crossover frequency with rate control added is 2 cpm or an ultimate period of 0.5 min. Thus, adding rate control has reduced the period to less than one half the value for proportional control. Note that at the open loop crossover frequency, open loop gain is only 0.27. This value is well below the 0.5 value for quarter decay, and thus controller gain could be materially increased. In fact, in many cases Equation (9-34) is noted as follows:

$$G_O = (0.6 \text{ to } 0.9)G_U \tag{9.34a}$$

This range of settings is typical of the two examples used in this discussion. To be on the safe side, the 0.6 value is recommended for initial settings. Gain values can be easily increased if system response is sluggish.

With increased controller gain as demonstrated in Figure 9-21, there is a marked reduction in initial overpeaking. Time response records for both proportional and proportional plus rate control of this process are show in Figure 9-22. These results clearly show influence of rate control on both period and initial overpeaking. The conclusion is that system time responses will show major improvement with rate action if the "process" phase curve is decreasing slowly. Under these conditions phase lead of rate or derivative control can cause a major change in crossover frequency as shown in Figure 9-21. This type of "process" is demonstrated by Equation (9-39), which is dominated by two, unequal time constants and has almost negligible dead time. Many thermal systems have this type of transfer function and are, therefore, prime candidates for rate control.

A secondary advantage of rate action that often occurs due to increased controller gain is a reduction of offset. This result is clearly shown in Figure 9-22. Increasing controller gain results in larger closed loop zero frequency gain as shown by Equation (7-10) and a reduction in offset. Offset will, however, not be eliminated, and if this is a requirement, reset control should be used. Increased controller gain will have a similar effect on load changes by increasing feedback element gain as noted in Chapter 8.

From a theoretical viewpoint it is easy to conclude that the application of rate control would improve most control systems. In actual practice this conclusion does not hold true. Even with a real controller as shown in Figure 9-19, high frequency noise can cause problems. Derivative or rate control amplifies this noise and can cause excessive valve wear as well as loss of control. For this reason rate control should not be applied to any process having a known noise history. Flow control systems are a typical example of this type of system.

For some unknown reason manufacturers have failed to find an alternate unit for rate time. All controllers, to the best of the author's knowledge, have rate time calibrated in minutes. This may be a concession on the part of manufacturers, assuming that rate action causes enough confusion without providing multiple units for rate time.

9.9. The PID Controller

It should be obvious to the reader by this time that many "processes" will require both reset and rate control. Reset is needed to eliminate offset and rate action to decrease period and initial overpeaking. This combination of actions may be accomplished as shown in Figure 9-23. The equation for this ideal PID controller is

$$\Delta P_o = Ke + (K/T_r)\int edt + KT_d\, de/dt \tag{9-45}$$

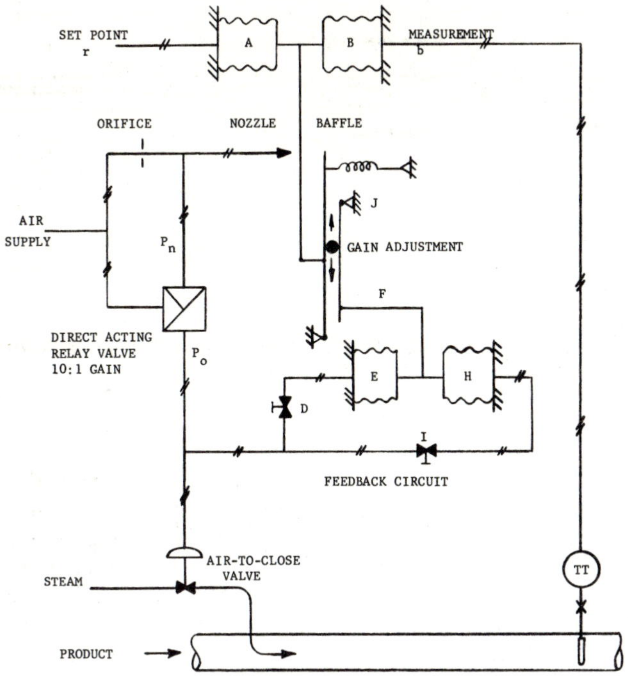

Figure 9-23. Proportional plus reset plus rate controller.

Equation (9-45) may be transformed as follows:

$$\Delta P_o/e = K(1 + [1/(T_r s)] + T_d s) \quad (9\text{-}46)$$

By allowing s to take on the value of $j2\pi f$ as previously noted, a frequency response may be obtained for Equation (9-46). A typical real controller frequency response with limited high and low frequency gains is shown on Figure 9-24. The envelope for a high-gain or on-off controller is also shown in Figure 9-24 for comparison. Plots in Figure 9-24 are intended to demonstrate basic concepts, and as such are not drawn to scale.

Frequency response plots of Figure 9-24 are most helpful in reconsidering basic objectives of reset and rate control. There should be no doubt in the reader's mind that the optimum controller is a high-gain or on-off controller. From considerations of simplicity, maintenance, initial cost, and offset, there can be no argument. A problem, of course, exists with stability. As "process" time constants decrease and gains increase, due to reduced mass and increased throughputs, controller gain must be reduced to obtain stability. The result is proportional control and resultant offset. As shown in Figure 9-24, reset "puts back" high gain at low frequency to eliminate offset. Associated phase lag of reset, however, causes a reduction in controller gain and crossover frequency. This leads to an increase in initial overpeaking and period. To overcome these disadvantages and reduce initial overpeaking and period, rate control is added. As noted in Figure 9-24, rate "puts back" a portion of the high-frequency characteristics of on-off control. The end result is an effort to use best features of on-off control and still maintain a stable system. Another way of describing this overall picture is to say that the controller compensates for "process" conditions resulting from design limitations. In fact in electrical terminology, the controller is often referred to as a compensating network. Reasons for design limitations may be due to physical or cost considerations, or just to poor design. These considerations will be considered later in the text. The problem at this time is to look at the PID controller.

In Sections 9.5 and 9.8 controller adjustments were determined according to the Ziegler-Nichols method using ultimate gain and period. Values for G_U and P_U were obtained from a "process" frequency response. As most operators and instrument engineers are aware, "process" frequency response plots are about as easy to obtain as a bald eagle. While they are known to exist, few people have seen one. Thus, it is obvious that some other technique for obtaining G_U and P_U must exist. In their original paper,[6] Ziegler and Nichols presented a method using a process reaction curve. This method, or a similar modification, is presented in the instructions supplied with many controllers. It can also be found in the original paper and other references.[1,2,6] Another method which gives G_U and P_U directly will be examined at this time.

In Chapter 5 it was noted that a system would cycle continuously in the closed loop, when open loop gain and phase angle were 1 and −180°, respectively. This analysis was made with a proportional controller such that open loop gain equals "process" gain times controller gain. At the cyclic condition, however, controller gain is equal to G_U and period is equal to P_U. The method under consideration makes use of these fundamentals to determine G_U and P_U experimentally. In theory it is possible to convert the PID controller of Figure 9-23 to a proportional controller as follows. Assume an average pressure (9

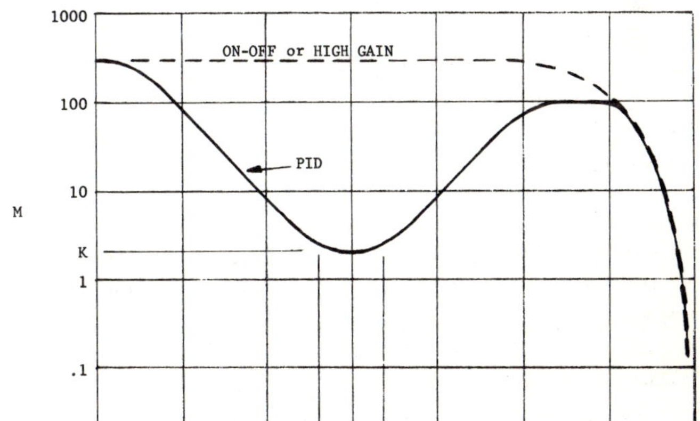

Figure 9-24. The PID and on-off controller envelope.

psig) is sealed in bellows H by closing valve I when controller output is steady or in equilibrium. Under this condition bellows H acts like a spring as previously noted. If valve D is now fully opened, the result is a proportional controller. In actual practice it is sometimes impossible to completely eliminate all rate action by opening valve D, and to lock out reset by closing valve I. The very presence of even an open valve D in the feedback circuit will produce additional flow resistance and cause some minimal rate time. Likewise, it may be impossible to completely seal valve I. In most cases, however, a satisfactory result can be obtained by setting T_r to its maximum setting (in minutes) and setting T_d to the minimum rate time. With electronic controllers it is often possible to switch rate action to an "off" setting. In the "off" setting, feedback resistance is shunted and rate action is completely eliminated.

With reset and rate control basically by-passed, proportional gain is reduced to a low dial setting. A set point or reference disturbance is now introduced into the control loop. Magnitude and duration of this disturbance will depend on system dynamics, and the reader will have to develop his own technique and experience. A typical set point disturbance and reaction are shown in Figure 9-25a. Measurement response of Figure 9-25a is obviously overdamped; thus, increase proportional gain and repeat the same procedure. A second test response is shown in Figure 9-25b. Repeat the procedure until the system cycles continuously as shown in Figure 9-25c. This result is identical to what was achieved in Chapter 5, and thus, G_U is equal to the proportional gain setting. A value of P_U may be obtained from the chart record as shown in Figure 9-25c. It is now possible to calculate controller settings for PID control.

For a PID controller, Ziegler and Nichols suggest the following settings:

$$G_O = (0.6 \text{ to } 0.9)G_U \tag{9-47}$$

$$T_r = P_U/2 \tag{9-48}$$

and

$$T_d = P_U/8 \tag{9-49}$$

Values for gain and rate time are identical to those for a PD controller. Reset time is decreased from $P_U/1.2$ to $P_U/2$ as a result of phase lead produced by rate control. Here again, it should be recognized that these settings are only first approximations, and tuning will normally be required. Typical results using settings calculated from G_U and P_U for proportional, proportional plus reset, proportional plus rate, and PID are shown in Figure 9-26. While records of Figure 9-26 do not show ideal

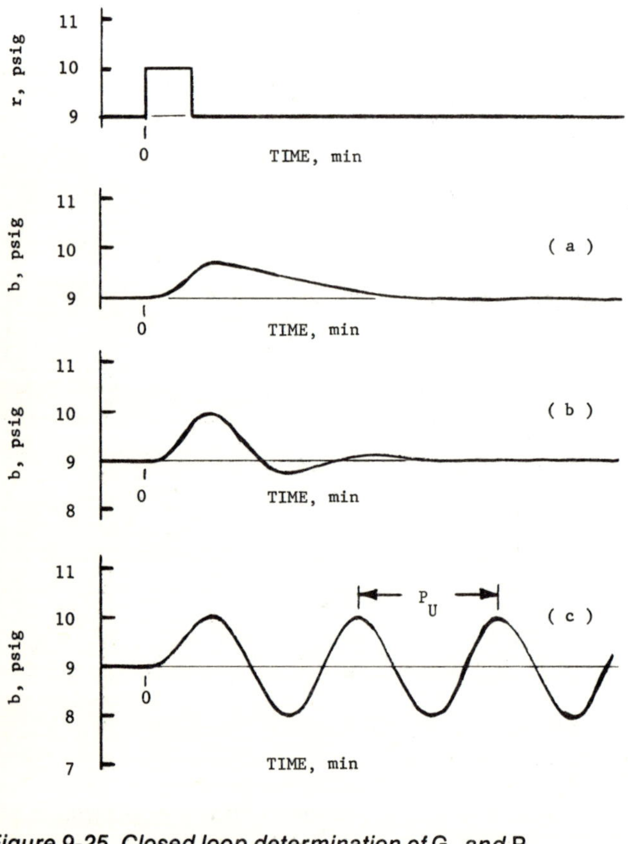

Figure 9-25. Closed loop determination of G_U and P_U.

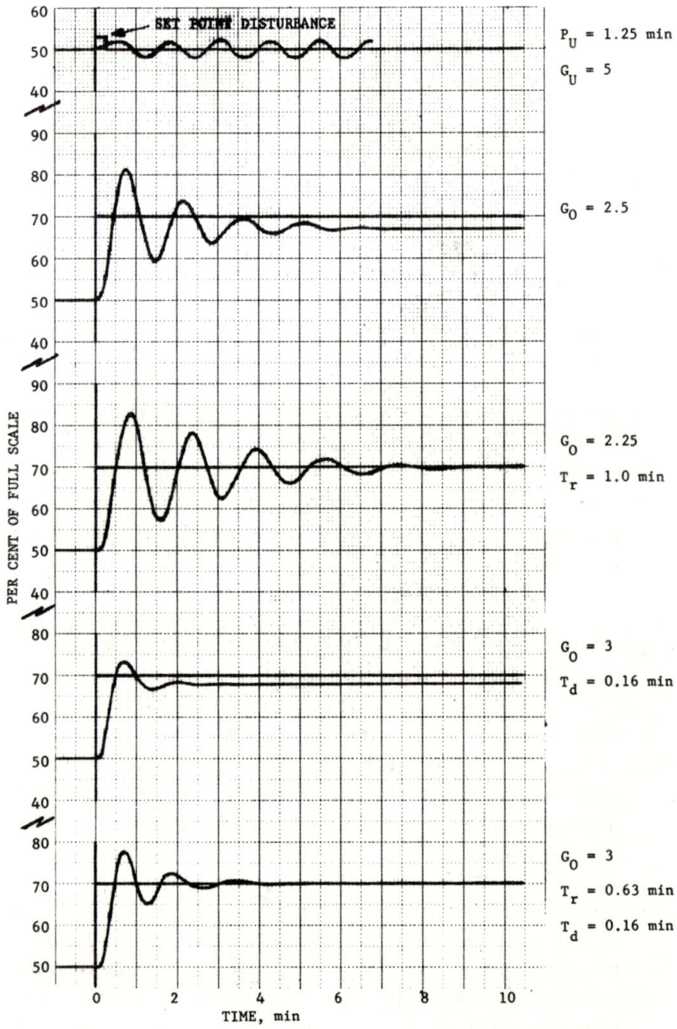

Figure 9-26. Control responses using calculated settings.

quarter decay, they do emphasize objectives of each control mode and validity of calculated settings.

A word of caution must be noted concerning reset and rate settings for the PID controller shown in Figure 9-23. Controllers using circuitry similar to what is shown are referred to as "parallel feedback" controllers. The name originates from parallel feedback paths to bellows E and H. It is not difficult to recognize that this type of circuit can present a problem if needle values I and D are set for equal flow resistance. This produces equal flow rates to bellows E and H, with the result that level F remains fixed. The obvious result is a high-gain proportional controller. This will occur when T_d/T_r equals 1. Using Ziegler-Nichols recommendations, T_d/T_r is equal to 1/4.

$$T_d/T_r = (P_U/8)(2/P_U) = 1/4 \tag{9-50}$$

Although not all controllers use parallel feedback, many do, and for this reason care should be exercised in selecting T_r and T_d. This is not a common problem and will normally not be encountered if techniques such as the Ziegler-Nichols method are used for adjusting settings. Other specific controller problems will be analyzed in Chapters 13 and 14.

Chapter 9—Study Questions and Problems

(Answers appear on pages 160-162)

1. In the pseudo proportional controller of Figure 9-2, bellows A and B are similar to a block diagram summing junction in which the __________ is subtracted from the __________ to produce the error signal.

2. Every proportional controller includes a central component such as the orifice, nozzle, and baffle combination of Figure 9-2. These central components are transducers or amplifiers which have a common characteristic of __________.

3. Although the orifice, nozzle, and baffle combination has a high gain, this combination like other high-gain transducers or amplifiers normally requires a __________ or __________ to achieve satisfactory controller response.

4. Is the primary function of feedback in a proportional controller circuit to reduce or increase gain?

5. The zero adjustment in a proportional controller should not __________ controller gain. Its function is to provide a controller __________ when reference and measurement are equal that will maintain the controlled variable at the desired value for a given load condition.

6. It is interesting to note that a proportional controller is actually a closed loop system consisting of an error detector, high-gain transducer or amplifier, power amplifier, and feedback gain adjustment. Sketch a block diagram of the proportional controller circuit labeling each of the above-noted components.

7. Using the three controller schematics and associated data shown below, fill in the blanks with the equivalent proportional controller values. Assume a 4-in. chart width, and input and output ranges of 3 to 15 psig if not otherwise specified.

a)

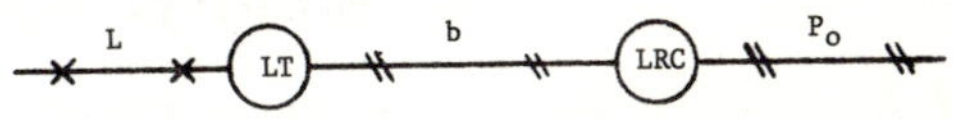

LT - Level Trans.
L: 0-24 in H_2O, input
b: 3-15 psig, output

LRC - Level Recording Cont.
b: 3-15 psig, input
P_o: 3-15 psig, output
Prop. Cont. Unit: Gain

b)

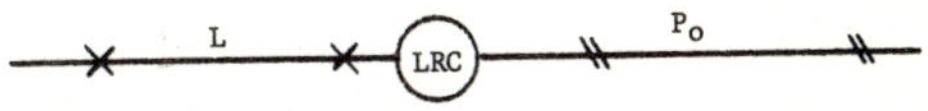

L: 0-24 in H_2O, input
P_o: 3-15 psig, output
Prop. Cont. Unit: Proportional Band, %

c)

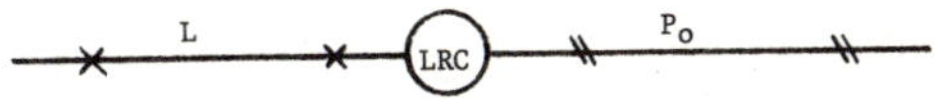

L: 0-24 in H_2O, input
P_o: 3-15 psig, output
Chart Width: 4 inches
Prop. Cont. Unit: Sensitivity, psi/in

Case	Gain	Prop. Band, %	Sensitivity, psi/in
a	5		
b		150	
c			10

Figure 9-P7. Controller schematics and associated data.

8. An electronic controller using the proportional band unit has an input range of 4 to 20 ma and an output range of 1 to 5 v. Calculate the output change for an input change of 4 ma and a proportional band setting of 40%.

9. Offset may be eliminated or reduced to a negligible value by high or low gain?

10. The initial response of a proportional plus reset controller corresponds to __________ controller action.

11. A proportional plus reset controller has a gain $K = 3$ and a reset time $T_r = 1.5$ min. Sketch the open loop time response of controller output pressure change for an error input change of unity. Assume a direct-acting controller and an initial equilibrium pressure of 9 psig.

12. The reset time in minutes is actually a measure of the time required for reset response to __________ the change in controller output resulting from proportional response.

13. Plot the frequency response for a proportional plus reset controller having a gain of 3 and a reset time of 0.5 min.

14. Show on your Bode plot for Problem 13 what happens if the controller gain of Problem 13 is changed from 3 to 5.

15. Show on your Bode plot for Problem 13 what happens if the controller reset time of Problem 13 is changed from 0.5 to 2.0 min.

16. For a given "process" frequency response, gain and frequency at −180° phase are 2.5 and 12 cpm, respectively. Calculate ultimate gain and ultimate period.

17. Based on ultimate gain and period obtained in Problem 16, calculate settings for a proportional plus reset controller.

18. Reset or integral controller action is used to eliminate ________ following set point or load changes.

19. Adding reset action has the adverse effect of increasing or decreasing period and initial overpeak?

20. If the controller of Problem 17 had been calibrated for reset rate, what would be the equivalent dial setting?

21. The objective of rate or derivative control is to reduce ________ and ________.

22. A proportional plus rate controller has a gain of $K = 4$, and a rate time $T_d = 0.4$ min. Assume an error change of $e = 0.5t$ and sketch the open loop time response of controller output. Use a direct-acting controller, an output range span of 3 to 15 psig, and an initial equilibrium pressure of 9 psig.

23. Rate time is defined as the time by which output of a PD controller leads output of a ________ controller.

24. Rate controller action will not eliminate ________.

25. Plot the frequency response for a proportional plus rate controller having a gain of 4 and a rate time of 0.5 min.

26. Show on your Bode plot for Problem 25 what happens if the controller gain of Problem 25 is changed from 4 to 2.

27. Show on your Bode plot for Problem 25 what happens if the controller rate time of Problem 25 is changed from 0.5 to 2.0 min.

28. Rate control action will provide very little help in controlling a "process" that is dominated by ________.

29. Application of rate control to a "process" with how many unequal time constants and negligible dead time lag will provide a major improvement in control?

30. Rate control should not be applied to a "process" known to have a high degree of ________ in the measurement.

31. For a given "process" frequency response, gain and frequency at −180° phase are 1.6 and 0.7 cpm, respectively. Assuming a proportional plus rate controller, calculate initial gain and rate time settings.

32. Rate or derivative control when properly applied will reduce ________ and ________.

33. Although rate action may tend to reduce ________, this is not its prime function.

34. The PID controller uses high gain at low frequency to eliminate ________, and high gain at high frequency to reduce ________ and ________.

35. The effects of reset and rate action can normally be removed from a PID controller by setting reset time to its (maximum or minimum) value and rate time to its (maximum or minimum) value.

36. If a system cycles continuously using only proportional control, controller proportional gain is equal to ________.

37. If a system cycles continuously using only proportional control, period of cycling is equal to ________.

38. A system using only proportional control cycles continuously at a proportional gain of 3 and has a cycling period of 2 min. Calculate settings for proportional gain, reset and, rate times.

39. In a PID controller having a parallel feedback circuit, the ratio of T_d/T_r should be approximately equal to ________. If the ratio of T_d/T_r equals 1, will the controller act as a high- or low-gain controller?

References

1. Eckman, D. P., *Automatic Process Control*, New York: John Wiley, 1958.
2. *Process Instruments and Controls Handbook*, ed. by D. M. Considine, Section 9, New York: McGraw-Hill, 1957.
3. LaJoy, M. H., *Industrial Automatic Control*, Chapter 7, Englewood Cliffs, NJ: Prentice-Hall, 1954.
4. Caldwell, W. I., Coon, G. A., and Zoss, L. M., *Frequency Response for Process Control*, New York: McGraw-Hill, 1959.
5. Harriott, P., *Process Control*, New York: McGraw-Hill, 1964.
6. Ziegler, J. G., and Nichols, N. B., "Optimum Settings for Automatic Controllers," *Trans. ASME*, Vol. 64, p. 759, 1942.

10 Associated Control Components

Thus far, the closed loop system has been described as consisting of a "process" and controller. The concept of controller is easily recognized and can be defined as noted in Chapter 9. The "process," however, is much more complex and is even hard to define. As previously noted in Chapter 4, "process" was defined as including every component in the control loop except the controller. Using this definition, measuring elements, transmitters, valve operators and valves, and transmission lines all become a part of the "process." While it is not the author's intention to negate the "process" term, it is convenient to divide the "process" into two parts. Part 1 will be defined as associated control components and will include measuring elements, transmitters, valve operators and valves, and transmission lines. Part 2 will be defined as actual process hardware, namely, heat exchangers, distillation columns, reactors, rotating machinery, furances, and the endless list of industrial equipment for which control is required. Process hardware will be considered in Chapters 11 through 14.

10.1. Measuring Elements and Transmitters

An attempt to discuss all possible types of measuring elements and transmitters is obviously beyond the scope and objectives of this text. For general information there are numerous good measurement texts available,[1,2,3,4,5,6] and for specific information the reader may obtain detailed company literature. Objectives of this section will be to examine basic measurement and transmitter characteristics such as accuracy, precision, sensitivity, dynamics, and overall effect on controllability. Some specific measuring elements will be used as examples.

The combination of measuring element and transmitter is usually considered as one component for control system analysis and may be mathematically described by a transfer function. This procedure was used to define a thermal sensing element and transmitter for the simple heating system of Figure 4-1. A transfer function, as the reader will remember from Chapter 3, is a differential equation having two solutions, namely, the steady-state solution and the characteristic or transient solution. To demonstrate this more clearly, the measuring or sensing element and transmitter of Chapter 4, as expressed in Equation (4-6), will be reexamined:

$$\Delta P_3/\Delta \Theta_2 = h = k_5/(1 + T_5 s) \qquad (4\text{-}6)$$

As s approaches zero, the steady-state solution is k_5. This is the steady-state or zero frequency magnitude, which may also be referred to as sensitivity. Since the importance of "process" gain has already been shown in Chapter 5, 7, and 8, it is fairly obvious that selection of measurement and transmitter sensitivity is an important consideration. The characteristic or transient solution is defined by $(1 + T_5 s)$ and determines measurement and transmitter dynamics. The importance of "process" dynamics has been one of major consideration throughout the text and must certainly be considered for measuring elements and transmitters.

Static characteristics of measuring elements and transmitters are often characterized by such terms as accuracy, precision, sensitivity, and linearity. Accuracy in measurement is defined as "the degree of correctness with regard to the true value of the variable being measured."[5] Usually accuracy is expressed in terms of full-scale deflection. Thus, an instrument having a range span of 100°F and an accuracy of 0.5% is capable of determining true temperature within 0.5°F. There is no easy way to define required accuracy for control. Accuracy required for a particular system must be individually determined for each application. It is the author's opinion that as a rule, accuracy specifications for control purposes are set much too stringently. Emphasis here is on control, and this statement does not apply to metering for billing. What is normally more important for control purposes is precision. Precision is defined as "the degree of reproducibility among several independent measurements of the same true value under specified conditions."[5] For control purposes, ability to reproduce or repeat a given condition can often insure product quality in spite of accuracy.

Linearity is another important characteristic of measuring elements and transmitters. Linearity is defined as "conformity to a straight line."[5] This characteristic is important from the viewpoint of readout and system stability. With a linear output from a measuring element and transmitter, a uniform indicating scale or recording chart may be used to cover the range span. A simple example is a recording or indicating thermometer having a range from 0-220°F. If a linear or straight line relation exists between temperature and change in output signal, it is only necessary to manufacture a common scale or chart and calibrate at two points.

If the relationship is nonlinear, many more calibration points are necessary. A linear measuring element and transmitter has a constant sensitivity throughout the instrument range span. With constant sensitivity or magnitude, closed loop stability will remain fixed for any set point or load change within the given range span. This assumes that all other system components are equally linear. Linearity, however, is not an absolute necessity, and there are several exceptions of measuring elements and transmitters which function using nonlinear relationships. The most notable of these are obstruction-type flow meters such as the orifice plate, flow nozzle, and venturi. For these devices flow is proportional to the square root of pressure drop across the obstruction. A flyball speed transmitter is another example of a square root device. A vapor pressure thermometer is another example of a nonlinear measuring element. In this instrument the functional relationship between pressure and temperature depends on the vapor pressure curve of the liquid employed. There is in all these examples one common consideration, namely, although the relationship is nonlinear, it is readily definable and reproducible. From a stability viewpoint a nonlinear instrument can produce control problems, since sensitivity or magnitude varies with set point and load. The extent of this problem, however, depends on the degree of nonlinearity, size of set point and load changes, and other nonlinear components in the control loop. The effect of other nonlinear components will be examined in more detail in the section of valves.

One potential advantage of a nonlinear instrument is increased sensitivity for a particular operating point. Both linear and square-root flow element curves are shown in Figure 10-1. Assuming an operating or set point flow of 75%, sensitivity of the linear element is

Sensitivity = output change/input change (10-1)
= 20%/ 20% = 1

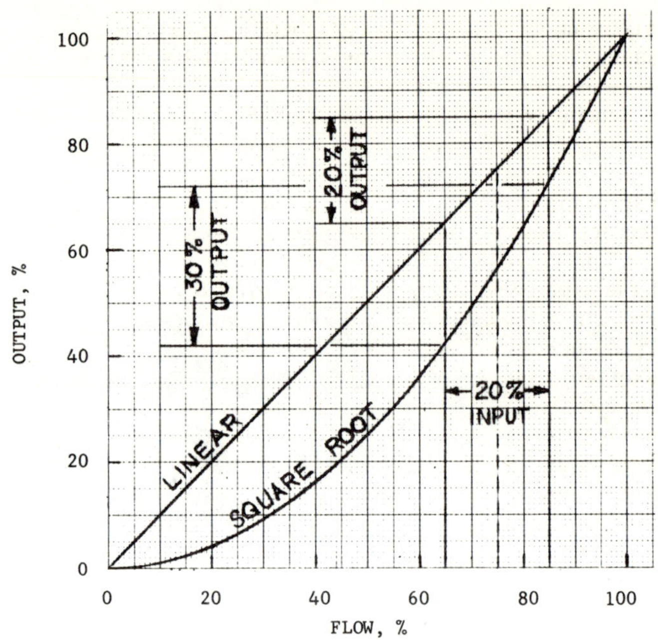

Figure 10-1. Linear and square-root flow element curves.

Using the same 20% input change and linearizing the square root curve in this region as shown in Figure 2-3, the snsitivity is

Sensitivity = output change/input change (10-2)
≅ 30%/20% = 1.5

Figure 10-1 shows that although both elements cover the same range span, the square root element is 1.5 times as sensitive to changes at a flow set point of 75%. In control systems where changes of set point or load are not expected to vary over wide ranges, this increased sensitivity could be an advantage. Figure 10-1 also shows how sensitivity can change with operating point. If the operating set point is changed to 25% of flow, sensitivity for the linear element remains at 1. Even assuming a linear curve for the square root element, the sensitivity changes as follows:

Sensitivity = (12.25-2.25)% Output/(35-15)% Flow (10-3)
= 10%/20% = 0.5

It is easy to see how a sensitivity or magnitude change from 1.5 or 0.5 can influence system stability. This is a prime disadvantage of nonlinear elements.

Range span selection is another factor which must be considered. Range span defines the signal limit in which the measuring element will function. In Chapter 4, for example, the range span was selected as 100°F and operated between limits of 100° and 200°F. Another way to define this would be to state a range zero along with range span. In the example of Chapter 4 both range zero and range span would be 100°F. To introduce variety assume a range zero of 200°F, a range span of 100°F, and a transmitter output of 4 to 20 ma. The instrument in question could be a conventional thermocouple and thermocouple transmitter. Assuming linearity, sensitivity or magnitude of this combination is

$$\text{Sensitivity} = (20\text{-}4)\ \text{ma}/(300\text{-}200)°\text{F} = 16/100 = 0.16\ \text{ma}/°\text{F} \quad (10\text{-}4)$$

Part of the problem in selecting a range span is involved with transfer function gain and time responses as discussed in Chapters 7 and 8, and particularly in Section 8.4, In Chapter 8 transfer functions were developed for b/r, b/U_2, b/U_3, and b/U_4 (see Problem 3, Chapter 8 and its solution, and Equations (8-20), (8-21) and (8-22)). To minimize offset it was noted that closed loop zero frequency gain for b/r should be 1, and that for load changes feedback element zero frequency gain should be very much greater than 1. The conclusion of Section 8.4 to insure these results under all conditions was to have a large controller gain. While this is theoretically a valid conclusion, it does have some practical limitations as will now be shown.

The measured variable b is not the controlled variable c! The importance of this statement may be observed by determining the transfer function for c/r based on the figure in Problem 3, Chapter 8.

$$c/r = (1/h)\left[g_1g_2g_3g_4h/(1 + g_1g_2g_3g_4h)\right] = (1/h)(b/r) \quad (10\text{-}5)$$

Equation (10-5) clearly shows that c is not equal to b but is proportionally related to the reciprocal of the measuring element and transmitter transfer function. As h approaches 1, which will occur when the transfer function sensitivity or gain is 1 and the time constant is zero, c does equal b. These are the conditions assumed in many control texts. In practice these assumptions are very seldom valid since a measuring element and transmitter will exist, steady-state sensitivity or gain is seldom 1, and the time constant will normally have some finite value. Sensitivity shown in Equation (10-4) is a typical example, and time constants of 0.05 to 8 min are typical for thermal sensing elements.[7] The practical solution is to minimize the measuring element and transmitter time constant and select a range span that will provide a steady-state sensitivity of as close to 1 as possible. Under these condtions, b will represent c as close as physically possible. In the example of Equation (10-4) this would mean selecting a range span of 16°F to obtain a sensitivity of 1 ma/°F. Normal reaction to this conclusion is "What about controller gain and noise problems with this narrow range?"

The answer to the problem of controller gain can be easily explained by reference to Equation (10-5) and load change equations c/U_n.

$$c/U_2 = \left[1/(hg_1)\right]\left[b/r\right] \quad (10\text{-}6)$$

$$c/U_3 = \left[1/(hg_1g_2)\right]\left[b/r\right] \quad (10\text{-}7)$$

$$c/U_4 = \left[1/(hg_1g_2g_3)\right]\left[b/r\right] \quad (10\text{-}8)$$

Equations (10-6) through (10-8) are also based on the system of the figure in Problem 3, Chapter 8. If it is assumed that g_2, g_3, and g_4 have fixed magnitude or gain, any change in h must be accounted for in g_1 in order to maintain a fixed stability. Thus, if sensitivity of h is increased from 0.16 to 1 ma/°F, or a factor of 6.25, controller gain must be decreased by this same factor. The immediate reaction is to view this as a contradiction of conclusions expressed in Section 8.4. An examination of Equation (10-5), however, shows that as long as the product of g_1 and h remains constant, b/r will have the same value. This means that control stability will remain constant. The only change will be that the measured variable b will give a truer picture of what is actually happening to the controlled variable c. A similar effect results from load changes. Equations (10-6) through (10-8) show that the product of h and g_1 exists in the feedback path of all load changes. Here again, b/r remains constant and provides equivalent stability. If h is increased, g_1 must be decreased and net offset is identical. The difference is that b presents a clearer picture of what is actually happening at c as it did for set point changes. The increase in the measured variable is what causes problems with both noise and operating personnel.

To show what happens when the measuring element and transmitter sensitivity is changed from 0.16 to 1 ma/°F or from a range span of 100° to 16°F, it is necessary to compare control records. Assuming a 4-inch strip chart recorder, a set point change of 5°F, and a controller having reset action, results are shown in Figure 10-2a. Similar results are shown for a load change in Figure 10-2b. While control results are identical, it is obvious what reaction would be obtained from operators reviewing these charts. The effect of signal noise of plus or minus 1°F will also cause considerable operator reaction. Actually, the only difference is that with a 16°F range span a more truthful picture of what is actually happening to the controlled variable is presented.

In actual practice range spans are seldom selected according to the procedure presented in this chapter and, more often than not, have a very wide range. Noise problems and a need to know the controlled variable value during start-up are usually given as reasons. If control is critical, this approach is like an ostrich with its head in the sand. While a nonlinear element that will provide a wide range with increased sensitivity at the operating point is often helpful, it is not the ultimate answer. *A better solution for critical control is to use a wide range indicator for start-up and a shorter range for control purposes*. When small range spans are used, it is usually a good idea to provide operator training to minimize the shock of seeing records such as shown in Figure 10-2.

Dynamics of a sensing element are also very important, as shown by Equations (10-5) through (10-8). In general it is desirable to have the measuring element and transmitter time constant as small as possible. As noted earlier in this section, if the time constant of h is zero, c and b differ only by the sensitivity or gain of h. While it is desirable to have a small time constant, it is not always an absolute necessity. Assume, for example, a "process" in which g_2, g_3 and g_4 of the figure in Problem 3, Chapter 8, have time constants of 0.8, 1.6, and 8 min, respectively. A Bode plot using gains of 1 for each transfer function is shown in Figure 10-3, along with product curves of gain and phase angle for $g_2g_3g_4$. If a measuring element and transmitter time constant of 0.2 min is assumed with a gain of 1, the result will be as shown in Figure 10-3. By observing gain and phase of

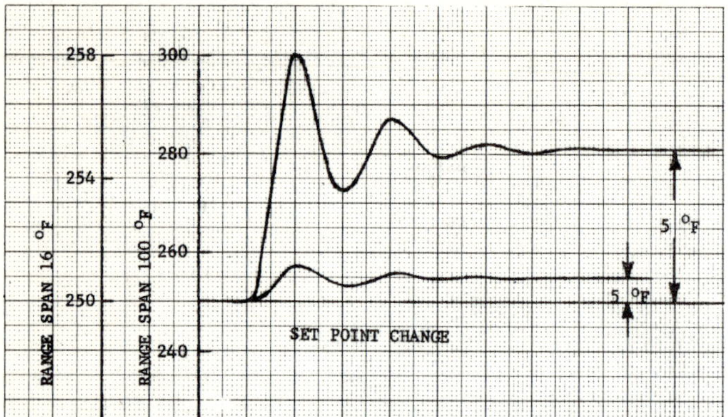

Figure 10-2. Range span variations, set point, and load change responses.

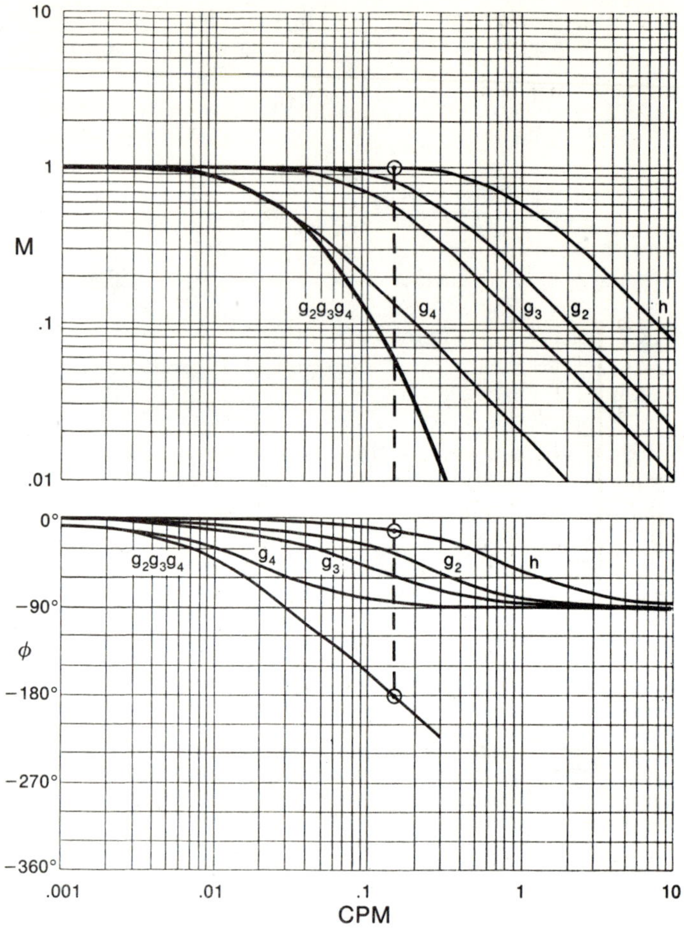

Figure 10-3. Measurement dynamics.

h at the frequency where the combined phase lag of $g_2g_3g_4$ is 180°, it is easy to see that the effect of h's time constant is minimal. Another way to look at this is to assume a step input to g_4 as shown in Figure 10-4. Output at c would be an exponential with a response of 0.63 at 8 min. Measured output b, however, will have an additional lag and respond as a second-order S-shaped reaction curve. Results for both c and b are shown in Figure 10-4. The difference between b and c, as previously noted, is almost negligible because the time constant h is much smaller than any other time constants in the control loop. If, however, the measuring element and transmitter time constant is 1.2 min, a much different result is obtained. As could be shown in Figure 10-3, the curve for h would move to the left of g_2 and have a greater influence on the process than g_2. This effect is also shown by the second b curve in Figure 10-4. In this case there is a decided difference between c and b, and this can cause problems in control.

The assumption of a measuring element and transmitter being represented by a single exponential lag is frowned on by the purist. Actually most measuring element and transmitters will exhibit more complex transfer functions. While most will have at least second-order transfer functions, there are some that can be of even higher order and also include dead time lag. An analytical measuring device that functions on a sampled basis is a typical example of a device with dead time lag. The approximation of a single lag, however, is normally valid to about 30° phase lag.[7] Thus, the single lag representation is acceptable as long as the measuring element and transmitter time constant is smaller than other major component time constants as shown in Figure 10-3. Whenever the measuring element and transmitter time constant becomes one of the principal components or time lags, a more detailed analysis would be required.

10.2. Transmitter Mechanics

In the very brief discussion of Section 10.1, the transmitter was considered an integral part of the measuring element. While this can often be a valid assumption as noted, there are times when a more detailed analysis is required. Usually this information can be supplied by the manufacturer. As a rule manufactur-

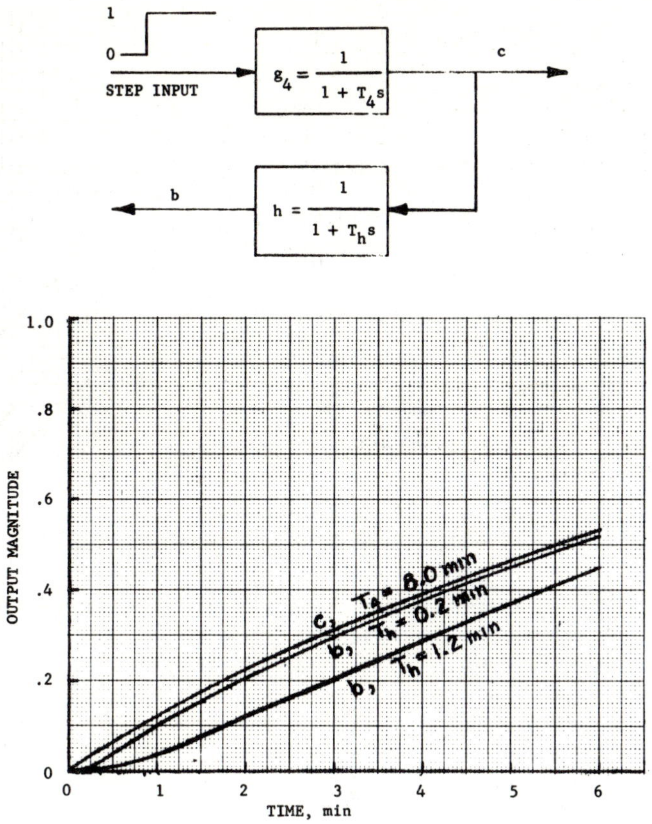

Figure 10-4. Measurement deviation as a function of T_h.

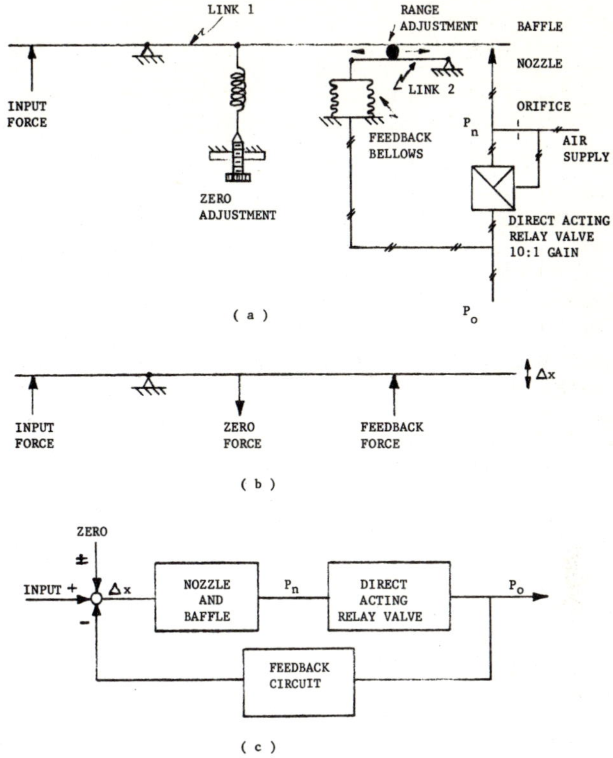

Figure 10-5. Pseudo transmitter circuit.

ers' data will not match the particular application at hand. To solve this problem, some manufacturers will supply a series of test results covering a range of applications.[8] From these data, results for a particular application can generally be obtained by interpolation. To see why this is necessary and obtain a clearer picture of the function of a transmitter, a pseudo transmitter will be examined.

In most original instrument systems, the measuring element was directly connected to the controller. Through some type of amplifying device, sufficient energy was developed to drive a pen or indicator. This arrangement was applicable when most control systems were field mounted and close-coupled to equipment under control. With advent of the control room and centralized control, transmitters became a necessary control component. The prime function of a transmitter is to reproduce the low-energy measurement signal with sufficient energy that it may be transmitted over an appreciable distance. In process control this distance can vary from 10 to 1000 ft. There are exceptions to the above statement, such as systems employing telemetry equipment. However, in these cases some additional component or transducer is used to convert to electrical transmission. The general process problem being described at this time is demonstrated in Figures 1-5 and 1-6.

The pseudo transmitter that will be used for demonstration is shown in Figure 10-5a. The reader should recognize similarity between this transmitter and controllers of Chapter 9. Similarity is intended and is commonly found in commercial instruments. To help explain how the transmitter operates, Figures 10-5b and c are also shown. The input force could be developed by any sensing element and applied to link 1 by a Bourdon spring, diaphragm, bellows, or torque tube. As shown in both Figures 10-5a and b, link 1 acts as both a summing junction and a baffle. The objective is to obtain a 3 to 15 psig change in output P_o for one range span change of input. As an example, assume a thermal tube system, with a Bourdon spring or bellows, having a range span of 100° to 200°F. When the temperature is 100°F and the measurement signal is connected to link 1, P_o should produce an output of 3 psig. If P_o is not equal to 3 psig, the zero adjustment of Figure 10-5a is changed to achieve this result. The force balance of input, feedback (output), and zero is shown in Figure 10-5b and as a summing junction in Figure 10-5c. As temperature is increased from 100° to 200°F, link 1 acts as a baffle to increase nozzle pressure P_n. Nozzle pressure acting through the relay valve or power amplifier should increase P_o to 15 psig. If P_o does not equal 15 psig, a range adjustment is required. As shown in Figure 10-5a, the effect of P_o on link 1 can be varied by position of the range adjustment ball between links 1 and 2. In an actual transmitter several cycles of range and zero adjustment are required to achieve the proper force balance on link 1.

The basic components in a transmitter are some form of summing junction (link 1), error detector (nozzle-baffle), power amplifier (relay valve), and feedback mechanism (feedback bellows and link 2). The result as shown in Figure 10-5c is a closed loop circuit, which can usually be defined by a second-order transfer function. With good design, however, a response can be obtained over a wide range of frequencies that justifies assuming a constant gain and phase angle for the transmitter. A typical plot is shown in Figure 10-6. For results of Figure 10-6 to have any real value, complete test data should be attached. This is especially true for transmission line size and length, and termination volume, since changes in loading can change response results. The important point to note, however, is that gain and phase angle, as shown in Figure 10-6, are constant for all practical purposes out to a cut-off frequency of approximately 100 cpm. While the plot in Figure 10-6 is for a particular instrument and load, it is typical of many commercial transmitters. Since most systems in the process control field operate with natural frequencies of less than 60 cpm (1 cps), the assumption of constant gain and phase angle is, therefore, very realistic.

While a pneumatic transmitter was chosen for the example of this section, note that the same considerations apply to all transmitters. A simple circuit for an electronic transmitter using an operational amplifier is shown in Figure 10-7. By selecting correct polarities for e_i and e_z, an equation may be written for this circuit as follows:[10]

$$e_o = (R_f/R_r)\,e_i - (R_f/R_z)\,e_z, \tag{10-9}$$

where e_o = output emf signal,
e_i = input emf signal,
e_z = constant emf source,
R_f = feedback resistance,
R_r = input range resistance,
R_z = zero resistance.

Assume that e_i is generated by a thermocouple and that a range of 100° to 200°F is desired. By adjusting R_z it is possible to null out the input signal e_i for 100°F. As the temperature increases to 200°F, e_i will change and cause an output to be generated by the high-gain amplifier at e_o. The output signal is fedback through R_f to summing point P for comparison with e_i. If the desired output is not obtained, range resistance R_r is adjusted to achieve this result. Here again, several cycles of range and zero resistance adjustment will be required to achieve desired results. This system requires that amplifier gain A be very much greater than 1 (values of 10^6 are not uncommon) such that current changes in the microamp range at P will drive e_o through the desired range. In this transmitter, node P acts as the summing junction for input, feedback, and zero adjustment current flows. The amplifier detects microamp changes in the same manner that the nozzle-baffle of Figure 10-5 senses position change. These small changes are then amplified to provide the necessary power for transmission, performing the same function as the relay valve in the pneumatic transmitter of Figure 10-5. The ratio of R_f/R_r provides feedback comparison between e_o and e_i and also determines steady-state transmitter gain as shown in Equation (10-9). Although an actual transmitter using this exact circuit is unknown to the author, it does present fundamental components of commercial electronic transmitters in a very simple form. Actual commercial transmitters are usually somewhat more complicated in their circuitry to minimize range span and zero interaction, and compensate for line voltage variations, ambient temperature changes, and load impedance matching. Basic functions as shown in Figure 10-7 will, however, be found somewhere in the circuitry.

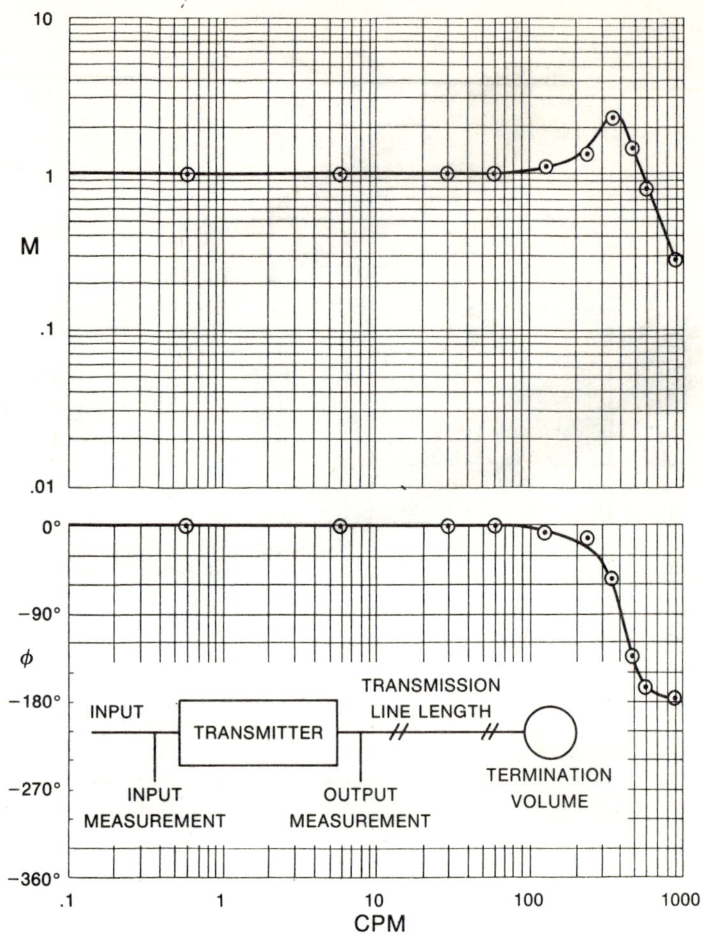

Figure 10-6. Typical transmitter frequency response.

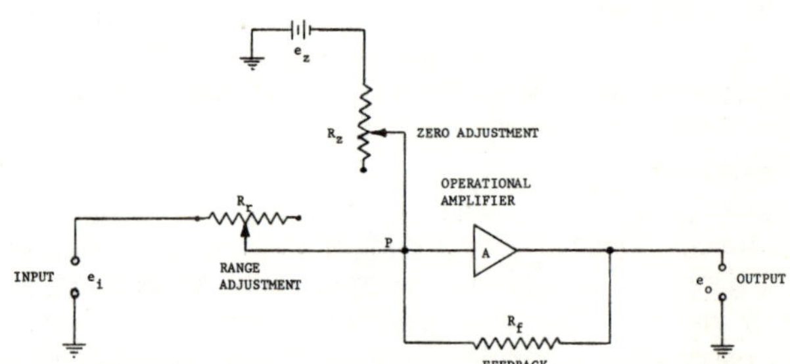

Figure 10-7. Operational amplifier used as an electronic transmitter.

10.3. Valve Operators and Valves

The subject of valve operators and valves, or final control elements, has to be one of the more mystical phases of practical control considerations. There are undoubtedly an infinite number of applications, and to the beginner it must appear that if operator and valve combinations are not infinite, they must at least approach numbers similar to the national debt. For this reason, no attempt will be made to present an exhaustive treatment of this subject. There are many texts either partially or completely devoted to this subject,[5,9,11,12,13] and for detailed information the reader is referred to these texts and control valve suppliers. The intent of this section is to describe the function of a final control element, note some actual elements, explain certain basic terminology, and examine common problems of selection.

A final control element in most process control applications consists of a valve operator, or actuator, and a valve. The basic function is to convert the controller output signal p to a manipulated variable m, as shown in Figure 1-10. Another analogy would be a transducer, which converts from some input energy media to some other type of energy media at its output. The problem is that the combination of input signals coming from pneumatic, electric, and hydraulic controllers, along with the variety of manipulated variables used in industry, leads to an almost limitless number of applications. To minimize problems at this stage of the reader's training, the more common pneumatic diaphragm actuator and conventional globe-type valve will be used as specific examples.

A typical pneumatic diaphragm actuator is shown in Figure 10-8. The controller output signal normally has a range of 3 to 15 psig, and this signal is applied to chamber A. Input force resulting from pressure acting on the diaphragm area is opposed by the range spring. Resultant output is stem motion. This type of actuator is normally referred to as direct-acting. For a reverse-acting actuator the input air signal is connected to chamber B. In the design of a particular actuator it is necessary to consider such characteristics as load capacity, dynamic characteristics, hysteresis, dead band, reliability, safety, and cost. Some of these characteristics, namely, those which have a direct influence on control, will be examined later in this chapter. There are many other types of actuators using mechanical, electrical, and hydraulic principles. Many also exist using combinations such as electro-pneumatic, electro-hydraulic, and pneumatic-hydraulic. Actuator selection is a function of power, speed, and cost. From a viewpoint of power, hydraulic and pneumatic actuators are usually used where power demands are large. Where high speed is required, the hydraulic actuator has the inherent advantage of an incompressible fluid. There are undoubtedly advantages and disadvantages for each actuator, and the examples noted are not intended as a complete picture. As a general rule, however, the pneumatic actuator of Figure 10-8 is the best compromise for process control and is the most common type of valve operator found in these applications.

The number of valve types, such as ball, butterfly, globe, gate, and Saunder's, along with construction variations as previously noted, is almost endless. For this reason the student should seek out one or more of the referenced texts on valves for detailed study. For the example of this section, a globe-type valve will be used since this is one of the most common types found in process control applications. A simple schematic for this type of valve is shown in Figure 10-9, along with titles for some control valve components. The valve plug shown in Figure 10-9 is a quick opening plug. Other plug types, such as linear, equal percentage, V-port, parabolic and square root, may be obtained. The type of plug determines what is known as the flow characteristic. This is the relation between flow and stem travel as the stem is varied from zero to 100% of its lift. Another term would be valve sensitivity, namely, output flow as a function of input stem travel. Since is is part of the "process" gain, this characteristic will be examined in more detail.

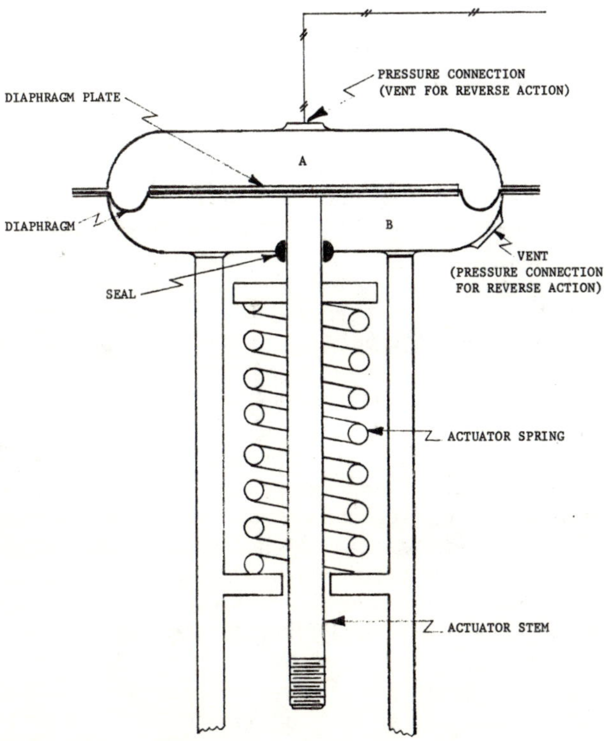

Figure 10-8. Schematic of a pneumatic diaphragm actuator.

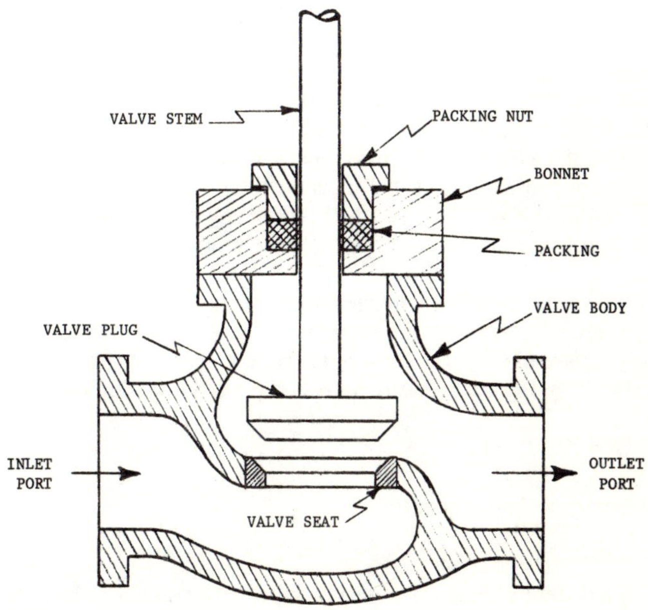

Figure 10-9. Single-seat globe valve schematic.

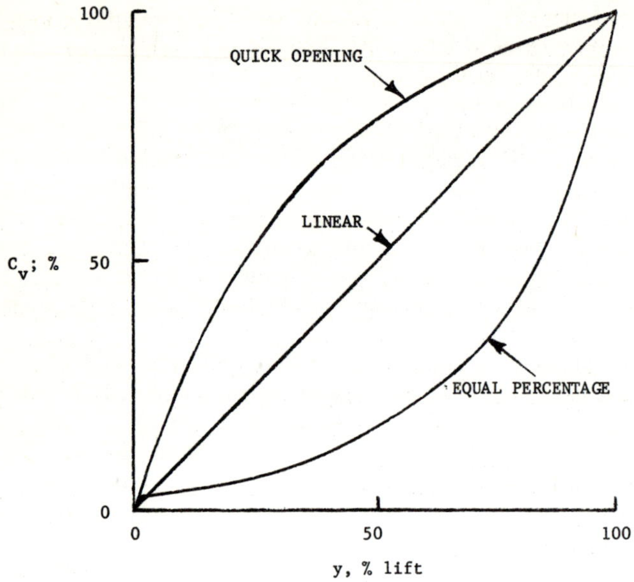

Figure 10-10. Inherent flow characteristics for constant pressure drop.

The typical valve flow equation for incompressible liquids as derived from fluid mechanics principles is

$$Q = C_v (\Delta P/\text{sp. gr.})^{1/2}, \tag{10-10}$$

where Q = liquid flow, gpm,
ΔP = valve pressure drop, psi,
C_v = valve size coefficient, number of gal/min of 60°F water that will flow through a fully open valve with 1 psi pressure drop,
sp.gr. = specific gravity relative to water.

For each type of valve plug configuration there is a specific relationship between C_v and valve lift, or opening, called the inherent flow characteristic. This relationship is identical to the relationship between flow and lift for a constant pressure drop across the valve. If, for example, a linear plug were used at constant pressure drop, the inherent flow characteristic would appear as shown in Figure 10-10. While any type of characteristic is possible, most valves fall into the following three characteristics: quick opening, linear, and equal percentage. Quick opening and equal percentage characteristics for constant pressure drop are also shown in Figure 10-10. Since a valve is actually a variable orifice, flow through the valve is a function of valve stem lift and pressure drop as shown in Equation (10-11):

$$Q = f(\Delta P, y), \tag{10-11}$$

where y = valve stem lift, in.

By differentiating Equation (10-11) it can be seen that the total change in flow is the sum of changes due to both pressure drop and lift.

$$dQ = [\partial Q/\partial(\Delta P)]_{y = \text{const}}\, d(\Delta P) + [\partial Q/\partial y]_{\Delta p = \text{const}}\, dy \tag{10-12}$$

If the pressure drop ΔP is assumed constant, then Equation (10-12) reduces to

$$dQ/dy = [\partial Q/\partial y]_{\Delta P = \text{const}} \tag{10-13}$$

The change of flow dQ with respect to change in lift dy is a measure of valve sensitivity or gain. Since a constant system gain at all operating conditions is desirable to maintain stability, the reader is probably wondering why any characteristic other than the linear plug should be used. Unfortunately, the answer is very complex and presents a common problem that is continually debated by experts in the field of control.[9,11,12,13,14]

The original idea of a characterized valve was to compensate for some nonlinear gain term within the system such that total "process" gain would remain constant at any process load or operating point. If, for example, some process component exhibits high gain at low flow and low gain at high flow, the reciprocal gain characteristics of an equal percentage valve would in theory be ideal for this process. Ideally, with this kind of a match, "process" gain would remain constant throughout the complete operating range of the system. The problem, however, is seldom this simple as will be shown.

In determining the curves of Figure 10-10 a constant pressure drop was assumed for the control valve. While it is possible to have this condition in an industrial application, a variable pressure drop is probably a more common occurence. As an example, consider the system shown in Figure 10-11 where flow is controlled from a centrifugal pump by a valve in series with the line and process equipment load. A typical pressure versus flow curve for this system is also shown in Figure 10-11. While the story told by Figure 10-11 is very straight forward, the results can often be troublesome. For example, assume that a linear control valve is selected based on designed flow Q_d and allowable pressure drop as shown in Figure 10-11. Obviously, to provide control, the control valve must change flow. However, as shown in Figure 10-11, as flow changes on either side of Q_d, pressure drop across the control valve also varies. The end result can be shown by plotting a series of flow versus lift curves, each at a different pressure drop. A curve may then be plotted using the locus of flow and pressure drop intersections. The resultant curve of flow versus lift as shown in Figure 10-12 represents the actual installed valve characteristic. As can be seen, the installed characteristic is no longer linear, but has a characteristic that is more like that of a quick-opening plug. If the same type of analysis is made for an equal percentage plug, it is found that the installed characteristic approaches that of a linear plug. Thus, it can be seen that selection of a valve characteristic to compensate for nonlinear gain components within a control loop can be a very complex problem. In fact, the only way to be completely sure is to make a dynamic analysis of the control loop.[11,14] This procedure, however, is normally very time-consuming and too costly to be used on most applications. To overcome this problem, manufacturers and user companies have developed valve-sizing nomographs, slide rules, and specific regulations

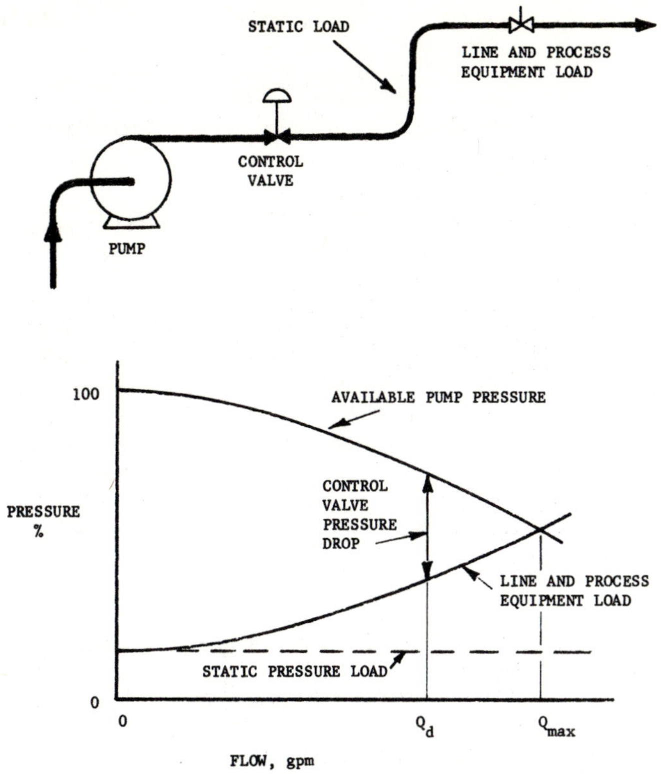

Figure 10-11. Typical control valve application and system pressure characteristics.

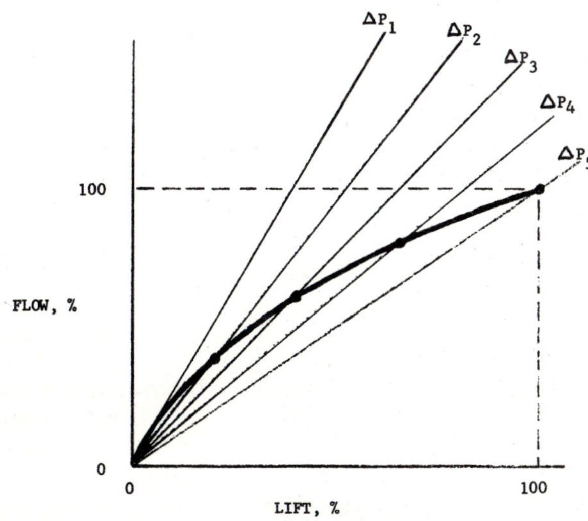

Figure 10-12. Installed flow characteristics.

regarding selection of both valve size and flow characteristics. Many of these are noted in the References.

In addition to the above it also necessary to consider problems of rangeability, compressibility (for gas and vapor flow), flashing,[15] cavitation, choked flow,[11] and materials of construction. While these problems are noted, no attempt will be made to indicate solutions. They are mentioned only to create an awareness on the part of the reader of the many problems that can be encountered in control valve selection. For a specific problem the reader should refer to the References and solicit help from control valve manufacturers.

10.4. Transmission Line Dynamics

In a modern control installation it is common practice to operate with field-mounted transmitters and locate controllers in a control room which may be some distance from process components. Under these conditions signal transmission is required from transmitter-to-controller and from controller-to-valve as shown in Figure 10-13. The example shown in Figure 10-13 is a typical column bottoms level control system. Level transmitter LT-105 transmits the measured variable, over some given distance, to level controller LRC-105 located in a unit control room. Controller output is in turn conveyed over a comparable distance to the final control element CV-105. For most control systems the problem of transmission line dynamics is negligible. This is particularly true if signals are transmitted electrically.[13] When long pneumatic transmission lines are used, however, it is possible that transmission line dynamics can influence controllability. While this problem of pneumatic transmission lines is generally overemphasized, it is helpful to examine basic concepts. Results not only show when to be suspicious, but also help to explain other accepted practices in control applications.

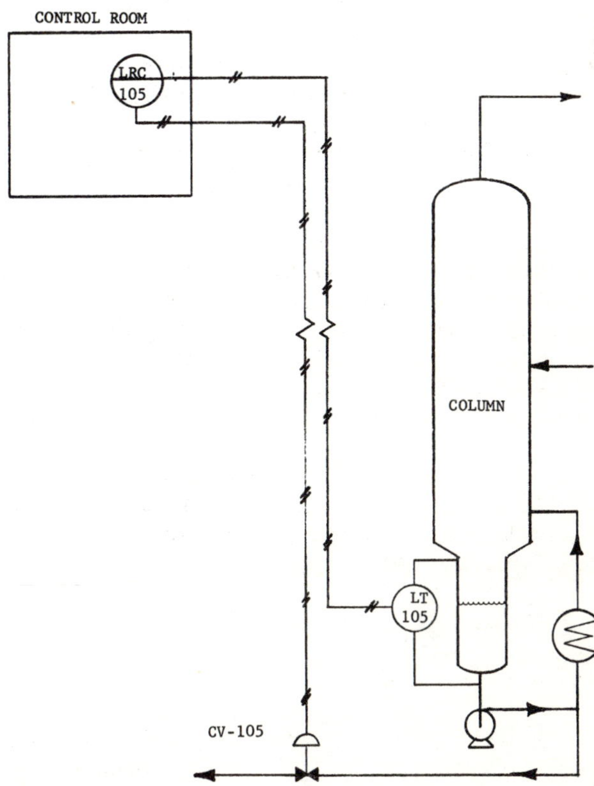

Figure 10-13. Typical column level control system.

Analysis of transmission line dynamics is a very complex process that must take into account numerous variables, such as line length and diameter, air temperature, pressure and velocity, and load or receiver volume at the line termination. Since these calculations are difficult and time-consuming, they are seldom carried out in detail.[15] In some cases experimental data are plotted to show results for particular line sizes and load volumes. One such example, for 3/16 in. I.D. tubing and negligible end volume, shows that transmission line dynamics may be approximated by dead time and a single exponential lag.[16] From these data it can be shown,

> "...pneumatic transmission may be employed up to 350 ft distance if a time constant of about 1 sec and a dead time lag of 1/3 sec are not important in automatic control, and up to about 1000 ft distance if a time constant of about 7 sec and a dead time of 1 sec are not important."[13]

Even though the above statement and referenced works are correct (and very well done), there is an inherent problem in this type of statement. While the authors are very specific about conditions of tube size and receiver volume, the user is prone to forget everything but line length. This can cause problems.

Consider, for example, the level control system of Figure 10-13. If both lines from transmitter-to-controller and from controller-to-valve are 350 ft, it is not a valid conclusion that each line can be represented by a time constant of 1 sec and a dead time of 1/3 sec. The reason is receiver volume. For the transmitter-to-controller line this conclusion could very well be valid, since most controllers have negligible input or receiver volume. The controller-to-valve line, however, is another story. Valve operator volumes can vary from a few up to 800 cu in. or more, and this load can have a very pronounced influence on overall dynamics. Since in most pneumatic transmission lines it can be assumed that air is the gas, pressure is 3 to 15 psig, temperature is close to ambient, and line diameter is 3/16 in. I.D., receiver volume and line length become the only real variables. Under these conditions it would be desirable if a simple formula existed to check transmission line dynamics in questionable applications. Such a formula does indeed exist and may even be simplified for most applications.

A reasonable approximation for pneumatic transmission lines up to phase lags of about 150° may be obtained using the following formula:[17]

$$\Delta P_i/\Delta P_r = 1/\left\{\left[1+\zeta_1 s/(\pi f_1)+s^2/(4\pi^2 f_1^2)\right]\left[1+\zeta_2 s/(\pi f_2)+s^2/(4\pi^2 f_2^2)\right]\right\} \quad (10\text{-}14)$$

where

$$f_1 = (10.42\,\Theta^{1/2}/L)\left\{1/\left[1+V/(5Ld^2)\right]\right\}^{1/2} \quad (10\text{-}15)$$

$$f_2 = (31.26\,\Theta^{1/2}/L)\left\{\left[1+V/(5Ld^2)\right]/\left[1+0.45V/(Ld^2)\right]\right\}^{1/2} \quad (10\text{-}16)$$

$$\zeta_i = 7410\left[\mu\Theta/(f_i P_o d^2)\right], \quad (10\text{-}17)$$

and
ΔP_i = pressure change in a transmission line inlet, psi,
ΔP_r = pressure change in receiver volume, psi,
f_i = undamped natural frequencies of quadratic lags, cps, where $i = 1,2$,
ζ_i = damping ratio for quadratic lags, where $i = 1,2$,
Θ = absolute temperature, °R,
L = line length, ft,
V = receiver volume, cu in.,
d = I.D. of line, in.,
μ = absolute viscosity of air, (lb-sec)/(sq ft),
P_o = line absolute pressure level, psia.

Although Equation (10-14) is much less complicated than most transmission line equations, considerable time is required to obtain results. A much simpler method to approximate transmission line dynamics has been suggested using Equation (10-14) as a base.[18] In this technique, only the first quadratic term is used. While this degrades the approximation, it is not the problem that might be expected. To understand this more clearly it is helpful to review some basic concepts. To determine stability using a Bode plot it is only necessary to determine a phase lag of 180°. If one large time constant or several small time constants exist in process equipment, as would be expected, only a few degrees of lag remain to be contributed by control components and transmission line lags. If, on the other hand, a large portion of the phase lag is contributed by transmission line dynamics, it is obvious that a control problem will exist. Since there are normally two lines to consider, any approximation that will give a reasonable indication up to 30° phase lag (a total of 60° for both lines) is sufficient to indicate if a problem exists. Once a problem is located, more complex equations can be used to increase accuracy and help in determining a solution.

Using only the first quadratic term of Equation (10-14) and assuming air at a temperature and pressure of 70°F and 10 psig, respectively, and 3/16 in. I.D. tubing gives the following:

$$\Delta P_i/\Delta P_r = 1/\left[1+\zeta_1 s/(\pi f_1)+s^2/(4\pi^2 f_1^2)\right] \quad (10\text{-}18)$$

where

$$f_1 = (238/L)\left\{1/\left[1+5.7\,V/L\right]\right\}^{1/2} \quad (10\text{-}19)$$

$$\zeta_1 = 1.1/f_1 \quad (10\text{-}20)$$

Results for a line of 350 ft and negligible receiver volume, using Equations (10-18) through (10-20), are shown in Figure 10-14. The noted experimental values of a 1-sec time constant and 1/3-sec dead time for this same line are also shown for comparison in Figure 10-14. Agreement shown in Figure 10-14 is sufficient to warn the designer of any problems.

When damping ratio ζ is greater than 1.6 which is true for the 350-ft line calculation, Equation (10-18) may be approximated by two linear lags. This result is shown in Equation (10-21) and was used to calculate values for Figure 10-14:

$$\Delta P_i/\Delta P_r \cong 1/\left\{\left[1+\zeta_1 s/(\pi f_1)\right]\times\left[1+s/(4\pi f_1\zeta_1)\right]\right\} \quad (10\text{-}21)$$

In this format calculations and plots can be quickly produced with very little trouble.

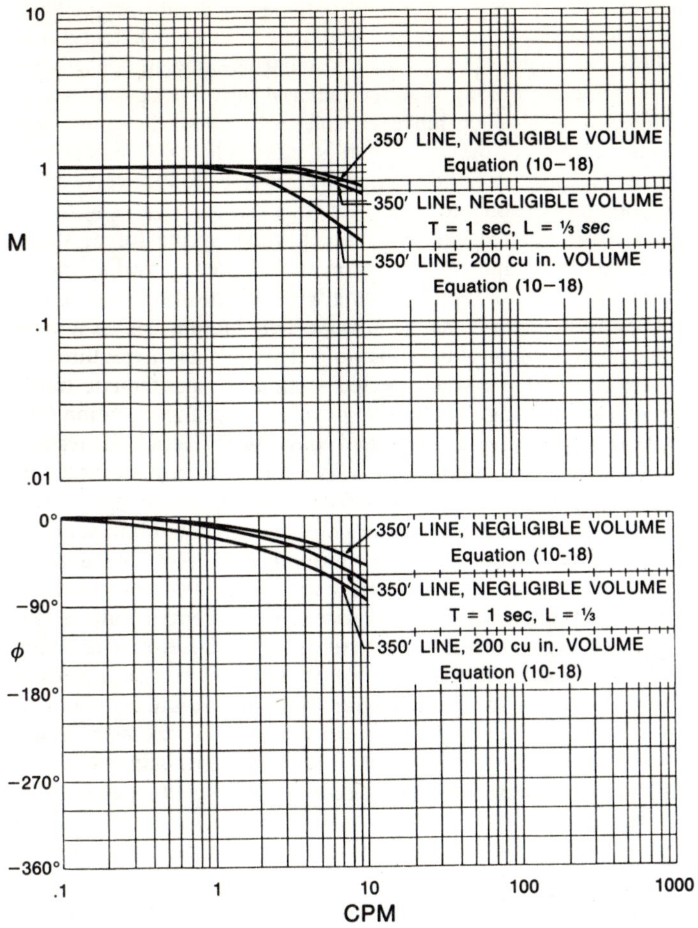

Figure 10-14. Examples of pneumatic transmission line dynamics.

A third response in Figure 10-14 shows the effect of receiver volume. In this calculation a receiver volume of 200 cu in. was assumed. From a practical viewpoint it could be assumed that results for 350 ft and negligible volume represent the transmitter-to-controller line of Figure 10-13, and results for 350 ft and 200 cu in. the controller-to-valve operator line. The important concept, however, is to recognize influence or receiver volume.

At this point the reader may question why the author has devoted such detail to transmission line dynamics, after an earlier statement that the "problem of pneumatic transmission lines is generally over emphasized." The reason is that many so-called standard control techniques, used with pneumatic instrumentation, have resulted from a need to reduce transmission line dynamics. To explain these techniques it was necessary to first examine transmission line dynamics in more detail. As an example, consider an industrial control process adjusted for quarter decay and having a period of 15 min. A control response would show a damped sinusoidal oscillation with a frequency of 0.067 cpm. If controller gain is increased to the point where the system continuously oscillates, frequency of oscillation will probably increase to 0.075 or 0.08 cpm. If the reader doubts this, a review of Chapters 6 and 7 should be undertaken. When the closed loop cycles continuously, gain and phase angle in the open loop are 1 and −180°, respectively. In other words, process and control components have produced 180° of phase lag at a frequency of 0.08 cpm. In order to be more specific, assume that these conditions are representative of the level control process of Figure 10-13. The problem is to determine influence of transmission line dynamics on this system. Results in Figure 10-14 clearly show that in this particular system transmission line gain is 1 and phase lag is negligible to 0.1 cpm. Thus, it may be concluded that in this system two 350-ft transmission lines with receiver volumes of approximately zero and 200 cu in. act as simple unity gain terms and have no dynamic influence on control. With different process and control components, however, the possibility of transmission line influence does exist.

As a second example consider a flow control loop. The transfer function for a flow process is for all practical purposes equal to unity. Stated another way, no exponential time lag exists since there is no capacity or storage term. Also, dead time is zero or negligible. For liquid flow these conclusions are based on incompressible flow concepts. If a pipe is filled, no additional material can be stored within the system, and a change in flow at the control valve is immediately observed at the measuring element. For gas flow the result is similar and is based on pressure ratios. For storage to take place in gas flow there must be a change in pressure with respect to time as noted in Chapter 2. Since pressure drop across a measuring element is normally small in comparison to absolute line pressure, the storage term is also small. As an example, consider a pressure drop of 100 in. of water for an orifice plate and a line pressure of 100 psia. From zero to 100% flow the pressure change of 100 in. of water is approximately 3.5 psi, or only 3.5% of total line pressure. With this small change there will be little storage, and results are similar to those for liquid flow. Thus, with flow control the process transfer function is unity. This means that dynamic considerations in flow loops involve only control components such as the measuring element, transmitter, controller, valve, and possibly transmission lines. In actual practice, flow loop periods will normally vary in a range of from 1 to 15 sec. Assuming cyclic conditions and a period of 6 sec yields a frequency of 10 cpm. If the same line lengths and receiver volumes are applied as in the previous example, some interesting results can be noted. From Figure 10-14 the combined phase lag of both transmission lines at 10 cpm is approximately 130°, and gain is reduced by the two lines from 1 to approximately 0.25. The result is that control action is limited by transmission line dynamics.

The above conclusion has been recognized and is usually corrected by one or more of the so-called standard control techniques, which will now be examined.

If signal transmission is absolutely necessary over this long distance, the easiest solution is to select electronic instrumentation. This simply eliminates the transmission line problem. If electronic instrumentation is unacceptable, a common solution is to reduce transmission line length by using a close-connected loop and field mounting the controller. This is an accepted practice with flow control loops and the reason should now be obvious. If an indication of flow rate is required in the control room,

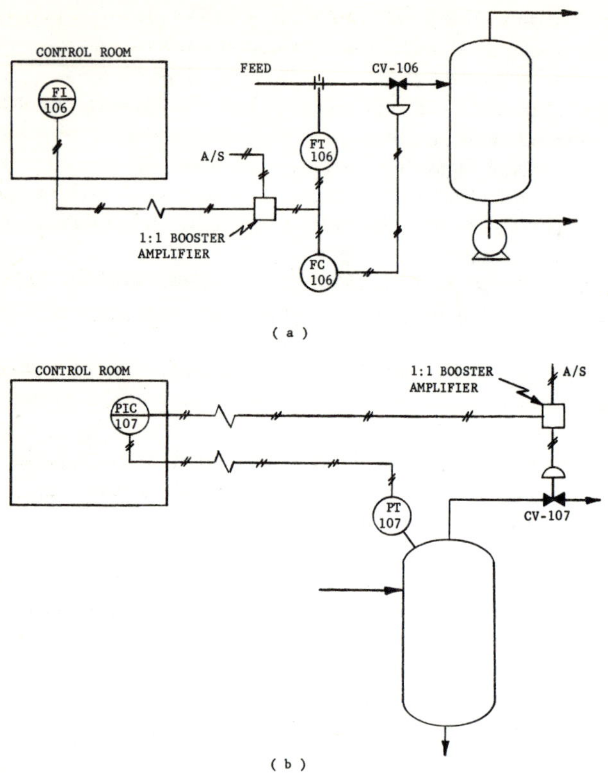

Figure 10-15. Booster amplifier applications.

this can be achieved by placing a tee in the transmitter line and adding a 1:1 relay or booster amplifier. A close-coupled control loop with a booster amplifier is shown in Figure 10-15a. A booster amplifier has a small or negligible inlet volume and, therefore, does not create any appreciable load on the transmitter. Its main advantage results from increased port size in the pilot valve that allows greater air flow or capacity. This increased capacity provides "power" to transmit the measurement signal without loading the control loop. Mechanical construction of a booster amplifier is similar to a reducing valve. In a booster amplifier the input pressure signal acting on a diaphragm area replaces the handwheel or screw adjustment.

Booster amplifiers are also used in many applications to effectively eliminate or reduce receiver volume. Effect of receiver volume on dynamics is shown by the transmission line equations and plots of Figure 10-14. A marginal system can often be improved by adding a booster amplifier ahead of the valve operator as shown in Figure 10-15b. In this application, controller-to-valve transmission line receiver volume is for practical purposes eliminated, or at least reduced to a negligible quantity. This has the effect of providing two transmission lines with negligible receiver volume and reduces total phase lag. If data shown in Figure 10-14 are taken as an example, reduction in phase lag is approximately 30°. While this does not seem to be appreciable, it is equivalent to 1/6 of the total 180° phase lag. In many cases this technique has provided an inexpensive solution to a control problem.

Before concluding this chapter it is necessary to consider one additional control component that is found in many industrial control loops. This component is the valve positioner. A very simple schematic for a positioner along with a block diagram is shown in Figure 10-16. This particular component could have been discussed along with the material on valve operators and valves. A positioner, however, can serve several functions and in the opinion of the author can be more fully explained at this time. The prime function of a positioner is to produce valve stem motion proportional to controller input signal. A valve operator is a force balance device in which valve spring force, static friction forces, and valve plug forces are balanced by diaphragm force. In a control loop operating without a positioner any force change will cause a change in valve stem motion, and in turn, a change in the manipulated variable. This change travels through the process loop and is detected as an error, and the controller then acts to change valve stem position. If change in controller signal to the valve is small and valve stem friction forces are greater than this change, the valve stem will remain fixed. A continued or increased error will cause additional changes to the diaphragm chamber, and eventually the valve will break loose. When it does change, it will normally move too far, and the sequence will repeat in a reverse direction. On a chart record, the result may look like a square wave generator output. As noted earlier, the basic function of a positioner is to reduce effect of these dynamic and static force problems.

As shown by the block diagram of Figure 10-16, a positioner is a simple fixed-gain proportional controller in a feedback loop. Any change in controller input signal to the input bellows causes a change in nozzle-baffle separation, Δx. Resultant change in

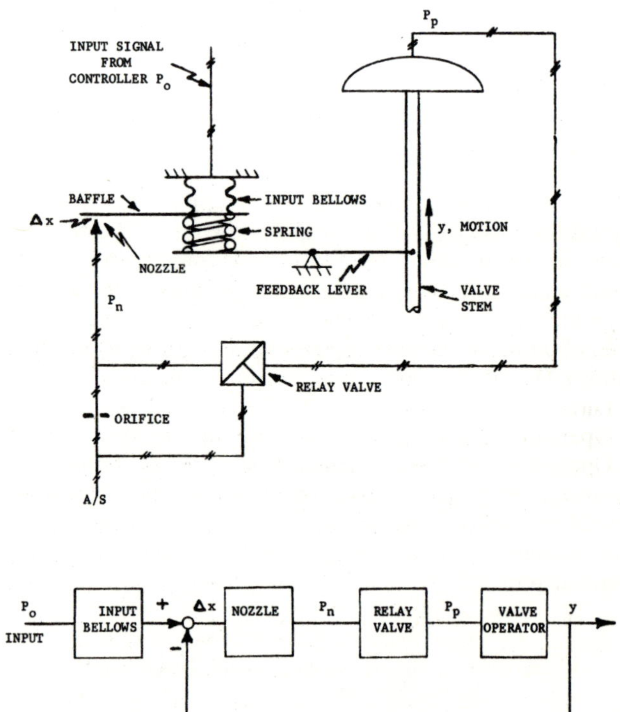

Figure 10-16. Valve positioner.

nozzle pressure P_n is amplified by the relay valve. Relay valve output P_p acts on the diaphragm to move the valve stem. If the valve stem moves, a feedback motion is generated through a feedback lever and spring. Assuming a stable loop, the system will come to balance with valve stem position change being proportional to change in input signal. In the case of excessive static friction, P_n and P_p will continue to change until the valve stem moves to the correct position. This same type of action also works to reduce hysteresis and improve linearity.

A change in line pressure at the valve plug can unbalance forces and move the valve stem even though a constant controller input signal is applied. This type of change will be detected through the feedback lever and spring, and the positioner will function to return the valve stem to the required position. Thus, a positioner is an important control component that can reduce effects of static and dynamic force changes, reduce hysteresis, and improve linearity. It can also work to reduce transmission line lag.

In a control loop without a positioner, controller input signal P_o is directly connected to the valve operator. With large valves an appreciable receiver volume can exist. One way to reduce the effect of this volume was to use a booster amplifier as shown in Figure 10-15b. Using a positioner is another way to reduce valve operator receiver volume. As shown in Figure 10-16, input bellows volume of a positioner is small and for most applications may be neglected. This assumption is similar to neglecting controller receiver volume for the transmitter-to-controller transmission line of Figure 10-14. Since most manufacturers tend to use common components, positioner relay valves are usually the same as those used in transmitters and controllers. One objective of controllers and transmitters is to minimize air flow without any sacrifice of dynamic response. The reason is simple: instrument air is a utility and, in a large control complex, can be costly. Thus, while a positioner will serve to reduce effect of receiver volume and improve dynamic response, it will normally not produce the same improvement as a booster amplifier. The main difference is that the booster amplifier, which is designed for increased air flow capacity, will outperform a positioner in filling valve operator receiver volume. A positioner, however, in addition to earlier-noted advantages can also work to improve system dynamics. Relative response times for one type of operator having a diaphragm area of 50 sq. in. are as follows:[5]

Direct operation from controller	15 sec
Operation with positioner	3 sec
Operation with booster amplifier	1 sec

In view of the many advantages of positioners, their application should be seriously considered on all critical control loops. For more details the reader should check noted references[5,11,13] and manufacturer's literature.

Chapter 10—Study Questions and Problems

(Answers appear on pages 162-163)

1. Accuracy in measurement defines how close the measured value corresponds to the ________ value.
2. Accuracy for an instrument is usually defined in terms of ________ of full-scale deflection.
3. For control purposes, ________ may be more important than accuracy.
4. Precision is a measure of the ________ of several independent measurements under identical conditions.
5. Name one advantage of a linear measurement throughout the range span.
6. Sensitivity of a nonlinear measuring element varies with the ________.
7. Range ________ defines the signal limit in which a measuring element will function, while range ________ defines its starting point.
8. An ideal measuring element and transmitter system would have a sensitivity of ________ and a time constant of ________ minutes.
9. Does reducing the range span of a measuring element and transmitter increase or decrease the instrument's sensitivity?
10. If in a given system measurement sensitivity is increased, will controller magnitude or gain have to be increased or decreased to maintain the same stability?
11. Increased measurement sensitivity has the advantage of showing a more truthful picture of the ________.
12. The assumption of describing a measuring element and transmitter as a single exponential lag is considered valid as long as the time is (less or greater) than time constants of other major components in the control loop.
13. The prime function of a transmitter is to reproduce low-energy measurement signals with sufficient ________ for transmission.
14. A transmitter will have some form of summing junction where what three types of signals are compared?
15. The function of a relay valve or amplifier in a transmitter is to provide ________ for transmission.
16. Feedback in a transmitter provides the means for checking to see that the correct ________ is developed between output and input signals.
17. Since the frequency spectrum of most process components are less than the transmitter cut-off frequency, transmitter gain or sensitivity can generally be considered to be ________.

18. An electrical transmitter having a force or motion input and using differential transformers is shown below. Explain how this system functions and define the basic functions of summing, detection, amplification, and feedback. This may be done using a block diagram with individual components defined.

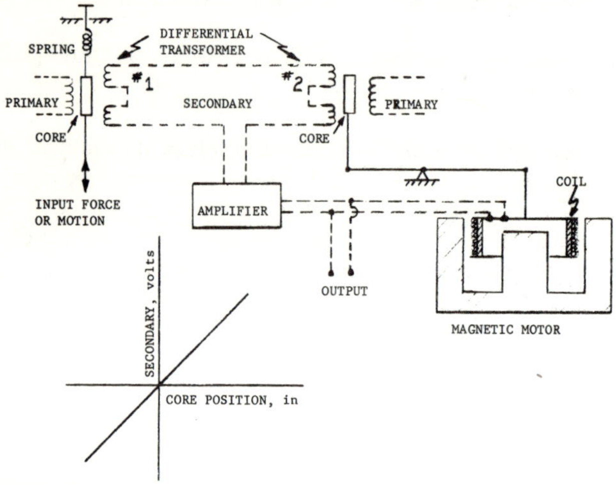

Figure 10-P18. Electrical transmitter using differential transformers.

19. For a simple pneumatic actuator as shown in Figure 10-8, the input signal is received from the __________ and the output signal is __________ motion.

20. Hydraulic and pneumatic actuators usually have an advantage in systems where __________ demands are large.

21. Are hydraulic actuators more suited for systems which require high-speed or low-speed valve travel?

22. Each plug type used in a globe valve has an __________ based on a constant pressure drop across the valve.

23. Actual flow through a valve is a function of __________ and __________.

24. The installed flow characteristic of a valve shows how flow varies as a function of __________ and actual __________.

25. The original purpose of nonlinear valve characteristics was to compensate for __________ process characteristics.

26. In the consideration of transmission line dynamics it is necessary to consider such variables as line __________ and __________, air __________, __________ and __________, and load or receiver __________.

27. Based on experimental data, it is possible to define transmission line dynamics in terms of an __________ lag and a __________ lag.

28. Assuming air at a temperature and pressure of 70°F and 10 psig, respectively, and a tubing diameter of 3/16 in. I.D., calculate the two time constants that can be used to represent a line of 500 ft with a receiver volume of 100 cu in.

29. In systems with large process equipment time constants, will transmission line dynamics be significant or negligible?

30. One method to reduce transmission line length is to __________ the controller.

31. The effect of valve operator receiver volume on a transmission line can be overcome by using a __________.

32. The main function of a valve positioner is to overcome __________ and __________ forces that can exist in a valve operator. Is process response increased or decreased by reducing receiver volume load on the transmission line?

References

1. Doebelin, E. O., *Measurement Systems: Application and Design,* New York: McGraw-Hill, 1966.
2. Beckwith, T. G., and Buck, N. L., *Mechanical Measurements,* Reading, MA: Addison-Wesley, 1969.
3. Ambrosius, E. E., Fellows, R. D., and Brickman, A. D., *Mechanical Measurement and Instrumentation,* New York: Ronald Press, 1966.
4. *Flowmeter Computation Handbook,* A.S.M.E. Research Committee on Fluid Meters, New York: Am. Soc. Mech. Engr., 1961.
5. Considine, D. M., *Process Instruments and Controls Handbook,* New York: McGraw-Hill, 1957.
6. Johnson, E. F., *Automatic Process Control,* New York: McGraw-Hill, 1967.
7. Looney, R., "Method for Presenting the Response of Temperature Measuring Systems," *A.S.M.E. Trans.,* Nov. 1957.
8. "Dynamic Response Testing of Process Control Instrumentation," I.S.A. Standard—S26, Pittsburgh, PA: Instrument Society of America, 1968.
9. Lloyd, S. G., and Anderson, G. D., *Industrial Process Control,* Marshalltown, IA: Fisher Controls Co., 1971.
10. Stice, J. E., and Swanson, B. S., *Electronic Analog Computer Primer,* Waltham, MA: Blaisdell, 1965.
11. Hutchison, J. W., ed., *I.S.A. Handbook of Control Valves,* Pittsburgh, PA: Instrument Society of America, 1971.
12. Buckley, P. S., *Techniques of Process Control,* New York: John Wiley, 1964.
13. Eckman, D. P., *Automatic Process Control,* New York: John Wiley, 1958.
14. Ziegler, J. G., and Nichols, N. B., "Valve Characteristics and Process Control," *Instruments,* Vol. 22, No. 1, Jan. 1949.

15. Rohmann, C. P., and Grogan, E. C., "On the Dynamics of Pneumatic Transmission Lines," *A.S.M.E. Trans.*, Vol. 79, pp. 853-874, May 1957.
16. Moise, J. C., "A Theoretical and Analytical Investigation of Long Pneumatic Transmission Lines," Ph.D. thesis, Case Institute of Technology, 1952.
17. Caldwell, W. I., Coon, G. A., and Zoss, L. M., *Frequency Response for Process Control,* New York: McGraw-Hill, 1959.
18. Edelman, R. I., "Equations for Pneumatic Transmission Line Transfer Functions," notes used in preparation of material for Reference 17, 1957.

11 Experimental Determination of Process Transfer Functions

In basic control theory of Chapter 1-8 it was shown how exponential and dead time lags were related to very simple physical systems. As a rule, however, components were assigned transfer functions and given values with little or no indication as to how these values were obtained. (This is a common procedure among textbook authors in control.) Actual determination of process transfer functions by either analytical or experimental means is probably the most difficult step in control system analysis. Since there is normally enough difficulty in explaining basic theory to the beginner, it is easy to see why process transfer functions and values are assumed. There is also another reason, namely, that it is seldom necessary to obtain actual process transfer functions. Out of every 1000 control loops, 990 will probably provide satisfactory control with conventional equipment and proper adjustment of controller settings. Of the 10 remaining systems, 9 can usually be salvaged by replacing equipment. Examples are wrong type of sensing element, over- or undersized valve, corrosion or temperature problem, or some other physical cause that was not evident at the design state. The one remaining loop in this group of 1000 processes is a *possible* candidate for analysis or test.

In examining the last sentence of the previous paragraph, pay particular attention to the word "possible." To begin, the author would be hard-pressed to give a factual breakdown of figures used in the example. They represent opinion only and could be difficult to prove to an operator or instrument maintenance man after a bad day. There are, however, two basic concepts that should be considered. First, many problems are a result of some abnormal process condition or may be considered under the heading of routine maintenance. Second, analytical and/or experimental analysis is a costly venture and should be undertaken only when absolutely necessary. The problem is to determine when analysis is necessary.

At this point the reader may wonder why the author is devoting a chapter to a problem that may seldom occur. The reason is very simple: a faulty control loop can often totally inhibit or limit product production. When either of these conditions occurs, economics of profit will usually dictate the determination of a solution. Under this condition the problem which seldom occurs becomes "the" problem, and techniques to be presented in this chapter become very important. Thus, production of product, increase in production rate or quality, reductions in energy charges or processing agents, and any other reasons that will produce a savings equal to or greater than study costs will determine when analysis is required.

Once a problem is defined and economics are justified, the question is how to proceed. This chapter will cover two methods that have been used successfully by the author. Assumptions are that a physical process exists, normal troubleshooting techniques have been used, and control is unsatisfactory.

11.1. Incomplete Response Technique

In Section 2.2 the incomplete response technique was described and applied to a first-order or exponential lag system. It was noted that this technique was trivial when applied to a first-order system, but would be used later on a more complex system. That promise will now be fulfilled. As noted in Chapter 5 and used throughout the remaining chapters, stability, period, and initial overpeak are directly related to a "process" phase lag of 180°. Although a "process" can have many exponential and dead time lags, it should be evident that the two largest exponential lags will contribute a majority of this 180°. The exception to this statement is a dominant dead time lag. This exception requires control techniques beyond which the author intends to cover in this basic text.[1] The remaining phase lag required to reach a total of 180° may be produced by the third largest time constant as shown in the solution to Problem 8, Chapter 7. It may also be produced by a dead time lag or by a combination of small exponential and dead time lags. No matter what the source, the magnitude will usually be about 30° and can be approximated by a dead time lag. As a result, most "processes" can be approximated dynamically by a transfer function having two, main time constants and a dead time lag. To complete the approximation it is also necessary to add a steady-state gain term. When combined the result is

$$\text{out/in} = Ke^{-Ls}/[(1 + T_1 s)(1 + T_2 s)], \qquad (11\text{-}1)$$

where K = steady-state "process" gain or magnitude ratio,
L = equivalent dead time lag,
T_1, T_2 = major process time constants.

Having assumed a model, the problem is to test the process and evaluate variables.

To demonstrate the incomplete response technique for determining a process transfer function, assume the process shown in Figure 11-1. Although it is sometimes possible to collect experimental data using a standard recorder, conventional chart speeds of from 3/4 to 1 in. per hr usually leave much to be desired. One alternative is to use a conventional two-pen recorder having a chart speed in the range of 3/4 to 1 in. per min. The faster chart speed is an option offered by most suppliers and can be connected as shown in Figure 11-1. To obtain the most meaningful data, the process should be tested at its normal operating point. For the example of Figure 11-1 assume a

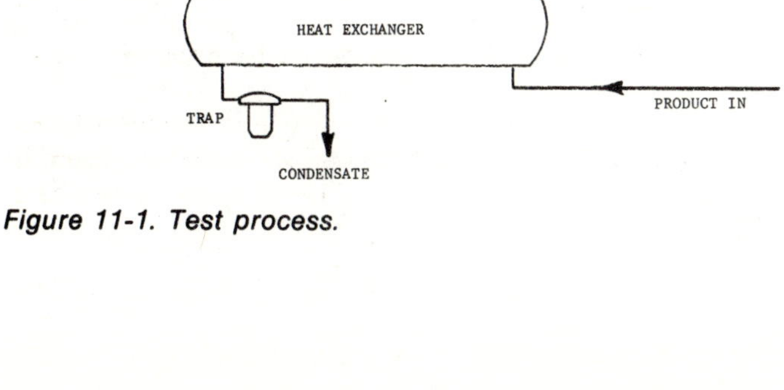

Figure 11-1. Test process.

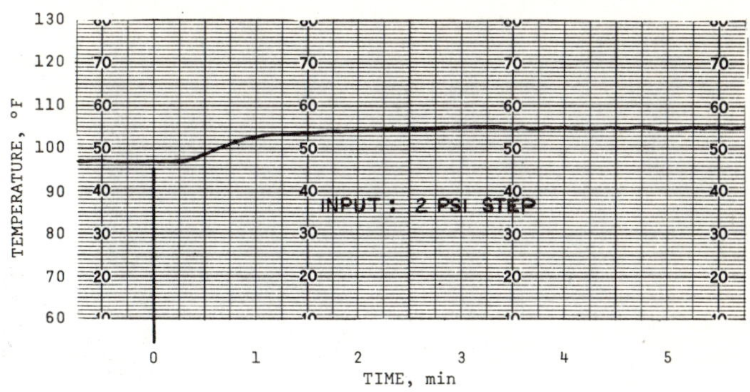

Figure 11-2. Actual data, heat exchanger step response.

temperature-sensing element range of 50 to 150°F, with a desired control point or reference value of 100°F. Since control is the problem, TRC-120 will probably be operating in the manual mode of operation. Most commercial controllers have a switch that allows an operational change from Automatic-to-Manual. In the manual position the set point signal is switched to feed directly to the control valve. This opens the control loop and allows the operator to regulate control valve position. If the controller is not already in "Manual," the first step in the test procedure is to make this conversion. A suggested method for this change can be found in the instructions supplied with each instrument. If the instrument does not have an Automatic-to-Manual switch, the instrument output line to the control valve must be disconnected and connected to an auxiliary regulated supply. A simple reducing valve and gauge will usually serve this purpose. Having achieved manual operation, adjust the signal to CV-120 to produce a product output temperature of approximately 95°F. A step change may now be introduced to the system by quickly changing pressure to the control valve. Magnitude of this pressure change can be determined experimentally, and should produce a 10 to 20% change in the measured variable or output. A change of this magnitude will give reasonable test accuracy and may still be considered as a small signal for linear theory as noted in Chapter 2. If a change of less than 10% of range is necessary, some form of amplification of the measured variable will usually be required to obtain test accuracy. For this particular test, valve pressure was changed from 8.5 to 10.5 psig. A record of the measured variable is shown in Figure 11-2. Although the output change is only 8% of range, namely, 97° to 105°F, there is little or no noise on the record, and it will be used for analysis. The problem is how to fit these results to the process transfer function of Equation (11-1).

Before attempting to analyze results shown in Figure 11-2, the incomplete response technique will be examined for a second-order system. The classical solution for a second-order differential equation is[2]

$$P_3 = 1 - \left[T_1/(T_1 - T_2)\right]e^{-t/T_1} + \left[T_2/(T_1 - T_2)\right]e^{-t/T_2} \qquad (11\text{-}2)$$

for a unit step input and assuming that $T_1 > T_2$. A physical example of such a system with typical input and output records is shown in Figure 11-3. If each side of Equation (11-2) is now

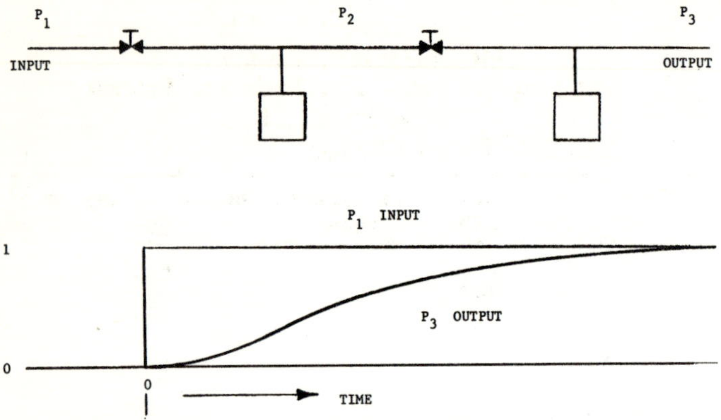

Figure 11-3. Second-order system with output response to a unit step input.

subtracted from 1, as was done in Chapter 2, the equation for the *IR* becomes

$$IR = \left[T_1/(T_1 - T_2)\right] e^{-t/T_1} - \left[T_2/(T_1 - T_2)\right] e^{-t/T_2} \quad (11\text{-}3)$$

An incomplete response plot for Equation (11-2) is shown in Figure 11-4. It is easy to see from Figure 11-4 that, as time increases, the *IR* curve becomes a straight line. This is the same result that was obtained for a first-order system in Chapter 2. What actually happens may be explained by examining Equation (11-3). Since $T_1 > T_2$, e^{-t/T_2} aproaches zero much quicker than e^{-t/T_1} for increasing values of time t. Or stated in another way, after a certain time $e^{-t/T_2} \rightarrow 0$, and Equation (11-3) reduces to

$$IR = \left[T_1/(T_1 - T_2)\right] e^{-t/T_1} \quad (11\text{-}4)$$

Taking the logarithm of Equation (11-4) gives

$$\log_{10} IR = -(0.4343/T_1)t + \log_{10}\left[T_1/(T_1 - T_2)\right] \quad (11\text{-}5)$$

If Equation (11-5) was plotted on semi-log paper, the result would be a straight line with a slope of $-0.4343/T$ and a y-axis intercept of $T_1/(T_1 - T_2)$. This result can actually be obtained, as shown in Figure 11-4, by extending the straight line portion of *IR* curve back to the *IR*- or *y*-axis. For reference this is labeled as Curve A. Since Curve A is a straight line, it must, as noted in Chapter 2, represent a first-order system and be defined by a single time constant. The time constant is T_1 and may be found from Curve A by taking 37% of the *IR* value at $t = 0$. Using Figure 11-4 gives

$$IR \text{ for } T_1 = 0.37\,(1.50) = 0.55, \text{ or } T_1 = 1.45 \text{ min.} \quad (11\text{-}6)$$

If Curve A is now defined as

$$\text{Curve A} = \left[T_1/(T_1 - T_2)\right] e^{-t/T_1} \quad (11\text{-}7)$$

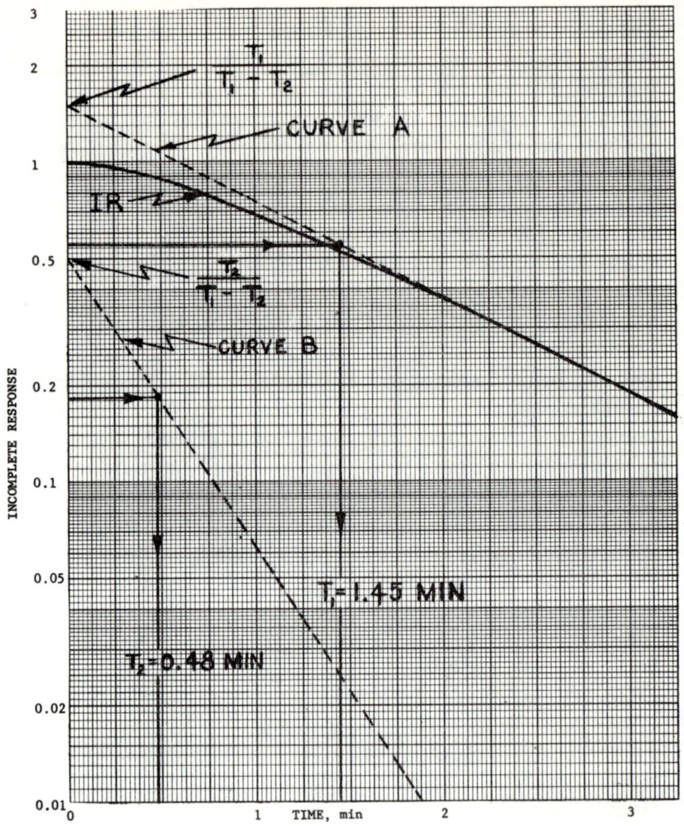

Figure 11-4. Incomplete response plot for a second-order system.

and substituted in Equation (11-3),

$$IR = \text{Curve A} - \left[T_2/(T_1 - T_2)\right] e^{-t/T_1}, \quad (11\text{-}8)$$

the result may be arranged to provide

$$\left[T_2/(T_1 - T_2)\right] e^{-t/T_2} = \text{Curve A} - IR \quad (11\text{-}9)$$

Equation (11-9) shows that subtracting numerical values of *IR* from Curve A should produce a second straight line, which represents time constant T_2.

For example, at $t = 0$, Curve A = 1.5 and $IR = 1.0$, giving a difference of 0.5. Note that in this case even though a log scale is used, actual numerical values are subtracted, and not distances. The reason is that Equation (11-9) shows an actual subtraction, and not a division that would result if distances were subtracted. This point is important since most readers, after "adding" or "subtracting" distances on Bode plots for eight or more chapters, will want to subtract distances in this method also. The result of course is meangingless. As shown in Figure 11-4 the difference of Curve A and *IR* is a second straight line, which will be defined as Curve B. If Curve B is substituted in Equation (11-9) and the log of both sides is taken, the result is

$$\log_{10} \text{Curve B} = -(0.4343/T_2)t + \log_{10}\left[T_2/(T_1 - T_2)\right] \quad (11\text{-}10)$$

Equation (11-10) confirms the straight line concept. The value of T_2 may now be found at 37% of Curve B value at $t = 0$. The result is

$$\text{Curve B for } T_2 = 0.37\,(0.5) = 0.18, \quad \text{or } T_2 = 0.48 \text{ min.} \tag{11-11}$$

A simple check on T_1 and T_2 may be made by substituting values in $T_1/(T_1 - T_2)$ and $T_2/(T_1 - T_2)$. This does not validate T_1 and T_2, but will give an indication if the graphical technique has been used with reasonable care. For example,

$$T_1/(T_1 - T_2) = 1.45/(1.45 - 0.48) = 1.5 \tag{11-12}$$

$$T_2/(T_1 - T_2) = 0.48/(1.45 - 0.48) = 0.49 \tag{11-13}$$

These values are extremely close, and much higher values will be noted in actual problems. As a rule, however, the error should be less than 20%.

Before this theory can be applied to the actual response of Figure 11-2, data must be read and converted to an incomplete response. One of the first things to do, however, is to determine "process" gain K for Equation (11-1). The reaction curve of Figure 11-2 was taken directly from TRC-120, as shown in Figure 11-1. Controller TRC-120 was in Manual mode and acted as a recorder only. The input step to CV-120 was 2 psi as noted in Figure 11-2, and total temperature change was 8°F, from 97° to 105°F. To obtain the "process" gain it is necessary to convert this 8°F, steady-state change to psi. Since a 3 to 15 psig change at the recorder input causes a chart travel from 50 to 150°F, the equivalent pressure change for 8°F is

$$P_{\text{out}} = 8°\text{F}\left[(15 - 3)\ \text{psi}/(150\text{-}50)°\text{F}\right] = 0.96\ \text{psi} \tag{11-14}$$

The value of K is then

$$K = P_{\text{out}}/P_{\text{in}} = 0.96\ \text{psi}/2.0\ \text{psi} = 0.48 \tag{11-15}$$

The next step is to get the incomplete response curve.

Actual data from Figure 11-2, using a time increment of 0.05 min, are shown in Table 11-1. Temperature change is found by subtracting the initial value of 97°F from each reading. Normalized values are obtained by dividing each delta value by the total change. For example at $t = 0.80$ min,

$$\text{value} = 101.1 - 97.0 = 4.1 \quad \text{normalized value} = 4.1/8.0 = 0.51 \tag{11-16}$$

The normalized value is equivalent to a complete response value. Thus, subtracting the normalized value from 1 gives the incomplete response value. Data from Table 11-1 are plotted in Figure 11-5. It is fairly obvious that Figure 11-5 is different from the simple theory expressed previously and demonstrated using Figure 11-4. In spite of this, it is possible to get a realistic process transfer function.

The two prime differences between Figures 11-5 and 11-4 exist at the beginning and end of the curve in Figure 11-5. The first 0.25 min of Figure 11-5 represent a dead time lag L as shown in Equation (11-1). While this is important in determining the process transfer function, it cannot be considered as part of the incomplete response curve. An easy way to eliminate dead time lag is to sketch in a pseudo zero time line, which is equal to dead time lag. A second problem is encountered when the reader tries to sketch in a straight line to fit the *IR* curve. While a straight line is evident between 0.25 and 1.0 min, values above 1.25 min leave much to be desired. The answer to the problem is fairly simple. An 8% change in range as shown in Figure 11-2 is too small a change to provide good output data. Amplification of the output signal should have been used. In this case amplification was omitted by intent to demonstrate this problem. In the time range from 0.25 to 1.25 min, where output changes are pronounced, good data are obtained. Beyond a time of 1.25 min, data should be neglected due to errors in reading Figure 11-2. Using this assumption, Curve A intersects the pseudo time zero at 1.20. Thus,

$$IR \text{ for } T_1 = 0.37\,(1.20) = 0.44, \quad \text{or } T_1 = 0.87 - 0.25 = 0.62 \text{ min.} \tag{11-17}$$

Subtracting the *IR* from Curve A gives Curve B, which intersects the pseudo time zero at 0.20. Here again, reliability of data between 0.25 and 0.4 min is much greater than that taken be-

Table 11-1
Incomplete Response Data
Step Response Data—Heat Exchanger Process
Input Step: 2 psi
Total Output Change: 8°F (0.96 psi)

Time (min)	Value (°F)	Δ Value (°F)	Normalized Value	Incomplete Response Value
0.05	97.0	0.0	0.00	1.00
0.10	97.0	0.0	0.00	1.00
0.15	97.0	0.0	0.00	1.00
0.20	97.0	0.0	0.00	1.00
0.25	97.0	0.0	0.00	1.00
0.30	97.1	0.1	0.01	0.99
0.35	97.3	0.3	0.04	0.96
0.40	97.7	0.7	0.09	0.91
0.45	98.2	1.2	0.15	0.85
0.50	98.7	1.7	0.21	0.79
0.55	99.1	2.1	0.26	0.74
0.60	99.6	2.6	0.32	0.68
0.65	100.0	3.0	0.37	0.63
0.70	100.3	3.3	0.41	0.59
0.75	100.7	3.7	0.46	0.54
0.80	101.1	4.1	0.51	0.49
0.85	101.3	4.3	0.54	0.46
0.90	101.6	4.6	0.57	0.43
0.95	101.9	4.9	0.61	0.39
1.00	102.1	5.1	0.64	0.36
1.25	102.9	5.9	0.74	0.26
1.50	103.2	6.2	0.77	0.23
1.75	103.5	6.5	0.81	0.19
2.00	104.0	7.0	0.87	0.13
2.50	104.4	7.4	0.92	0.08
3.00	104.7	7.7	0.96	0.04

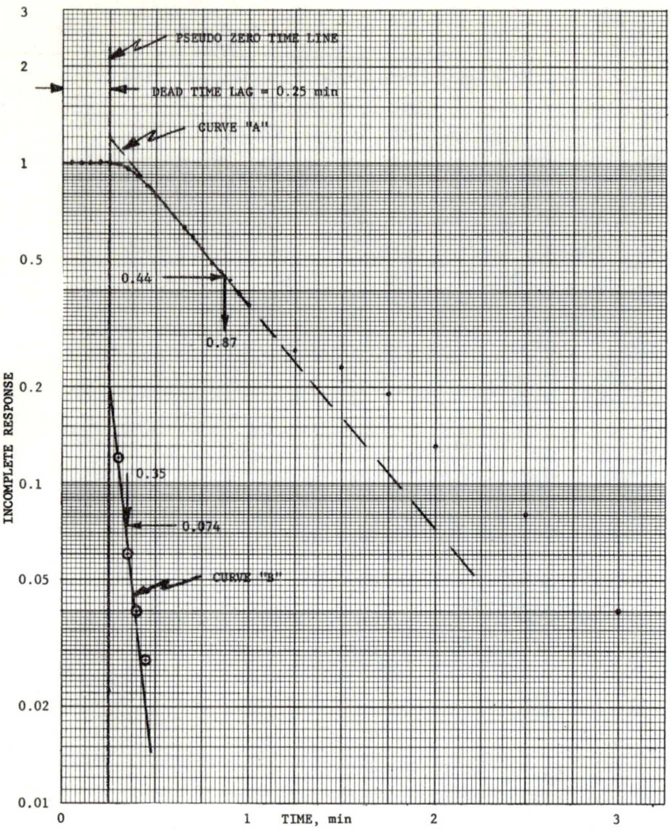

Figure 11-5. Heat exchanger incomplete response plot.

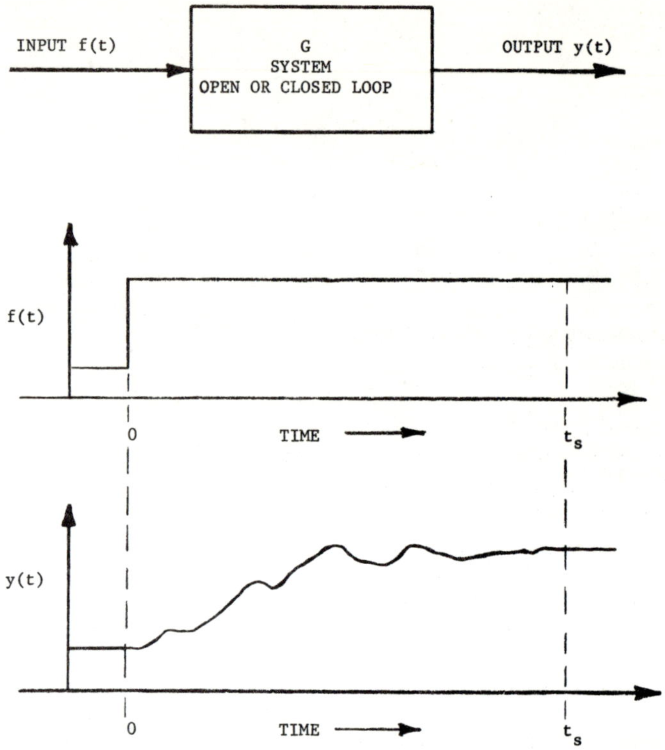

Figure 11-6. Model for step to frequency response conversion.

yond 0.4 min and should be given preference in determining Curve B. From Curve B,

$$IR \text{ for } T_2 = 0.37\,(0.20) = 0.074, \quad \text{or } T_2 = 0.35 - 0.25 = 0.10 \text{ min}. \tag{11-18}$$

It should be noted that in both Equations (11-17) and (11-18) the dead time lag or zero time shift of 0.25 min is subtracted from graph readings to get actual time constant values. If values of T_1, T_2, K, and L are now substituted in Equation (11-1), a "process" transfer function for the test process of Figure 11-1 results:

$$\text{out/in} = 0.48e^{-0.25\,s}/\left[(1 + 0.62s)(1 + 0.10s)\right] \tag{11-19}$$

If T_1 and T_2 are checked as in Equations (11-12) and (11-13), results are

$$T_1/(T_1 - T_2) = 0.62/(0.62 - 0.10) = 1.19 \tag{11-20}$$

$$T_2/(T_1 - T_2) = 0.10/(0.62 - 0.10) = 0.19 \tag{11-21}$$

These values show very favorable agreement with the intercepts of Figure 11-5.

The incomplete response technique is a very useful graphical method for obtaining a process transfer function model. The technique does, however, have a very definite limitation. In terms of application, the method can only be used on an overdamped system. In terms of results, values of T_1 and T_2 have no association with physical parameters of the process. Thus, the effect of changes in process parameters on T_1, T_2, L, and K cannot be predicted. *For these reasons the incomplete response technique is more useful for troubleshooting than design.*

11.2. Step to Frequency Response by Computer

While the incomplete response technique is a very useful tool, it does have certain limitations. A system must be overdamped, which in most cases rules out a closed loop, and it must have unequal time constants. In addition there is the problem of accepting and working with data as shown in Figure 11-5. It is very difficult for the scientific mind to accept what appears to be a total disregard for accuracy. Fortunately a solution is available which overcomes actual problems and alleviates most scientific reservations. The technique is a transformation from time to frequency and may be computer programmed with few problems.

To described the technique, assume a system as shown in Figure 11-6. Input $f(t)$ and output $y(t)$ are functions of time and may be shown in Figure 11-6. For this particular transformation it will be assumed that $f(t)$ is always a step function and that initial and final values of both $f(t)$ and $y(t)$ are constant. If $y(t)$ and $f(t)$ can be transformed to the frequency domain, it will be possible to obtain the frequency response. The transfer function may be obtained as follows:

$$G(s) = \text{out/in} = \mathscr{L}\{y(t)\}/\mathscr{L}\{f(t)\}, \tag{11-22}$$

where $G(s)$ = transfer function,
$\mathscr{L}\{y(t)\}$ = Laplace transform of $y(t)$,
$\mathscr{L}\{f(t)\}$ = Laplace transform of $f(t)$.

For the step input the transformation is

$$\mathscr{L}\{f(t)\} = 1/s \tag{11-23}$$

Transformation of the output may be obtained using the differential as noted in Chapter 3:

$$\mathscr{L}\{dy(t)/dt\} = s\,\mathscr{L}\{y(t)\} - y(0^+) \tag{11-24}$$

If $y(0^+)$, the bias value, is taken as zero, then,

$$\mathscr{L}\{y(t)\} = \mathscr{L}\{dy(t)/dt\}/s \tag{11-25}$$

Substituting Equations (11-23) and (11-25) into Equation (11-22) gives,

$$G(s) = \mathscr{L}\{dy(t)/dt\} \tag{11-26}$$

Equation (11-26) may be evaluated using the Laplace integral as follows:

$$G(s) = \int_0^\infty e^{-st}\left[dy(t)/dt\right]dt \tag{11-27}$$

Equations (11-22) through (11-27) require a knowledge of Laplace transforms that many readers may not have. To provide this background in detail would require additional chapters and will not be done. The transformations are standard and are described in most control texts.[1,3] The important result of Equation (11-27) is that the transfer function can be obtained through an integration involving slope of the output function.

To obtain the frequency response let $s = j2\pi f$, as has been done previously, and substitute in Equation (11-27):

$$G(j2\pi f) = \int_0^\infty e^{-j2\pi ft}\left[dy(t)/dt\right]dt \tag{11-28}$$

substituting

$$e^{-j\Theta} = \cos\Theta - j\sin\Theta,$$

gives

$$G(j2\pi f) = \int_0^\infty\left[\cos(2\pi ft) - j\sin(2\pi ft)\right]\left[dy(t)/dt\right]dt \tag{11-29}$$

In order to evaluate the integral from zero to infinity, rewrite Equation (11-29) as follows:

$$G(j2\pi f) = \int_0^{t_s}\left[\cos(2\pi ft) - j\sin(2\pi ft)\right]\left[dy(t)/dt\right]dt + \int_{t_s}^\infty\left[\cos(2\pi ft) - j\sin(2\pi ft)\right]\left[dy(t)/dt\right]dt \tag{11-30}$$

However, after t_s, $y(t)$ is a constant and

$$dy(t)/dt = 0,\ \text{for}\ t \geq t_s, \tag{11-31}$$

Thus,

$$G(j2\pi f) = \int_0^{t_s}\left[\cos(2\pi ft) - j\sin(2\pi ft)\right]\left[dy(t)/dt\right]dt \tag{11-32}$$

Equation (11-32) is a simple integration or summation that may be easily programmed for digital computation.

Assume a $y(t)$ output curve as shown in Figure 11-7. The slope or derivative of $y(t)$ for any incremental Δt may be defined as follows from Figure 11-7:

$$dy(t)/dt = \left[y(n\Delta t) - y(m\Delta t)\right]/\Delta t \tag{11-33}$$

The midpoint of an increment or time for which the derivative of Equation (11-33) applies is

$$t = k\Delta t = \left[(2n-1)/2\right]\Delta t \tag{11-34}$$

Substituting Equations (11-33) and (11-34) in Equation (11-32) and using the summation technique gives

$$G(j2\pi f) = \sum_{n=1}^{k}\{[\cos\{2\pi f[(2n-1)/2]\Delta t\} - j\sin\{2\pi f[(2n-1)/2]\Delta t\}] \times \left[(y(n\Delta t) - y(m\Delta t))/\Delta t\right]\Delta t\}, \tag{11-35}$$

which may be reduced to

$$G(j2\pi f) = \sum_{n=1}^{k}[\cos\{(2n-1)\pi f\Delta t\}\{y(n\Delta t) - y(m\Delta t)\} - j\sin\{(2n-1)\pi f\Delta t\}\{y(n\Delta t) - y(m\Delta t)\}] \tag{11-36}$$

or

$$G(j2\pi f) = A - jB, \tag{11-37}$$

where

$$A = \sum_{n=1}^{k}\cos\{(2n-1)\pi f\Delta t\}\{y(n\Delta t) - y(m\Delta t)\} \tag{11-38}$$

and

$$B = \sum_{n=1}^{k}\sin\{(2n-1)\pi f\Delta t\}\{y(n\Delta t) - y(m\Delta t)\} \tag{11-39}$$

Magnitude ratio, or gain, and phase angle may then be found as follows:

$$M = |G(j2\pi f)| = (A^2 + B^2)^{1/2} \tag{11-40}$$

$$\Phi = -\tan^{-1}(B/A) \tag{11-41}$$

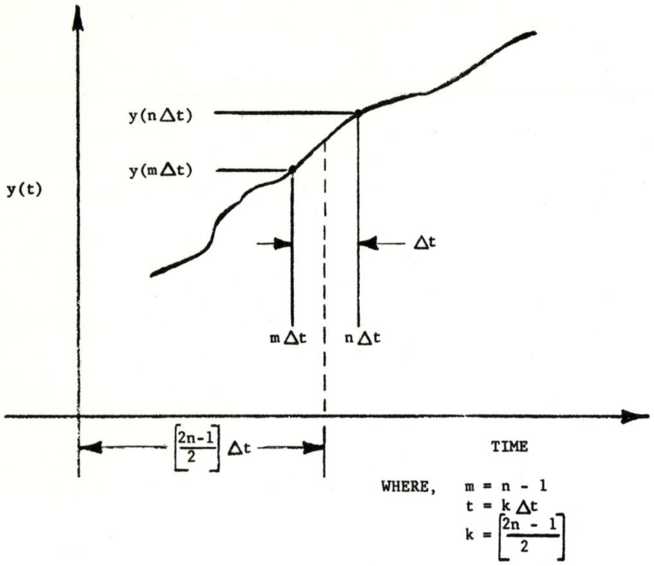

Figure 11-7. Slope determination at t = kΔt.

```
1       ZZJOB 5                               44100  STEP TO FREQUENCY     ZOSS
2       ZZFORX5
3       C      A PROGRAM TO COMPUTE FREQUENCY RESPONSE FROM STEP RESPONSE
4              DIMENSION H(90)
5              PUNCH 1
6            5 PUNCH 6
7              READ 7
8              PUNCH 7
9              READ 9,DELT,F,Q,MDATA,JOUT,STEP,BIAS
10             DO 11 I=1, MDATA
11             READ 12, H(I)
12          11 H(I) = (H(I)-BIAS)/STEP
13             PUNCH 13
14          12 FORMAT (F10.0)
15          13 FORMAT (/38HFREQUENCY              GAIN               PHASE)
16             DO 20 K=1, JOUT
17             THETA=3.1415927*F*DELT
18             A=H(1)*COS(THETA)
19             B=H(1)*SIN(THETA)
20             DO 14 N=2, MDATA
21             CN=N
22             ALPHA=(2.*CN-1.)*THETA
23             A=A+(H(N)-H(N-1))*COSF(ALPHA)
24          14 B=B+(H(N)-H(N-1))*SINF(ALPHA)
25             GSQED=A*A+B*B
26             GAIN=SQRTF(GSQED)
27             IF (A) 15, 16, 17
28          15 TANPH=-B/A
29             PHASE=ATANF(TANPH)*57.29578-180.
30             GO TO 18
31          16 PHASE=-90.
32             GO TO 18
33          17 TANPH=-B/A
34             PHASE=ATANF(TANPH)*57.29578
35          18 PUNCH 19, F, GAIN, PHASE
36          20 F=Q*F
37             GO TO 5
38           1 FORMAT (37HFREQUENCY RESPONSE FROM STEP RESPONSE)
39           6 FORMAT (/)
40           7 FORMAT(49H
41           9 FORMAT (3F10.0,2I5,F10.0)
42          19 FORMAT (F11.6, F13.4, F14.2)
43             END
```

Figure 11-8. Computer program to convert from step to frequency response.

Equations (11-38) through (11-41) show that, for any value of frequency, magnitude ratio and phase angle can be determined if values for $y(t)$ are known for time increments up to $k\Delta t$. A Fortran II program for this process is shown in Figure 11-8. To show how this program works, the heat exchanger reaction curve of Figure 11-2 will be analyzed.

Lines 1 and 2 of Figure 11-8 (which will be assumed as the reference for all "line" comments) are job identification numbers for a particular computer and will vary based on user conditions. Line 3 is a comment card to help identify the program. Line 4 is a dimension statement which provides storage capacity for up to 90 input values of $y(t)$. Line 5 punches the program title. Line 6 provides a blank card to skip a space. Line 7 reads the first input data card, which defines the job title. For this job the title is "Heat Exchanger Step Response Test." In general a more specific title with identifying number and date should be used. Line 8 provides instruction to punch job title on output. Line 9 requests all pertinent data and will require extended explanation.

The DELT of Line 9 is the value to Δt, and F is the first frequency. Selection of Δt and F involves process dynamics and is an interacting sequence. For example, Figure 11-2 shows a total temperature change of 8°F to a step input of 2 psi. If this system is assumed to be a single time constant, then at 0.63 of 8°F or approximately 102°F (97° + 5° = 102°F) a pseudo time constant for the system may be obtained. From Figure 11-2 this value is about 1 min. This would provide a corner frequency of 0.16 cpm. From past experience it is obvious that data are required at frequencies above and below the corner. A decade in either direction would yield a range of 0.016 to 1.6 cpm. For convenience only, F will be selected as 0.01 cpm. A value for DELT is then selected[2] on the basis that

$$2\pi f \Delta t << 1 \tag{11-42}$$

Using 0.5 as a value much less than 1, and f equal to 1.6 cpm, gives

$$\Delta t = 0.5/(2\pi f) = 0.5/[2\pi(1.6)] = 0.05 \text{ min} \tag{11-43}$$

This is the same Δt that was used in collecting data for the incomplete response, and the reason why it was used.

For a resonant or underdamped system a close approximation to resonant frequency may be obtained from the reciprocal of the period. Assuming a period of 1 min gives a resonant frequency of 1 cpm. From Figure 6-4 a range of frequencies from 0.03 to 3 cpm would be indicated. A first frequency of 0.01 could be used, but Δt must be reduced according to Equation (11-43) to meet the 3-cpm requirement.

The next input on Line 9 is Q, the frequency multiplier. When Equations (11-38) through (11-41) are solved, magnitude ratio and phase angle are obtained for a particular frequency. These results are punched according to instructions on Line 35. On Line 36, the first frequency is multiplied by Q to give a new F and the process of calculation is repeated. For an overdamped system such as the present heat exchanger example, a multiplier of 1.5 to 2 will supply adequate output information. For an underdamped or resonant system the multiplier should be between 1.2 and 1.5. This helps to clearly define the resonant peak. For the example at hand, Q will be assigned a value of 1.6.

The MDATA on Line 9 indicates the number of $y(t)$ inputs. In the present example, 60 inputs were used at 0.05-min increments. Basic data are in Table 11-1. Reading data from a recorder such as the one shown in Figure 11-2 can be a real problem. For example, between 2.5 and 3 min, 10 data points are required for a temperature change from 104.4° to 104.7°F. In the absence of noise or any major change in this type of time increment, the author has assumed a linear interpolation for data points. Total number of data points must meet two requirements. First, enough points should be used to meet the condition that $y(t)$ equals a constant, or $t_s = k\,\Delta t$. In the heat exchanger example a few more points should have been taken to reach the final temperature of 105°F. The result as will be seen in the final print-out is a reduction in steady-state or zero frequency gain. Second, the number of data points must be less than the dimension statement of Line 4. It is easy to change the dimension statement to accept more inputs if necessary. In most cases 60 to 80 data points are sufficient.

The value JOUT on Line 9 indicates the number of frequency values for which magnitude ratio or gain and phase angle should be calculated. With a frequency multiplier of 1.6, the required frequency range may be covered with 12 output points. This will give data from 0.01 to 1.75 cpm.

To provide a dimensional basis for determining magnitude ratio or gain, the inut step must be related to chart units of output. This is done through the STEP value on Line 9. It is defined as follows:

$$\text{STEP} = \text{chart equivalent input step} \tag{11-44}$$

$$= (\text{chart range/transmitter range}) \times \text{input step}$$

It is assumed in Equation (11-44) that transmitter range and input step have similar units, namely psi or ma. If units are mixed such as 3 to 15 psig and 4 to 20 ma, then an additional conversion factor will be required. In the heat exchanger example the following data apply:

Chart range	50° to 150°F
Transmitter range	3 to 15 psig
Input step	2 psi

Therefore,

$$\text{STEP} = (100°\text{F} \times 2\ \text{psi})/12\ \text{psi} = 16.7°\text{F} \tag{11-45}$$

A quick check shows that the 8°F change of Figure 11-2 divided by 16.7° gives a zero frequency gain of 0.48, which agrees with the result of Equation (11-15). In the case of Equation (11-15), the 8°F chart reading was changed to psi to conform to input step units. This shows that method is not as important as maintaining units.

The final item on Line 9 is BIAS. When $y(0^+)$ in Equation (11-24) is assumed to be zero, it becomes necessary to subtract the initial value of $y(t)$ from all input data points. This step is performed on Line 12. For the heat exchanger example the initial value or BIAS is 97°F. This is the temperature at the start of the test.

Details of format for input data are given on Lines 38 through 42. These are not critical and will vary depending on what computer is used. No attempt has been made to polish the program of Figure 11-8. It is used primarily as an educational tool, and when necessary on an actual process. The procedure of Lines 27 through 34 determines phase angle for values between +90° and −270°. This sometimes requires judgment in interpreting results, as will be noted shortly for the heat exchanger example. Based on reasonable input data, the program gives reliable results up to the limit of Equation (11-43). Input data

```
HEAT EXCHANGER STEP RESPONSE TEST
  0.05       0.01         1.6   60   12       16.7       97.0
 97.0
 97.0
 97.0
 97.0
 97.0
 97.1
 97.3
 97.7
 98.2
 98.7
 99.1
 99.6
100.0
100.3
100.7
101.1
101.3
101.6
101.9
102.1
102.3
102.5
102.7
102.8
102.9
103.0
103.1
103.15
103.2
103.25
103.3
103.35
103.4
103.45
103.5
103.55
103.6
103.7
103.85
104.0
104.04
104.08
104.12
104.16
104.2
104.23
104.26
104.29
104.32
104.35
104.38
104.41
104.44
104.47
104.5
104.54
104.58
104.62
104.66
104.7
```

Figure 11-9. Computer input data for heat exchanger step to frequency conversion.

and results for the heat exchanger example are detailed in Figures 11-9 and 11-10.

Obviously input data points could be combined and save input cards as shown in Figures 11-9 and 11-10. This, however, is a minor point and has the advantage of making it easy for students to correct key punch errors in input data. The phase limitation previously mentioned is evident in the number 12 output at 1.75 cpm. In development of the original program it was decided that phase angle determination would be from +90° to −270°. This took advantage of all four quadrants, allowed for phase lead of derivative-type networks, and carried phase lags at least 90° beyond the critical valve of −180°. While it is not needed for analysis, it is fairly obvious from the trend of previous results that the 62.34° phase value is actually −298°. This may be obtained by subtracting 62° from −360° and assuming a continuous function. This point is trivial, but is added to allay any reader fear. A comparison of incomplete response results using Equation (11-19) and computer results from Figure 11-10 are shown in Figure 11-11. Unfortunately actual frequency response data are not available for comparison.

11.3. Other Methods for Determining Process Dynamics

A review of available literature will quickly show a number of other methods for determining process dynamics on existing

FREQUENCY RESPONSE FROM STEP RESPONSE

HEAT EXCHANGER STEP RESPONSE TEST

FREQUENCY	GAIN	PHASE
.010000	.4607	-3.55
.016000	.4601	-5.68
.025600	.4586	-9.09
.040960	.4549	-14.51
.065536	.4455	-23.07
.104857	.4226	-36.31
.167772	.3721	-55.52
.268435	.2886	-78.20
.429496	.2417	-102.90
.687194	.1834	-156.87
1.099511	.0585	-217.58
1.759218	.0440	62.34

Figure 11-10. Computer output results for heat exchanger step to frequency conversion.

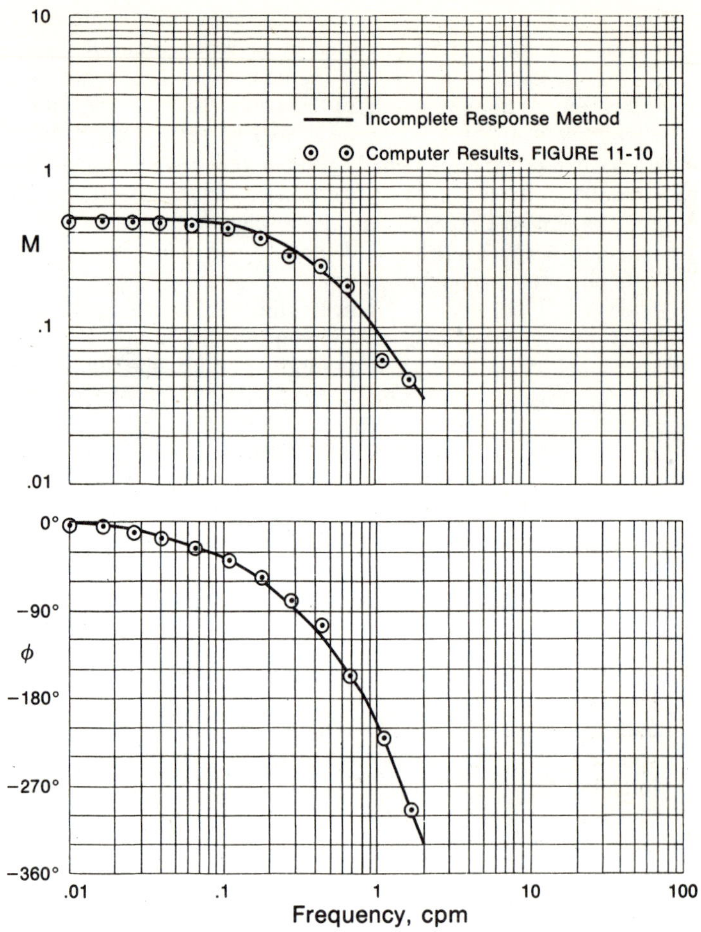

Figure 11-11. Frequency response comparison: incomplete response vs. computer results.

units. These methods range from an actual frequency response to correlation techniques.[4,5,6,7,8,9] There is no intention to cover these techniques in detail. Some advantages and disadvantages will be noted. Beyond this, the reader will have to explore the noted literature.

A *sine wave test* undoubtedly offers the most reliable method of obtaining process dynamics. It is also the most time-consuming from a viewpoint of collecting data. Although several tests were made in early studies on process plant dynamics,[4,5,9] the method has never found extensive use outside the laboratory.

Step response testing, using incomplete response and transform techniques described earlier in this chapter, provides one of the most common means of obtaining dynamic test data. The method is simple, results in minimum test time, and can often be conducted with conventional control equipment. It does have disadvantages of basing all information on one test, and an inability to obtain high-frequency results as noted in Equation (11-42). An impulse function may be used in the same manner as the step function to determine the frequency response. The transform technique is similar to the method noted in Section 11.2. Since the Laplace transform of the unit impulse is unity, the Laplace transform of the output is the transfer function.[1,3,9]

The *transform technique* of Section 11.2 may be extended beyond the concept of a step input.[6] It is possible to use any input $f(t)$ and transform both input and output functions in order to get the frequency response. The condition that initial and final values of $f(t)$ and $y(t)$ be constant still remains. The advantage of course is that $f(t)$ can take any random curve, similar to the $y(t)$ shown in Figure 11-6. The disadvantage is that input data for computer reduction, as shown in Figure 11-9, must also be supplied for the input record. In cases where a definite step input is undesirable, however, this technique is convenient.

Correlation techniques make use of a binary noise code as an input signal. For example a valve pressure of 9 psig may be selected as a "zero" and 9.2 psig as "one." The code may be generated from a computer or by means of digital shift registers. A specified number of pulses is used, and output is recorded for each pulse. By means of auto- and crosscorrelation techniques it is possible to obtain process dynamics.[10] Advantages are use of a small input signal, no limitations on initial and final conditions, and no need to wait for a complete output response to a given input signal. At a first glance, this technique appears to be a noise analysis method. This technique is probably the most expensive from a hardware consideration. It is a very powerful tool but has achieved only limited useage.[7,8,9,10]

All the above-noted methods suffer from noise, changes in loading during a test, and zero shifts in operating conditions. Generally these problems can be overcome by multiple tests and data averaging. There are, however, two important points that should be remembered regarding any dynamic data obtained from an operating unit:

—Results are valid only within a limited range of conditions that existed during the test;

—Results cannot be used to forecast effect of changes in equipment design parameters or operating conditions.

A simple example is dead time lag. If flow is doubled, dead time lag is halved. While it might be possible to forecast this change on dynamic data, the effect on heat transfer rates and mixing would be next to impossible to predict in any but the most trivial process. Thus, while experimental techniques provide useful information for troubleshooting, some additional method is needed to supply information for design considerations. This will be the subject of the next chapter.

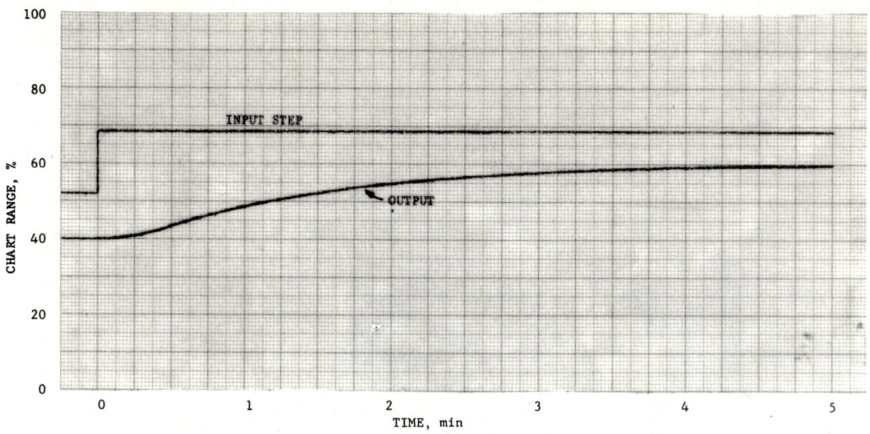

Figure 11-P1. Simulated process response data.

Chapter 11—Study Questions and Problems

(Answers appear on pages 163-166)

1. The graph above shows output response to a step input for a simulated "process" on which actual frequency response data are available. Using the graph and the incomplete response technique, find K, L, T_1, and T_2 for the process model of Equation (11-1). Plot the frequency response for Equation (11-1) and compare your result with actual frequency response data given in the appendix.

2. Using the figure in Problem 1 and the Step-to-Frequency conversion program of Figure 11-8, determine the frequency response. Compare computer results with actual frequency response data tabulated in the solution to Problem 1.

3. The __________ test offers one of the most reliable means for obtaining frequency response data.

4. A __________ test is probably the most common means for obtaining dynamic response data on actual processes.

5. Any experimental test will be influenced by what three factors?

6. In experimental testing it is often possible to overcome disturbances by averaging data and performing __________ tests.

7. Dynamic test data are valid only for __________ conditions that existed during the test.

8. In general, it is impossible to forecast from dynamic response data the change in response that will occur following a change in __________ conditions or __________ parameters.

References

1. Buckley, P. S., *Techniques of Process Control,* New York: John Wiley, 1964.
2. Caldwell, W. I., Coon, G. A., and Zoss, L. M., *Frequency Response for Process Control,* New York: McGraw-Hill, 1959.
3. Wilson, H. S., and Zoss, L. M., "Control Theory Notebook," reprinted from *I.S.A. Journal*, I.S.A., Pittsburgh, PA, 1964.
4. Hoyt, P. R., Stanton, B.D., and Union, D. C., "Dynamic Field Test of a Steam Turbine," A.S.M.E. Paper No. 57-IRD-15, 1957.
5. Zoss, L. M., Gollin, N. W., and Edelman, R. I., "Dynamic Response Analysis of Air Heater Temperature Control System," *Ind. and Engr. Chem.,* Vol. 48, pp. 1069-1073, 1956.
6. Wolfgang, G. H., Wagner, C. A., and Zoss, L. M., "Computer Techniques for Frequency Response Analysis," Paper No. 65.1.63, 18th Annual I.S.A. Conference and Exhibit, Chicago 1963.
7. Hutchinson, A. W., and Shelton, R. J., "Measurement of Dynamic Characteristics of Full-Scale Plant Using Random Perturbing Signals: An Application to a Refinery Distillation Column," *Trans. Instn. Chem. Engrs.,* Vol. 45, pp. T334-T342, 1967.
8. Briggs, P. A. N., Hammond, P. N., Hughes, M. T. G., and Plumb, G. O., "Correlation Analysis of Process Dynamics Using Pseudo-Random Binary Test Perturbations," *Proc. Instn. of Mech. Engrs.,* Vol. 179, Part 3H, pp. 37-51, 1964-65.
9. *Plant and Process Dynamic Characteristics,* Proceedings of a Conference at Cambridge, The Society of Instrument Technology, New York: Academic Press, or London: Butterworth's Scientific Publications, 1957.
10. Rex, R. L., and Roberts, G. T., " Correlation, Signal Averaging and Probability Analysis," *Hewlett-Packard J., Vol 21, No. 3, pp. 2-8, 1969.*

12 Analytical Determination of Process Transfer Functions

Development of a transfer function using techniques of Chapter 11 assumes an existing process and is usually limited to test conditions. As noted, this makes it almost impossible to predict influence of design changes on controllability. For design consideration the basic assumption of an existing process is not valid. The only solution under these conditions is to derive and solve the time-dependent or differential equations which describe the process. Since the words "differential equations" were used earlier in Chapter 2, they should not cause too much panic at this point. There is no question that development of process equations is probably the most difficult phase of control. In most applications it is a costly, time-consuming operation that should be justified in the same manner as for testing programs of Chapter 11. When justified, results can be very useful in predicting controllability. Results may also be used to explain in more detail the control theory of Chapters 1 through 8.

12.1. Energy or Mass Balance

The objective is to derive a transfer function for a given process that could be used to determine controllability as a function of "process" parameters. As a start, assume a control system design as shown in Figure 12-1. It is often helpful to sketch a preliminary block diagram in an effort to establish energy and/or mass flow through the system. A typical block diagram with product flow and inlet temperature loads is shown in Figure

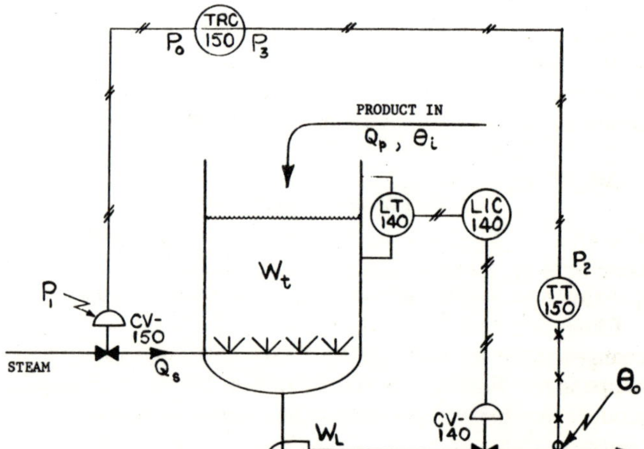

PRODUCT INLET FLOW, Q_p = 15 lb/min
PRODUCT INLET TEMP., θ_i = 70°F
SET POINT TEMP., r = 150°F
MAXIMUM STEAM FLOW, Q_s = 3 lb/min
PRODUCT IN TANK, W_t = 30 lb
PRODUCT IN PUMP & DISCHARGE LINES, W_L = 1.5 lb
VALVE MOTOR PRESSURE RANGE, P_1 = 3 to 15 psig
TEMP. TRANSMITTER SPAN, 100 to 200°F
TEMP. TRANSMITTER OUTPUT, P_2 = 3 to 15 psig

Figure 12-1. Example process for transfer function derivation.

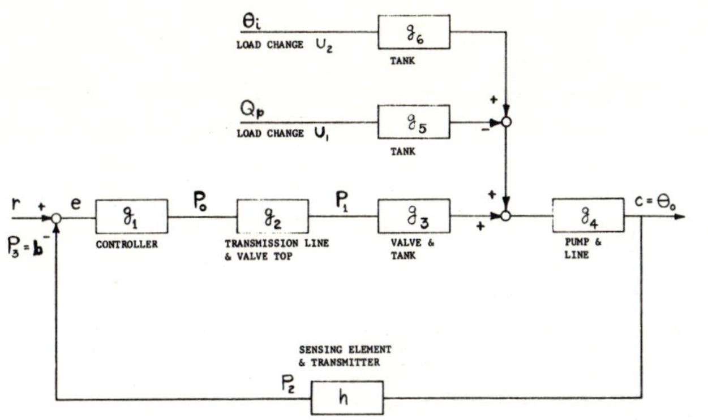

Figure 12-2. Block diagram of example process for transfer function derivation.

12-2. If loads are initially assumed to be constant as in Chapter 7, the problem reduces to determining output temperature change as a function of steam flow change, namely, $\Delta\Theta_o/\Delta Q_s$.

With steam flow and temperature as variables, an energy balance would appear to be a good starting point. For the process tank of Figure 12-1, the energy balance is

$$q_{in} = q_{out}, \tag{12-1}$$

where q = heat flow, B/min.

Equation (12-1) represents the steady-state conservation of energy principle. If a steady-state condition existed, there would be no need for control. What actually happens is

$$q_{in} + \Delta q_{in} = q_{out} + \Delta q_{out} + \Delta q_{storage} \tag{12-2}$$

Equation (12-2) includes the transient or changing condition as well as the steady state. The bias or steady-state condition, as previously noted, is of negligible interest and can be eliminated from consideration by subtracting Equation (12-1) from (12-2)

$$\Delta q_{in} = \Delta q_{out} + \Delta q_{storage} \tag{12-3}$$

The effect of Equation (12-3) is the same as neglecting initial conditions in Laplace transforms of Chapters 2 and 11. The question now is to relate Equation (12-3) with $\Delta\Theta_o/\Delta Q_s$.

Changes in heat flow to the process of Figure 12-1 result from changes in steam flow, product flow rate, and product inlet temperature. To keep this initial example as simple as possible, product flow and temperature changes have been assumed zero (*constant* flow and inlet temperature). Thus, steam flow provides the only source of heat flow change to the process. Steam flow through CV-150 of Figure 12-1 is a function of valve position, upstream steam pressure, and downstream steam pressure. This is shown as follows:

$$Q_s = f(P_1, P_u, P_d), \tag{12-4}$$

where Q_s = steam flow, lbs/min,
P_1 = valve motor pressure, psig,
P_u = upstream steam pressure, psia,
P_d = downstream steam pressure, psia.

To find total change in steam flow, Equation (12-4) is differentiated with respect to each variable. Stated in another way, total change in steam flow is equal to the sum of changes produced by each variable. This gives

$$\Delta Q_s = \partial Q_s/\partial P_1\,]_{P_u,P_d}\,\Delta P_1 + \partial Q_s/\partial P_u\,]_{P_1,P_d}\,\Delta P_u + \partial Q_s/\partial P_d\,]_{P_1,P_u}\,\Delta P_d \tag{12-5}$$

Equation (12-5) is a partial differential equation. The slope or rate of change for each variable, for example $\partial Q_s/\partial P_1$, is determined while maintaining associated variables, namely, P_u and P_d constant. By multiplying $(\partial Q_s/\partial P_1)$ by a change in valve diaphragm pressure ΔP_1, that portion of total steam flow due to valve diaphragm pressure can be determined. The sum of all individual or "partial" changes gives the total change. To simplify Equation (12-5) make the following substitutions:

$$\alpha = \partial Q_s/\partial P_1\,]_{P_u,P_d}$$

$$\gamma = \partial Q_s/\partial P_u\,]_{P_1,P_d}$$

$$\beta = \partial Q_s/\partial P_d\,]_{P_1,P_d}$$

which then gives

$$\Delta Q_s = \alpha\,\Delta P_1 + \gamma\,\Delta P_u + \beta\,\Delta P_d \tag{12-6}$$

From a practical viewpoint $(\partial Q_s/\partial P_1)$ represents the inherent valve characteristic described in Chapter 10. It gives flow as a function of valve opening for a constant pressure drop. It is helpful to observe that Equation (12-6) states that steam flow changes for any change in valve diaphragm pressure, upstream steam pressure, and/or downstream steam pressure.

A block diagram of Equation (12-6) is shown in Figure 12-3. It should become fairly obvious that ΔP_u and ΔP_d are additional load changes which were not defined in Figure 12-2. This original omission was intended to show how a detailed analysis can often reveal otherwise unconsidered problems. Actually in this

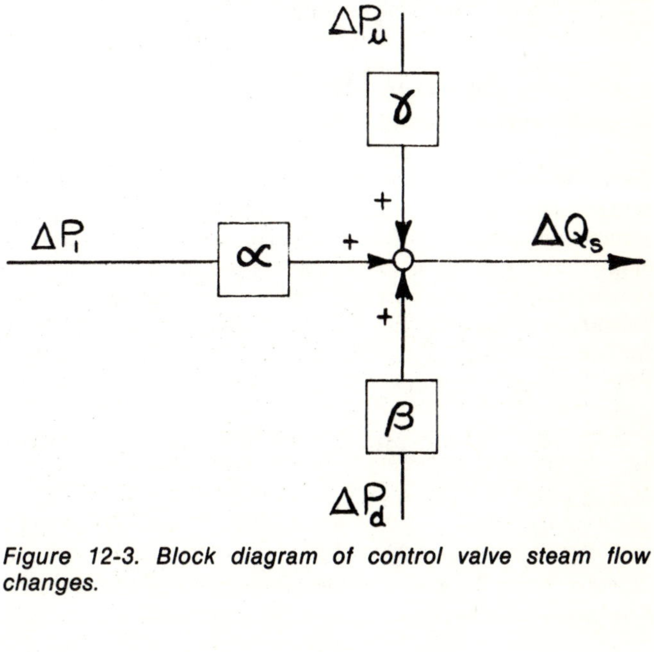

Figure 12-3. Block diagram of control valve steam flow changes.

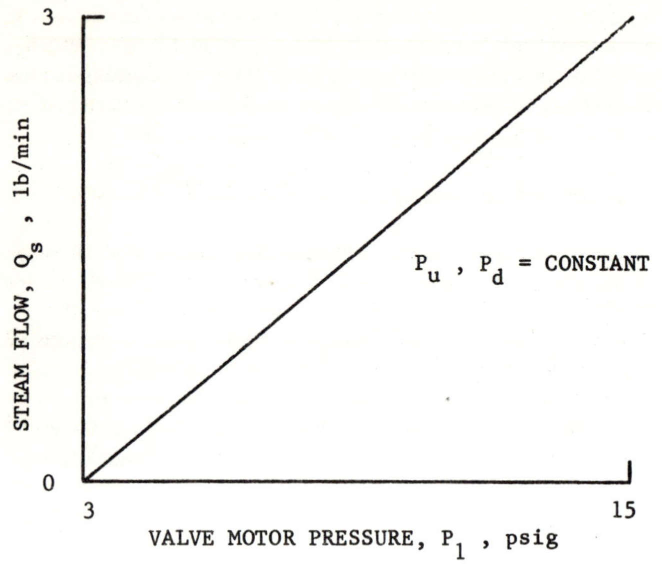

Figure 12-4. Linear characteristic valve curve.

case, ΔP_d can be neglected based on a constant downstream steam pressure. This results from the constant head produced by level control as shown in Figure 12-1. As a rule, upstream or supply steam pressure is more variable. This, however, depends on overall plant capacity and load cycles. To start, P_u will be assumed constant; thus ΔP_u = zero. Load changes merely present another input disturbance and may be handled as shown in Chapter 8. With these assumptions Equation (12-6) reduces to

$$\Delta Q_s = \alpha\, \Delta P_1, \tag{12-7}$$

and the problem is to evaluate α. In a real problem the value of α would be determined at the calculated operating point using the manufacturer's characteristic valve curves. For the present example a linear value will be assumed with a flow range of 0 to 3 lb/min as shown in Figure 12-1. The result is shown in Figure 12-4. The slope or value of α may then be calculated as follows:

$$\alpha = \left[3 - 0\,(\text{lb/min.})\right]/\left[15 - 3\,(\text{psi})\right] = 0.25\ \text{lb/(min} - \text{psi)} \tag{12-8}$$

As steam mixes with product and condenses, both latent and sensible heat are released. If saturated steam at 5 psig is assumed, latent heat is 960 B/lb, and condensing temperature is approximately 227°F (from steam tables). Since product temperature is 150°F, an additional 77 B/lb of sensible heat is available to give a total of 960 + 77, or 1037 B/lb of steam. To allow for line losses (nice excuse to get an even number) assume 1000 B/lb. Change in heat input may then be calculated as follows:

$$\Delta q_{in} = 0.25\ \text{lb/(min-psi)} \times 1000\ \text{B/lb} \times \Delta P_1\ \text{psi} \tag{12-9}$$

$$\Delta q_{in} = 250\, \Delta P_1\ (\text{B/min})$$

The next step is to determine change in heat output.

If level control is maintained as shown in Figure 12-1, product flow leaving the tank will equal the sum of product and steam flow into the tank. Assuming 1000 B/lb of steam, results in an average steam flow of 1.2 lb/min. This provides 1200 B/min, or the energy needed to heat 15 lb/min of product from 70 to 150°F. This calculation is based on a specific heat of 1B/(lb-°F). Thus, product and steam flow leaving the tank under normal conditions will be 16.2 lb/min. Steam flow will vary. If only small changes in steam flow are considered, such as plus or minus 0.5 lb/min, these changes can be neglected. The error which is introduced is less than 5% and is tolerated for two reasons:

1. Simplification of process equations;
2. Validity of process data does not normally justify any additional sophistication.

Change in heat output may then be determined as follows:

$$\Delta q_{out} = 16.2\ \text{lb/min} \times 1\text{B/(lb°F)} \times (\Theta_t - 150)\text{°F}, \tag{12-10}$$

where Θ_t = tank temperature at any time, °F.

The value $(\Theta_t - 150)$°F in Equation (12-10) shows the change in tank temperature from the 150°F set point. If Θ_t equals 150°F, there is no change in Δq_{out}. This is equivalent to being at the set point or desired value, which is the steady-state value q_{out}. This is the bias value which has been neglected. In defining Θ_t as the tank temperature at anytime, perfect mixing of steam and product is assumed. With steam sparging into this small volume this assumption appears justifiable. To complete Equation (12-3), $\Delta q_{storage}$ must be determined.

Whenever tank temperature changes, energy has been added or removed from the mass in the tank. The rate at which this energy may be added or removed is a function of product mass in the tank, tank mass, specific heat of materials, and the rate of change of tank temperature. If tank mass is neglected for this initial example, change in storage may be expressed as follows:

$$\Delta q_{storage} = MC_p(d\,\Theta_t/dt), \tag{12-11}$$

where M = product mass in tank, lb,
C_p = specific heat, B/(lb − °F).

Substituting values from Figure 12-1 gives

$$\Delta q_{storage} = 30\ \text{lb}\left[1\text{B/(lb°F)}\right](d\,\Theta_t/dt)\ (\text{°F/min}) \tag{12-12}$$

The next step is to substitute Equations (12-9), (12-10), and (12-12) into Equation (12-3):

$$250\, \Delta P_1 = 16.2(\Theta_t - 150) + 30\,(d\,\Theta_t/dt) \tag{12-13}$$

Before Equation (12-13) can be conveniently solved, it is standard practice to replace $(\Theta_t - 150)$ by $\Delta\Theta$. Thus, $\Delta\Theta$ is a temperature difference that represents any deviation from the 150°F set point. An equally convenient substitution may be made by differentiating as follows:

$$\Delta\Theta = \Theta_t - 150 \tag{12-14}$$

$$d(\Delta\Theta)/dt = d\,\Theta_t/dt \tag{12-15}$$

Equation (12-14) is a standard temperature difference substitution used in heat transfer studies. Equation (12-15) merely shows that slope of the difference, namely $\Delta\Theta$, is equal to the slope of Θ_t. Substituting Equations (12-14) and (12-15) in Equation (12-13) gives

$$250\,\Delta P_1 = 16.2\Delta\Theta + 30\left[d(\Delta\Theta)/dt\right] \quad (12\text{-}16)$$

Substituting $s = d/dt$ gives

$$250\,\Delta P_1 = 16.2\Delta\Theta + 30s\,\Delta\Theta \quad (12\text{-}17)$$

By factoring out $16.2\Delta\Theta$ from the right-hand side of Equation (12-17),

$$250\,\Delta P_1 = 16.2\Delta\Theta(1 + \left[30/16.2\right]s), \quad (12\text{-}18)$$

which may be resolved to

$$\Delta\Theta/\Delta P_1 = (250/16.2)/(1 + \left[30/16.2\right]s) \quad (12\text{-}19)$$
$$= \left[15.4/(1 + 1.85s)\right]\ (°\text{F/psi})$$

Equation (12-19) is the conventional exponential lag form and shows a magnitude ratio of 15.4°F/psi and a time constant of 1.85 min. If only symbols are carried through from Equation (12-17), the result would be

$$\Delta\Theta/\Delta P_1 = \left\{(\alpha h)/\left[(Q_p' + Q_s')C_p\right]\right\}/ \quad (12\text{-}20)$$
$$\left[1 + \left\{M/(Q_p' + Q_s')\right\}s\right],$$

where $\alpha = \partial Q_s/\partial P_1\big]_{P_u,P_d} = 0.25$ lb/(min − psi),
h = heating value of steam = 1000 B/lb,
Q_p' = average or design product flow = 15 lb/min,
Q_s' = average or design steam flow = 1.2 lb/min,
M = product mass in tank = 30 lb,
C_p = specific heat of product = 1 B/(lb − °F).

From Equation (12-20) it is possible to see the effect of design changes and predict controllability.

To complete the calculations it is necessary to examine product flow from the tank to the point of measurement. From Figure 12-1 it seems fairly evident that this could be treated as transportation or dead time lag. The lag would be

$$L = W_L/(Q_p' + Q_s') = 1.5\ \text{(lb-min)}/16.2\ \text{lb} \quad (12\text{-}21)$$
$$\cong 0.09\ \text{min},$$

which gives a transfer function as follows:

$$\Delta\Theta_o/\Delta\Theta = e^{-Ls} = e^{-0.09s}, \quad (12\text{-}22)$$

where $\Delta\Theta_o$ = change in product outlet temperature, °F.

Combining Equations (12-19) and (12-22) gives

$$\Delta\Theta_o/\Delta P_1 = 15.4e^{-0.09s}/(1 + 1.85s) \quad (12\text{-}23)$$

The overall "process" transfer function $\Delta P_3/\Delta P_0$ can be obtained by finding $\Delta P_1/\Delta P_0$ and $\Delta P_3/\Delta\Theta_0$. Techniques for transmitters and transmission lines, noted in Chapter 10, could be used along with manufacturer's data to yield the overall result. The important point to note is that transfer functions such as Equations (12-20) and (12-22) can be derived to predict influence of design changes on controllability.

12.2. Analysis and Test Results Compared

The textbook problems of Section 12.1 are necessary to demonstrate technique and develop proficiency. To develop a sense of reliability and recognize limitations, it is necessary to compare actual results with test data. There are numerous published papers which present both analysis and test results,[1,2,3] and many others that allude to application of noted techniques.[4] One such example will be detailed at this time based on the noted reasons.[5] The system, namely a concentric tube heat exchanger, is shown in Figure 12-5. Following the detailed analysis, predicted and actual control results will be compared. A block diagram of the system showing basic components and load changes is given in Figure 12-6. Transfer functions for each component will be determined. Loads will be assumed constant to limit complexity.

The first component to be considered in this system is the transmission line, positioner, and valve top combination. In most cases a manufacturer can supply dynamic data for applications of this type. Usually the manufacturer will provide data covering a range of conditions and a particular result is found by

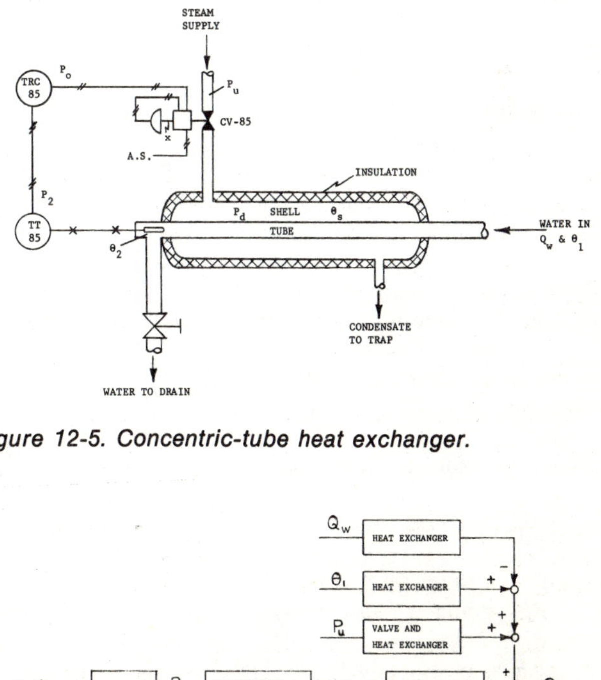

Figure 12-5. Concentric-tube heat exchanger.

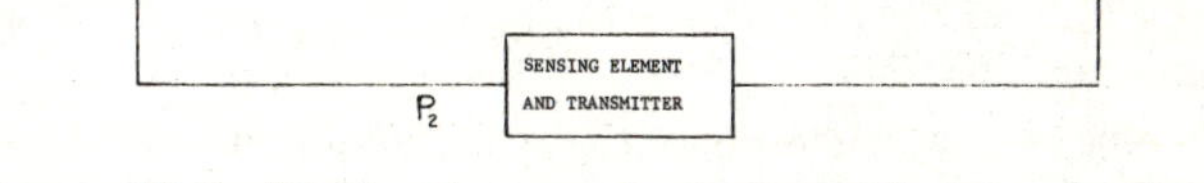

Figure 12-6. Block diagram for concentric-tube heat exchanger.

interpolation. In the example under study an exponential lag was assumed, and the time constant was determined experimentally. Zero frequency or steady-state gain was determined by dividing valve stem travel by pressure range. Results are

$$\Delta x / \Delta P_o = 0.047 \text{ in./psi}/(1 + 0.083s), \quad (12\text{-}24)$$

where x = valve stem travel, in. (9/16 in.),
P_o = control output pressure, psi (3 to 15 psig).

Thus k = 9/16 in./12 psi = 0.047 in./psi
T_v = 0.083 min, experimentally determined

The next component includes the valve and heat exchanger. Procedure for determining valve and heat exchanger transfer function is application of the energy balance concept shown in Section 12.1. Assuming that upstream pressure P_u is constant reduces Equation (12-5) to

$$\Delta Q_s = \partial Q_s/\partial x\,\big]_{P_d}\,\Delta x + \partial Q_s/\partial P_d\,\big]_x\,\Delta P_d, \quad (12\text{-}25)$$

where Q_s = steam flow, lb/hr,
x = valve stem travel, in.,
P_d = downstream steam pressure, psia.

It should be noted that in this application a positioner is used; thus, valve stem position rather than valve motor pressure is used for the input variable. Equation (12-25) can be reduced to

$$\Delta Q_s = \alpha \Delta x + \beta \Delta P_d, \quad (12\text{-}26)$$

where $\alpha = \partial Q_s/\partial x\,\big]_{P_d}$
$\beta = \partial Q_s/\partial P_d\,\big]_x$

To evaluate α and β it is necessary to obtain typical operating conditions and valve specifications. Data are

P_u = upstream steam pressure, 40 psia,
P_d = downstream steam pressure, 22 psia (design or normal operating condition),
x = valve stem travel, 9/16 in.,
Equal percentage flow characteristic.

The basic equation for valve capacity using steam is[6]

$$Q_s = 3C_v\left[(\Delta P)P_d\right]^{1/2}/K, \quad (12\text{-}27)$$

where Q_s = steam flow, lb/hr,
P_u = inlet pressure, psia,
P_d = outlet pressure, psia,
ΔP = pressure difference, psi
C_v = valve flow coefficient,
K = 1 + (0.0007 × °F superheat).

When P_d is less than $0.5P_u$, use $0.5P_u$ for $\left[(\Delta P)P_d\right]^{1/2}$ (this accounts for critial flow).

Using the normal values of P_u = 40 psia, P_d = 22 psia, and assuming K = 1 for saturated steam, it is possible to plot a curve of steam flow versus lift. This procedure also uses Equation (12-27) and data from Table 12-1. Results are shown in Figure 12-7. Since this curve is for a constant P_d, it is possible to determine α from this curve. To do this, however, it is necessary to determine the normal operating point for steam flow.

Table 12-1
Inherent Control Valve Specifications

Valve Opening (% of Total Travel)	C_v
10	0.226
20	0.332
30	0.488
40	0.710
50	1.04
60	1.46
70	2.06
80	2.91
90	3.97
100	5.03

Design conditions show water to be heated enters at 60°F and at a flow rate of 1080 lb/hr. Assuming a specific heat of 1 B/(lb − °F), the basic heat equation gives

$$q = Q_w C_w(\Theta_2 - \Theta_1), \quad (12\text{-}28)$$

where q = heat load, B/hr,
Q_w = water flow rate, lb/hr,
C_w = specific heat, B/(lb − °F),
Θ_2 = water temperature leaving exchanger, °F,
Θ_1 = water temperature entering exchanger, °F.

Using Equation (12-28) the operating heat load for a desired Θ_2 of 110°F is

$$q = 1080\,(1)\,(110\text{-}60) = 54{,}000 \text{ B/hr} \quad (12\text{-}29)$$

If sensible heat is neglected, enthalpy (or heat content) of steam at 22 psia may be taken from steam tables as 957 B/lb. Thus, designed steam load is

$$Q_{s\,\text{normal}} = 54{,}000\,(\text{B/hr})/\left[957\,(\text{B/lb})\right] = 56.4 \text{ lb/hr}, \quad (12\text{-}30)$$

and under these conditions the valve will be about 49% open as shown in Figure 12-7. For this condition, C_v is 0.94, as determined from Equation (12-27) using design pressure conditions.

$$C_v = KQ_s/\left\{3\left[(\Delta P)P_d\right]^{1/2}\right\} \quad (12\text{-}31)$$

$$C_v = 56.4/\left\{3\left[(18)\,22\right]^{1/2}\right\} = 0.94$$

By taking the slope of steam flow versus lift for the design condition of Q_s, it is possible to determine α. Using data from Figure 12-7 gives

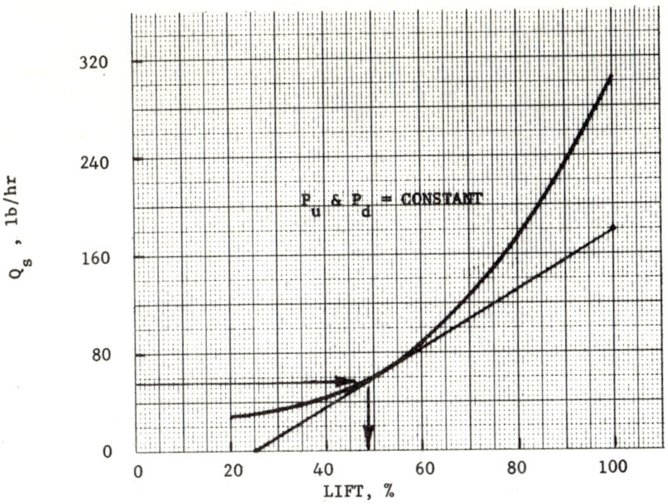

Figure 12-7. Steam flow vs. lift for constant P_u *and* P_d.

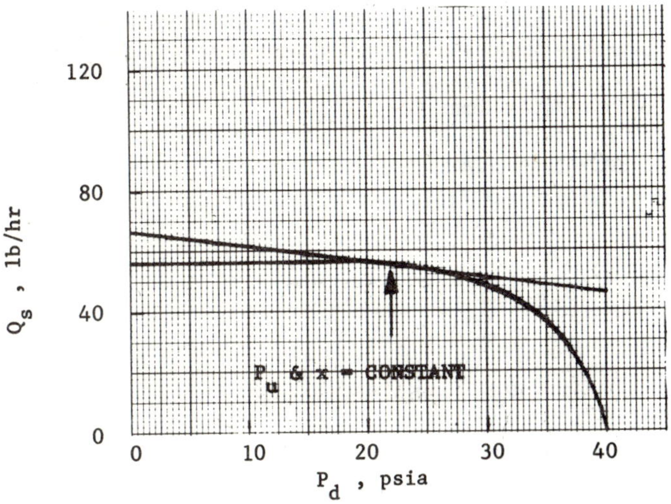

Figure 12-8. Steam flow vs. P_d *for constant* P_u *and x.*

$$\alpha = \left[(180\text{-}0)\ \text{lb/hr}\right] / \left[(100\text{-}25)\%\ \text{lift}\right] \times 100\%\ \text{lift}/(9/16\ \text{in.}) \tag{12-32}$$

$$\alpha = 180\,(100)/\left[75\,(0.562)\right] = 427\ \text{lb/(hr-in.)}$$

A curve of steam flow versus P_d may be calculated for design conditions using Equation (12-27) and C_v equal to 0.94. This result is shown in Figure 12-8. The value of β may be found using data from Figure 12-8 as follows:

$$\beta = \left[(46 - 66)\ \text{lb/hr}\right] / \left[(40 - 0)\ \text{psi}\right] = -\,0.5\ \text{(lb/hr)/(psi)} \tag{12-33}$$

The next step is to relate steam pressure in the exchanger shell to temperature.

If saturated steam is assumed in the shell, a unique relation exists between steam pressure P_d and steam temperature Θ_s.

$$\Delta P_d = \partial P_d/\partial \Theta_s\big] \, \Delta \Theta_s = z \, \Delta \Theta_s, \tag{12-34}$$

where $z = \partial P_d/\partial \Theta_s\big]$

In this application the slope may be found from steam tables at the operating shell pressure of 22 psia. A curve of P_d versus Θ_s, which may be used to find z, is shown in Figure 12-9. The result is

$$z = \left[(34\text{-}0)\ \text{psia}\right] / \left[(260\text{-}185)°\text{F}\right] = 0.45\ \text{psi/°F} \tag{12-35}$$

The next step is to apply the energy balance concept to the heat exchanger.

An energy balance on the heat exchanger shell shows that the change in heat flow from condensing steam must equal the rate of change of heat storage in shell and tube walls plus any change in heat flow to water in the tube.

$$\Delta q_{\text{in}} = \Delta q_{\text{storage}} + \Delta q_{\text{out}} \tag{12-36}$$

The change in input is

$$\Delta q_{\text{in}} = h \, \Delta Q_s, \tag{12-37}$$

where h = latent heat of steam, B/lb.

The value of h is assumed to be constant and is taken at 22 psia to be 957 B/lb. Sensible heat is neglected. The storage term is

$$\Delta q_{\text{storage}} = (W_t c_t + W_s c_s)\,(d\,\Theta_s/dt), \tag{12-38}$$

where W_t = tube weight, lbs,
c_t = specific heat of tube material, B/(lb-°F),
W_s = shell weight, lbs,
c_s = specific heat of shell material, B/(lb-°F),
Θ_s = steam temperature at any time, °F,
t = time in consistent units, sec, min, hr.

If $\Delta\Theta_s$ is set equal to $(\Theta_s - 233)$ and differentiated as in Section 12.1, Equation (12-38) reduces to

$$\Delta q_{\text{storage}} = (W_t c_t + W_s c_s)\, s \, \Delta \Theta_s \tag{12-39}$$

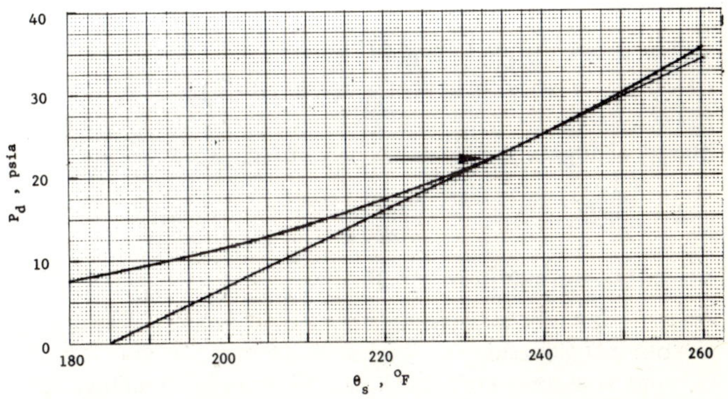

Figure 12-9. Pressure vs. temperature for saturated steam.

It should be noted that 233°F is the saturation temperature at 22 psia. In Equations (12-38) and (12-39) it is assumed that shell and tube walls are at the same temperature as steam in the shell. This assumption is based on a large value of h for condensing steam such that thermal resistance $1/(hA)$ is for practical purposes equal to zero.

Heat flow from condensing steam to water in the tube may be expressed as follows:

$$\Delta q_{\text{out}} = h_i A_i (\Delta\Theta_s - [\Delta\Theta_2/2]), \tag{12-40}$$

where h_i = heat transfer coefficient, tube-to-water, B/(hr-ft²-°F),
A_i = inside area of tube, ft²,
$\Delta\Theta_2$ = change in water temperature leaving heat exchanger, °F.

By assuming tube temperature equal to steam temperature, the shell-side film and tube-wall conductivity resistances are zero. Thus, the overall heat transfer coefficient reduces to the water-side film coefficient h_i. By assuming small temperature changes it is possible to neglect the log-mean-temperature difference and use numerical averages. For steady state this gives

$$\text{temperature difference} = \Theta_s - (\Theta_1 + \Theta_2)/2 \tag{12-42}$$

Since Θ_1 is assumed constant, the resultant difference in temperature changes is as shown in Equation (12-40). Substituting Equations (12-37), (12-39) and (12-40) into Equation (12-36) gives

$$h\Delta Q_s = (W_t c_t + W_s c_s) s\, \Delta\Theta_s + h_i A_i (\Delta\Theta_s - [\Delta\Theta_2/2]) \tag{12-42}$$

A similar energy balance may now be written for the water or tube side of the heat exchanger. In this case Δq_{out} of Equation (12-40) becomes the heat input. The storage term, in this case, is somewhat more difficult to see, since the water in the tube is flowing. The fundamental point, however, is that the tube is always full of water. Thus, at the end of a change that results in an increase in Θ_2, the energy level of water in the tube is higher than at the start of the change. Using the average tube water temperature as previously noted, this storage term is

$$\Delta q_{\text{storage}} = W_w c_w s\, (\Delta\Theta_2/2), \tag{12-43}$$

where W_w = weight of water in tube, lbs,
c_w = specific heat of water, B/(lb-°F).

Energy changes leaving the exchanger are

$$\Delta q_{\text{out}} = Q_w c_w \Delta\Theta_2, \tag{12-44}$$

where Q_w = water flow rate, lbs/hr.

Note that Equation (12-44) results from $q_{\text{out}} = Q_w c_w (\Theta_2 - \Theta_1)$ and assumes that both Q_w and Θ_1 remain constant. Substituting Equations (12-40), (12-43) and (12-44) into the heat balance of Equation (12-36) gives

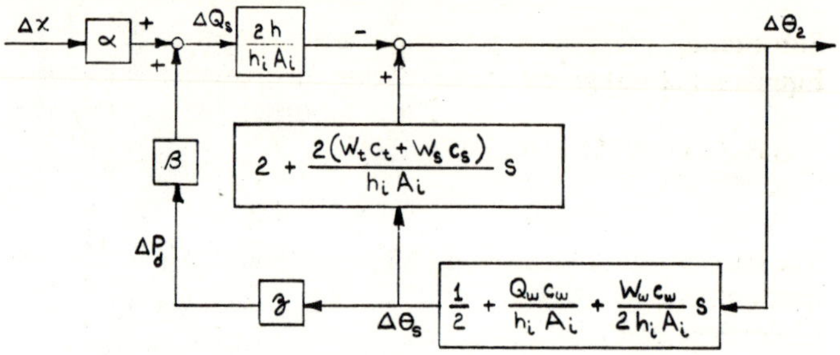

Figure 12-10. Block diagram of valve and heat exchanger transfer function, $\Delta\theta_2/\Delta x$.

$$h_i A_i (\Delta\Theta_s - [\Delta\Theta_2/2]) = W_w c_w s\, (\Delta\Theta_2/2) + Q_w c_w \Delta\Theta_2 \tag{12-45}$$

It is now possible to combine Equations (12-26), (12-34), (12-42) and (12-45) to obtain the valve and heat exchanger transfer function $\Delta\Theta_2/\Delta x$. This may be done by solving simultaneous equations or by block diagram analysis. A block diagram of the four equations is shown in Figure 12-10. Solution of the four equations is

$$\Delta\Theta^2/\Delta x = [(2h\alpha)/A]/[1 + (B/A)s + (C/A)s^2], \tag{12-46}$$

where
$$A = -\beta z h + 2Q_w c_w - (2Q_w c_w \beta z h)/(h_i A_i)$$
$$B = W_t c_t + W_s c_s + W_w c_w + 2Q_w c_w (W_t c_t + W_s c_s)/[h_i A_i] - [W_w c_w \beta z h]/[h_i A_i]$$
$$C = W_w c_w (W_t c_t + W_s c_s)/[h_i A_i]$$

Based on the following heat exchanger data, it is possible to calculate numerical values for Equation (12-46).

Shell data:

Length = 5.7 ft	Material—iron
O.D. = 0.292 ft	Specific heat = 0.1 B/(lb – °F)
I.D. = 0.256 ft	Specific weight = 485 lb/ft³

Tube data:

Length = 5.7 ft	Material—brass
O.D. = 0.07 ft	Specific heat = 0.092 B/(lb – °F)
I.D. = 0.052 ft	Specific weight = 560 lb/ft³

Average water velocity = 2.25 ft/sec

Using these values, results in the following:

Q_w = 1080 lb/hr
C_w = 1.0 B/(lb – °F)
h_i = 603 B/(hr – ft² – °F)
A_i = 0.93 ft²
W_w = 0.755 lb
W_t = 5.5 lb
W_s = 43.5 lb

Substituting these and previously calculated coefficients in Equation (12-46) gives

$$\Delta\Theta_2/\Delta x = (255°\text{F/in.})/(1 + 76.8 \times 10^{-4}s + 204 \times 10^{-8}s^2) \quad (12\text{-}47)$$

The denominator of Equation (12-47) may be factored using the quadratic equation to give two real, unequal, and negative roots. Under these conditions the denominator or characteristic portion of Equation (12-47) may be expressed as

$$\Delta\Theta_2/\Delta x = (255°\text{F/in.})/\left[(1 + T_1 s)(1 + T_2 s)\right] \quad (12\text{-}48)$$

From Equation (12-48) it can be shown that

$$(1 + T_1 s) = 0 \text{ and } (1 + T_2 s) = 0, \quad (12\text{-}49)$$

or $T_1 = -(1/s_1)$ and $T_2 = -(1/s_2)$

Values of s_1 and s_2 for Equation (12-47) using the quadratic equation are $s_1 = -135$ and $s_2 = -3630$. Substituting these values into Equation (12-49) and converting from hours to minutes gives

$$T_1 = -\left[1\text{ hr}/(-135)\right]\left[60\text{ min/hr}\right] = 0.44\text{ min} \quad (12\text{-}50)$$

$$T_2 = -\left[1\text{ hr}/(-3630)\right]\left[60\text{ min/hr}\right] = 0.017\text{ min}, \quad (12\text{-}51)$$

which results in

$$\Delta\Theta_2/\Delta x = (255°\text{F/in.})/\left[(1 + 0.44s)(1 + 0.017s)\right] \quad (12\text{-}52)$$

If steam supply pressure P_u, inlet water temperature, and water flow rate Q_w are allowed to vary, the equations become somewhat more complicated. Procedures, however, are identical and a block diagram, similar to Figure 12-10, including all three load changes, is shown in Figure 12-11. To obtain specific transfer functions shown in blocks in Figure 12-6, the equations of Figure 12-11 must be solved for each load variable while maintaining all other inputs constant. For example, $\Delta\Theta_2/\Delta P_u$ can be found if Δx, $\Delta\Theta_1$, and ΔQ_w are equal to zero (constant conditions). Procedure is identical to solving for $\Delta\Theta_2/\Delta x$. To complete the analysis it is necessary to find the sensing element and transmitter transfer function.

Based on manufacturer's data[7] the sensing element and transmitter transfer function may be approximated as an exponential lag. Gain or sensitivity is determined from transmitter input and output ranges as follows:

Input range = 50 to 150°F
Output range = 3 to 15 psig

Thus,

$$k_t = \text{output/input} = (15\text{-}3)\text{ psi}/(150\text{-}50)°\text{F} = 0.12\text{ psi/°F} \quad (12\text{-}53)$$

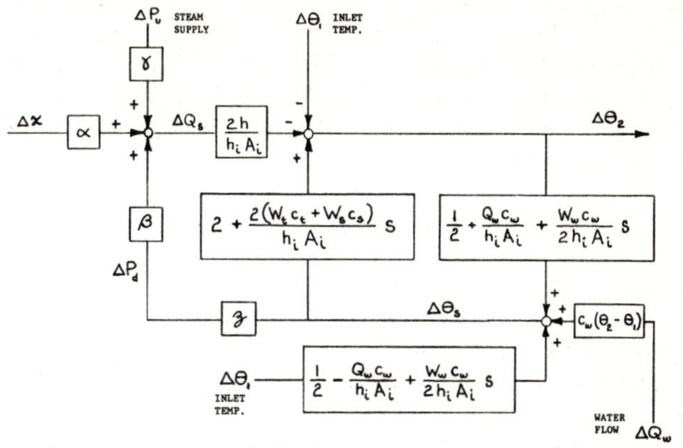

Figure 12-11. Block diagram of valve and heat exchanger transfer function, $\Delta\theta_2/\Delta x$, showing steam supply, inlet temperature, and water flow load changes.

The time constant is equal to thermal resistance times thermal capacitance of the sensing element. This same result was shown earlier in Table 2-1 as follows:

$$T_t = M_r C_p/(hA), \quad (12\text{-}54)$$

where M_r = equivalent weight of sensing element, lbs,
C_p = equivalent specific heat of sensing element, B/(lb-°F),
h = film coefficient, B/(hr-ft²-°F),
A = surface area of sensing element, ft².

For any particular sensing element, area, weight, and specific heat are fixed; thus, the time constant is a function of the film coefficient. If the manufacturer can furnish data for a given set of conditions, it is usually possible to predict the time constant for specific test conditions as a function of h, the film coefficient. For the thermal bulb used in this system, the time constant may be predicted over a limited range using the expression

$$T_t = 2.3/V^{0.58}, \quad (12\text{-}55)$$

where T_t = time constant, sec,
V = water velocity, fps.

Using product velocity of 2.25 fps as previously noted gives

$$T_t = 2.3/(2.25)^{0.58} = 1.34\text{ sec} \quad (12\text{-}56)$$

or $T_t = 1.34\text{ sec}/(60\text{ sec/min}) = 0.024\text{ min}.$

Substituting values from Equations (12-53) and (12-55) into an exponential lag transfer function gives

$$\Delta P_2/\Delta\Theta_2 = 0.12\,(\text{psi/°F})/(1 + 0.024s) \quad (12\text{-}57)$$

Note that Equation (12-55) is not a general equation. It is applicable only for the particular equipment used in this test. Specific recommendations should be obtained from the manufacturer for each component.

The overall "process" transfer function may now be determined by combining Equations (12-24), (12-48) and (12-57) as follows:

$$\Delta P_2/\Delta P_0 = 0.047\,(\text{in./psi})/(1 + 0.083s) \times 255\,(°\text{F/in.})/[(1 + 0.44s)(1 + 0.017s)] \times 0.12\,(\text{psi}/°\text{F})/(1 + 0.024s) \quad (12\text{-}58)$$

which may be combined to give

$$\Delta P_2/\Delta P_0 = 1.44\,(\text{psi/psi})/[(1 + 0.083s)(1 + 0.44s)(1 + 0.017s)(1 + 0.024s)] \quad (12\text{-}59)$$

If the two smallest time constants are combined and assumed as a dead time lag, the two time constants plus the dead time model of Chapter 11 are obtained. This approximation, which follows the practice of Ziegler and Nichols, simplifies the frequency response plot and is well within calculation accuracy. The result is

$$\Delta P_2/\Delta P_0 = 1.44\,e^{-0.04}\,(\text{psi/psi})/[(1 + 0.083s)(1 + 0.44s)] \quad (12\text{-}60)$$

A frequency response plot of Equation (12-60) is shown in Figure 12-12. From Figure 12-12 at $-180°$, "process" gain is 0.1 and frequency is 2.8 cpm. Using these values, ultimate gain and period are

$$G_U = 1/0.1 = 10 \quad (12\text{-}61)$$

$$P_U = 1/2.8\ \text{cpm} = 0.36\ \text{min} \quad (12\text{-}62)$$

Settings for a proportional plus reset controller may now be calculated according to Chapter 9:

$$G_O = 0.45G_U = 0.45(10) = 4.5\ \text{or}\ 22\%\text{P.B.} \quad (12\text{-}63)$$

$$T_r = P_U/1.2 = 0.36/1.2 = 0.3\ \text{min} \quad (12\text{-}64)$$

Records are shown in Figure 12-13 for the actual heat exchanger control system. Figure 12-13a shows results using proportional control with a gain of 10 (10% P.B.). The system cycles continuously and has a period of approximately 0.37 min. Proportional plus reset control is shown in Figure 12-13b using settings calculated in Equations (12-63) and (12-64). The variation in decay for increasing and decreasing set point changes indicates the potential of nonlinear characteristics. This could cause problems with large set point or load changes that exceed linear assumptions used in the calculations. Revised controller settings may be necessary to maintain satisfactory stability if large changes do exist. If such a problem does exist, the reader should check more advanced texts which treat the subject of nonlinear control.[8,9]

The important point to note from this example is that, *if necessary*, "process" transfer functions can be determined with an accuracy that is applicable to engineering design. Calculation of controller settings is a by-product and does not justify costs of analysis. The importance of this analysis is that gain, time constant, and dead time lag values, which determine stability, period, and overshoot, are defined in terms of process parameters. Thus, through the use of equations like Equation (12-46), it is possible to determine both qualitatively and quantitatively how variations in process parameters will affect control. If necessary, this technique can be a very useful tool.

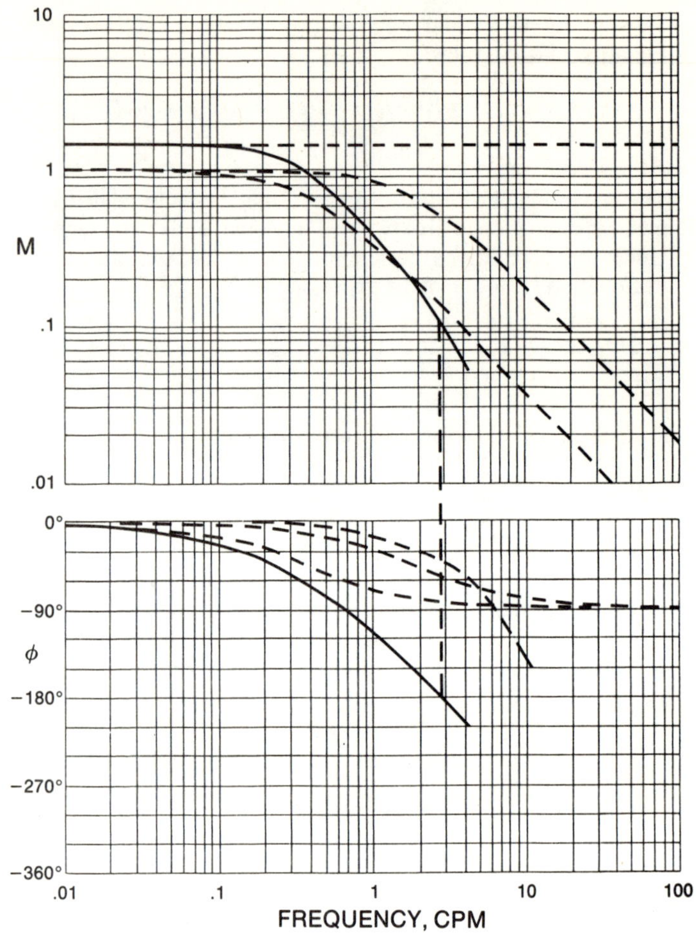

Figure 12-12. Heat exchanger frequency response from calculated transfer function.

12.3. Predicting the Time Response

In Chapters 6 and 7 closed loop time responses were predicted based on a second-order system analog. The technique is relatively simple and gives reasonable results for systems which have a characteristic second-order resonant-type frequency response.

When the frequency response deviates from the characteristic resonant response, a more sophisticated technique is required. A typical deviation from the characteristic response was shown for the load change b/U_4 in the figure obtained in the solution to Problem 2, Chapter 8. The time response for this load change could not be predicted using the analog technique of Chapter 6. Similar deviations can also occur when proportional plus reset controllers are used. For problems of this type, time-response

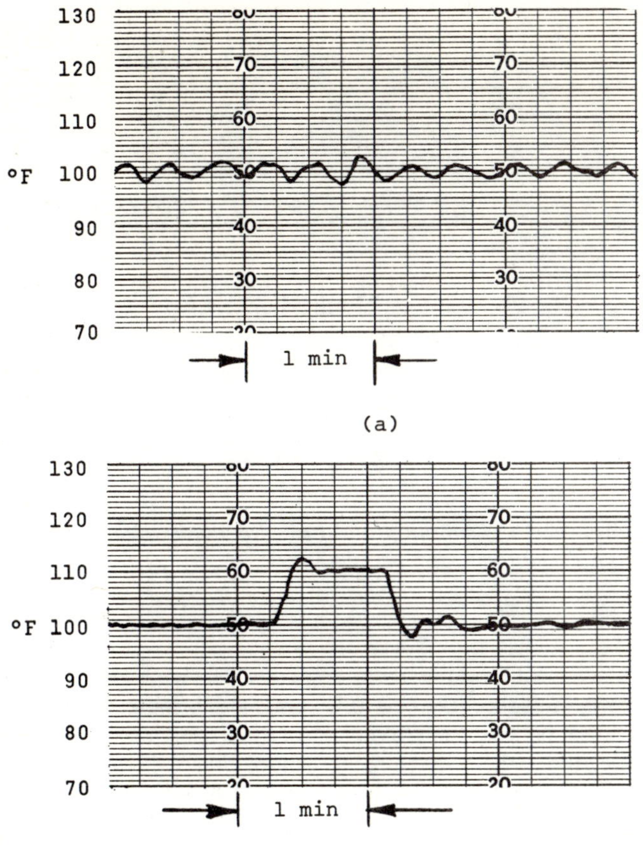

Figure 12-13. Heat exchanger control records: (a) proportional control only, 10% PB, P_U = 0.37 min; (b) proportional plus reset control, 22 % PB, T_r = 0.3 min.

results can be found using a computer program based on a Fourier series technique.[10]

The Fourier series for a square wave of amplitude A is[11]

$$y(t) = A\{(1/2) + (2/\pi)\sin(2\pi f_1 t) + (2/(3\pi))\sin(2\pi f_3 t) + (2/(5\pi))\sin(2\pi f_5 t) + \cdots + (2/(n\pi))\sin(2\pi f_n t)\}, \tag{12-65}$$

where $y(t)$ = output at time t,
A = amplitude,
f_n = odd multiples of initial frequency,
n = odd integers, 1, 3, 5,....

To prove a point, the first four terms are plotted and shown in Figure 12-14 for an amplitude of $A = 1$. Note that as more terms are added that a square wave will develop, the first half of which is equivalent to a step input. A consideration of what a frequency response reveals will lead to how the Fourier series may be used to obtain the time response.

The information on a frequency response or Bode diagram explains what happens to each sine wave as it passes through the particular component or process under study. For example, assume the calculated heat exchanger response of Figure 12-12. At a frequency of 2.8 cpm the Bode plot reveals that an input sine wave having unity amplitude will come out of the "process" with an amplitude of 0.1 and lag the input by 180°. In mathematical format this result may be expressed as follows:

$$\text{input} = 1\sin\{2\pi(2.8)t\} \tag{12-66}$$

and

$$\text{output} = 0.1\sin\{2\pi(2.8)t + (-180°)(2\pi/360)\} \tag{12-67}$$

The $2\pi/360$ in Equation (12-67) converts the $-180°$ to radians. Thus, the frequency response plot tells how each sine wave is altered as it passes through the process. This is the fundamental consideration in using the Fourier series for a square wave to predict the output time response. Figure 12-14 clearly shows that if sufficient terms of Equation (12-65) are used, a square wave or step input can be approximated. Since the frequency response shows how each of the terms in Equation (12-65) are altered as they pass through the "process," *the equivalent altered series of Equation (12-65) should give the output time response.* Using the format of Equations (12-66) and (12-67) the result is

$$x(t) = A\{(M_0/2) + (2M_1/\pi)\sin(2\pi f_1 t + [2\pi\Phi_1/360]) + (2M_3/(3\pi))\sin(2\pi f_3 t + [2\pi\Phi_3/360]) + (2M_5/(5\pi))\sin(2\pi f_5 t + [2\pi\Phi_5/360]) + \cdots + (2M_n/(n\pi))\sin(2\pi f_n t + [2\pi\Phi_n/360])\} \tag{12-68}$$

where $x(t)$ = output at anytime t,
M_0 = zero frequency or steady-state magnitude,
M_n = magnitude at f_n,
Φ_n = phase angle at f_n.

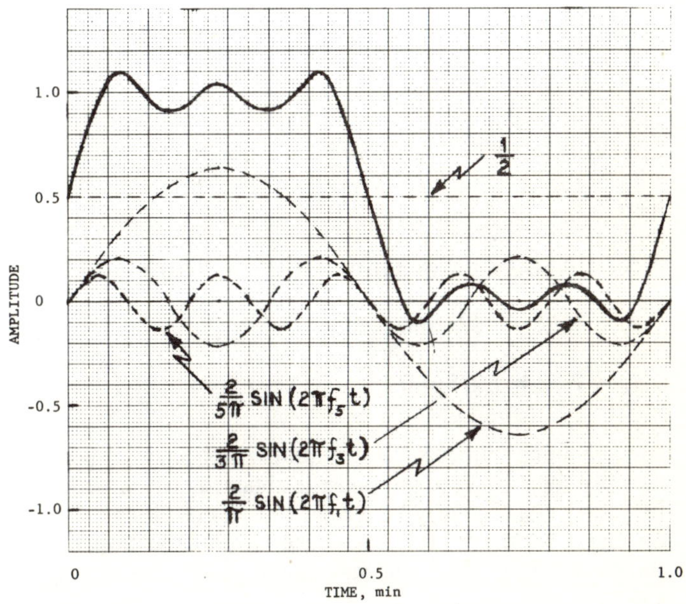

Figure 12-14. Square-wave development using first four terms of the Fourier series with A = 1.0.

Equation (12-68) is easily programmable and can be a very useful tool in predicting output time response. There are several details that must be considered, and these will be explained in an example solution. A Fortran program for Equation (12-68) is given as Figure 12-15.

As in Chapter 11, Lines 1, 2, and 3 show job classification numbers for a particular computer and a comment card to identify the program. Line 4 shows a dimension statement that provides capacity for up to 100 input values of M_n and Φ_n. Line 5 punches the program title. Line 6 provides a blank card to skip a space and start a second calculation if input data exist. Lines 7 and 8 read and punch the first input data card, which defines the particular problem. For the present example the title is "Heat Exchanger Transient from F.R." In general a more specific title with identifying number and date should be used. Line 9 requests all pertinent input data and requires extended explanation.

The N and KOUT of Line 9 define the number of input data values and desired number of outputs requested. Before these numbers can be selected it is necessary to determine U, the output time increment, and F, the initial frequency (f_1 of Equation (12-68)). To determine U and F it is necessary to examine the frequency response that will be converted. As an example, heat exchanger data of Figure 11-2 will be used. In Chapter 11 these data, as shown in Figure 11-9, were converted by both the incomplete response and step-to-frequency conversion to give the "process" frequency response as shown in Figures 11-10 and 11-11. Based on the frequency response of Figure 11-11, a PID controller was selected. Settings were calculated as follows:

From Figure 11-11,

$$P_G \text{ at } -180° = 0.12$$

$$f_u \text{ at } -180° = 0.85 \text{ cpm}$$

thus,

$$G_U = 1/P_G = 1/0.12 = 8.33 \text{ (P.B. = 12\%)}$$

$$P_U = 1/f_u = 1/0.85 = 1.2 \text{ min},$$

then

$$G_O = 0.6G_U = 0.6\,(8.33) = 5.0 \text{ (P.B. = 20\%)}$$

$$T_r = P_U/2 = 1.2/2 = 0.6 \text{ min}$$

$$T_d = P_U/8 = 1.2/8 = 0.15 \text{ min}$$

By combining "process" and controller frequency responses, the open loop frequency response is obtained. The closed loop frequency response may be obtained using open loop data and the Black-Nichols chart. The actual closed loop response is shown in Figure 12-16 and may now be used to determine U and F.

As demonstrated in Chapters 6 and 7, the period of the closed loop time response is directly related to resonant peak frequency

```
1        ZZJOB 5                                    44100  FREQUENCY TO STEP
2        ZZFORX5
3        C      PROGRAM TO COMPUTE STEP RESPONSE FROM FREQUENCY RESPONSE
4               DIMENSION GAIN(100),PHASE(100)
5               PUNCH 1
6             5 PUNCH 6
7               READ 2
8               PUNCH 2
9               READ 9, N, KOUT, U, GZERO, F, STEP
10              DO 11 I=1,N
11           11 READ 12, GAIN(I), PHASE(I)
12              PUNCH 13
13              DO 15 J=1, KOUT
14              H=.5*GZERO
15              DJ=J
16              T=DJ*U
17              DO 14 I=1,N
18              AI=I*2-1
19              THETA=AI*2.*3.1415927*T*F+PHASE(I)/57.29578
20           14 H=H+(GAIN(I)*2./(3.1415927*AI))*SINF(THETA)
21              H=H*STEP
22           15 PUNCH 16,T,H
23              GO TO 5
24            1 FORMAT (37HSTEP RESPONSE FROM FREQUENCY RESPONSE)
25            2 FORMAT(49H
26            6 FORMAT (/)
27            9 FORMAT(2I5,4F10.0)
28           12 FORMAT(2F10.0)
29           13 FORMAT (/20H    TIME      RESPONSE)
30           16 FORMAT (F10.5, F10.5)
31              END
```

Figure 12-15. Computer program for conversion from frequency response to step response.

of the closed loop frequency response. The peak frequency as shown in Figure 12-16 is 0.9 cpm, which indicates a time-response period of approximately 1/(0.9 cpm), or 1.1 min per cycle. Since the resonant peak gain of Figure 12-16 is greater than what would be expected for quarter decay, this system can be expected to be highly underdamped and take several more cycles than usual to decay to a final value. This type of result is not unusual with the Ziegler-Nichols method for determining settings. A need for field adjustment to trim these "ball park" settings was noted in Chapter 9. The point is not to trim at this time but to recognize that several cycles will be required for reaching a final value. As a start, assume that 5 to 6 cycles will be required for decay. Using 6 cycles and a period of 1.1 min per cycle results in a total time of 6.6 min for decay. Since only the first half of Figure 12-14 is to be used as the test step, the fundamental period of the square wave must be at least two times 6.6 min or 13.2 min. An approximation of this concept is shown in Figure 12-17. As can be seen from Figure 12-17, if the fundamental period is too small, the basic time response will not reach a steady-state value before the return step of the square wave occurs. If the fundamental period of 13.2 min is accepted, then the initial frequency is

$$F = 1/13.2 = 0.076 \text{ cpm} \tag{12-69}$$

While 0.076 cpm is mathematically correct, it does not provide a convenient base for reading data from Figure 12-16. For example, to solve Equation (12-68) it is necessary to read magnitude and phase angles from Figure 12-16 at odd multiples of F, namely $f_1, f_3, f_5, \cdots, f_n$. Thus, if $F = f_1 = 0.076$, then $f_3 = 3f_1 = 0.228$, $f_5 = 5f_1 = 0.38$ cpm, and so forth. This, of course, makes extra work and is very time-consuming. To avoid this complication, F is normally taken to be the next smaller unit frequency. In this case, F would be selected as 0.01 cpm. This results in a fundamental period of 100 min and gives 50 min for the time response to decay before square wave switching occurs. With F

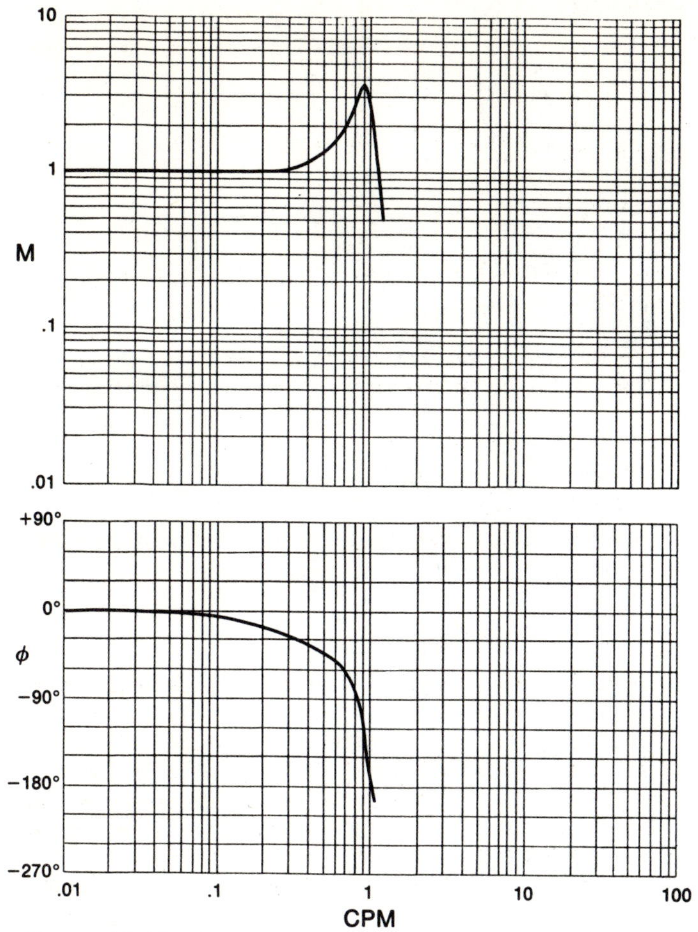

Figure 12-16. Heat exchanger example, closed loop frequency response.

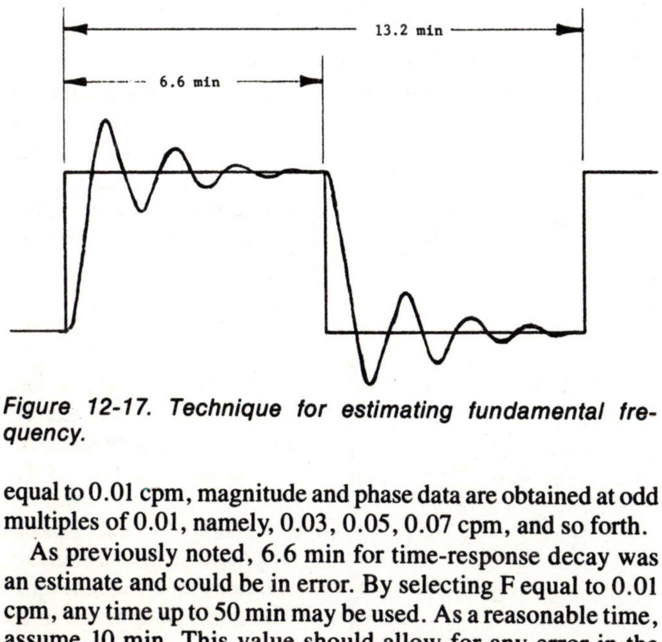

Figure 12-17. Technique for estimating fundamental frequency.

equal to 0.01 cpm, magnitude and phase data are obtained at odd multiples of 0.01, namely, 0.03, 0.05, 0.07 cpm, and so forth.

As previously noted, 6.6 min for time-response decay was an estimate and could be in error. By selecting F equal to 0.01 cpm, any time up to 50 min may be used. As a reasonable time, assume 10 min. This value should allow for any error in the 6.6-min approximation and still not require excessive computer time. Having selected a total time of 10 min, it is now possible to determine U and KOUT. Since this system is known to be underdamped, it will be necessary to select a small time increment U to define cyclic peaks. A value of 0.2 min for U is a reasonable selection and will be assumed for this example. To obtain a total time of 10 min with a time increment of 0.2 min, KOUT must be equal to 50:

$$\text{KOUT} = \text{total time/U} = 10 \text{ min}/0.2 \text{ min} = 50 \qquad (12\text{-}70)$$

The remaining inputs on Line 9 are N, number of inputs, GZERO, zero frequency gain, and STEP, input step. Determination of N can be accomplished once inputs are read from the frequency response, and will be explained shortly. The value of GZERO is fairly obvious and is also taken from the frequency response curve. In the present example GZERO as read from Figure 12-16 is 1.0, and is equal to M_0 in Equation (12-68). The STEP value depends on what set point or load change is desired. In the present example a 9°F set point change will be used, since experimental data are available for comparison. To complete Line 9, inputs for magnitude and phase angle will now be determined.

Lines 10 and 11 provide a DO LOOP with instructions to read all GAIN (magnitude) and PHASE (phase angle) inputs. These data are read from Figure 12-16 starting as previously determined at F = f_1 = 0.01 cpm. Some actual values are shown in Table 12-2.

Table 12-2
Input Data from Figure 12-16

Frequency (cpm)	Gain	Phase (degrees)
f_1 = 0.01	1.0	−2.0
f_3 = 0.03	1.0	−3.0
f_5 = 0.05	1.0	−5.0
f_7 = 0.07	1.0	−6.0
f_9 = 0.09	1.0	−8.0
f_{11} = 0.11	1.0	−10.0
•	•	•
•	•	•
•	•	•
f_{101} = 1.01	2.0	−166.0
f_{103} = 1.03	1.8	−172.0
f_{105} = 1.05	1.5	−180.0
f_{107} = 1.07	1.3	−185.0
f_{109} = 1.09	1.2	−188.0
f_{111} = 1.11	1.0	−193.0

The GAIN and PHASE inputs are punched in this program on individual cards and use the 12 FORMAT of Line 28. It should be noted that the negative sign must be used when applicable, since positive angles are possible. The F10.0 format allows the inputs to be placed anywhere within the field as long as the decimal point is included. The number of inputs N may now be found as follows:

$$N = (n - 1)/2, \qquad (12\text{-}71)$$

where n = number of final frequency.

```
HEAT EXCHANGER TRANSIENT FROM F. R.
  55   50       0.2         1.0        0.01          9.0
        1.0       -2.0
        1.0       -3.0
        1.0       -5.0
        1.0       -6.0
        1.0       -8.0
        1.0       -10.0
        1.0       -11.0
        1.01      -13.0
        1.01      -15.0
        1.02      -16.0
        1.03      -19.0
        1.03      -21.0
        1.04      -24.0
        1.04      -25.0
        1.05      -26.0
        1.06      -27.5
        1.07      -29.0
        1.08      -31.0
        1.09      -32.0
        1.095     -34.0
        1.10      -35.0
        1.105     -36.0
        1.12      -37.0
        1.125     -38.0
        1.18      -39.0
        1.20      -41.0
        1.22      -42.0
        1.25      -43.0
        1.3       -45.0
        1.33      -46.0
        1.37      -48.0
        1.41      -50.0
        1.47      -52.0
        1.52      -54.0
        1.60      -56.0
        1.71      -58.0
        1.82      -61.0
        2.02      -64.0
        2.20      -67.0
        2.40      -71.0
        2.65      -75.0
        2.90      -80.0
        3.15      -85.0
        3.40      -92.0
        3.45      -105.
        3.40      -120.
        3.15      -137.
        2.80      -148.
        2.30      -158.
        2.00      -166.
        1.80      -172.
        1.50      -180.
        1.30      -185.
        1.20      -188.
        1.00      -193.
```

Figure 12-18. Heat exchanger example, computer input data.

```
ZZJOB 5                    44100    FREQUENCY TO STEP
STEP RESPONSE FROM FREQUENCY RESPONSE

 HEAT EXCHANGER TRANSIENT FROM F. R.

   TIME       RESPONSE
    .20000     2.12796
    .40000     7.48401
    .60000    11.55320
    .80000    11.53320
   1.00000     8.44756
   1.20000     6.15377
   1.40000     7.12646
   1.60000     9.91242
   1.80000    11.18203
   2.00000     9.72549
   2.20000     7.57698
   2.40000     7.25103
   2.60000     8.79246
   2.80000    10.12048
   3.00000     9.77376
   3.20000     8.46242
   3.40000     7.83731
   3.60000     8.43704
   3.80000     9.33808
   4.00000     9.51750
   4.20000     8.97977
   4.40000     8.46907
   4.60000     8.48896
   4.80000     8.87662
   5.00000     9.17017
   5.20000     9.12246
   5.40000     8.85066
   5.60000     8.65185
   5.80000     8.72543
   6.00000     8.99385
   6.20000     9.16798
   6.40000     9.05100
   6.60000     8.78090
   6.80000     8.68578
   7.00000     8.87859
   7.20000     9.10581
   7.40000     9.07353
   7.60000     8.82942
   7.80000     8.68787
   8.00000     8.80955
   8.20000     9.01169
   8.40000     9.04024
   8.60000     8.89811
   8.80000     8.79523
   9.00000     8.84626
   9.20000     8.95286
   9.40000     8.98095
   9.60000     8.93209
   9.80000     8.89173
```

Figure 12-19. Heat exchanger example, computer time-response results.

In the present example,

$$N = (111\text{-}1)/2 = 55 \tag{12-72}$$

Although N has now been determined, the question is how and why? The answer of course is that the number of required inputs is a function of the frequency response. Input data must describe the frequency response. In the present example this includes the resonant peak, which determines or establishes cyclic behavior of the time response. As a general rule-of-thumb, 50 to 75 inputs will normally be required such that a frequency range of two or more decades is covered.

Combined input data are shown in Figure 12-18. The second line shows N, KOUT, U, GZERO, F, and STEP, according to the 9 FORMAT of Line 27. It should be noted that N and KOUT are integers and must be placed in specific columns. The remaining data, including GAIN and PHASE, use the more liberal F10.0 field. In this format it is necessary to include the decimal point on all values.

Remaining program steps in Figure 12-15 calculate the series value H, as shown in Equation (12-68). Line 21 multiplies the value of H by STEP, and punches both time T and response H according to Line 22. The program then returns to Line 13, advances time by one increment on Lines 15 and 16, and repeats the calculation. The "do loop" between Lines 13 and 22 is continued until the required number of outputs (KOUT) has been obtained. Results for this example are shown in Figure 12-19. Actual data of Figure 12-19 is limited to 49 outputs rather than the 50 requested. This results from a computer

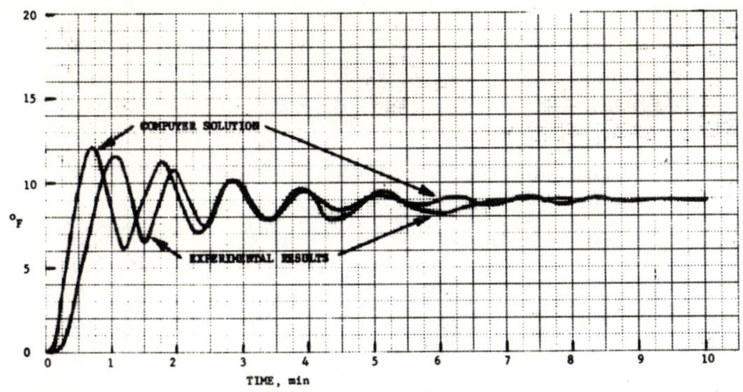

Figure 12-20. Computer and experimental time-response results.

room "stop" or limit of 49 output cards. This limit can be by-passed if necessary. It is set to avoid excessive output from unproven student programs. This information is provided for the critical reader who will note that only 49 KOUTs are shown.

A comparison of computer and experimental time-response results are shown in Figure 12-20. While the results are not identical, they do demonstrate that, *if necessary*, the time response can be adequately predicted for most engineering applications.

Chapter 12—Study Questions and Problems

(Answers appear on pages 166-170)

1. Given the temperature control process system shown below and noted data, develop the transfer function $\Delta\Theta_2/\Delta P_1$. Note and give reasons for any assumptions.

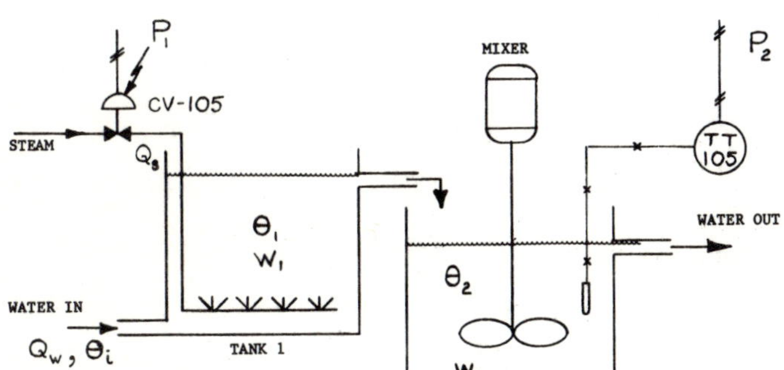

WATER FLOW RATE, Q_w = 30 lb/min
HOLD UP IN TANK 1, W_1 = 50 lb
HOLD UP IN TANK 2, W_2 = 25 lb
MAXIMUM STEAM FLOW, Q_s = 6 lb/min
WATER INLET TEMP., θ_i = 60°F
SET POINT TEMP., r = 160°F
VALVE MOTOR PRESSURE RANGE, P_1 = 3 to 15 psig
TEMP. IN TANK 1 AT ANYTIME = θ_1
TEMP. IN TANK 2 AT ANYTIME = θ_2
TEMP. TRANSMITTER SPAN, 100 to 200°F
TEMP. TRANSMITTER OUPUT, P_2 = 3 to 15 psig

Figure 12-P1. Temperature control process with two tanks.

2. Given the "process" (level control problem) and associated data shown below, determine the transfer function $\Delta P_t/\Delta P_a$. Sketch a block diagram of the transfer function $\Delta P_t/\Delta P_a$. Assuming an input step change of −1 psi for ΔP_a (from 9 to 8 psig) and a recorder having a range of 0 to 30 in. of water for an input of 3 to 15 psig in P_t, sketch the recorded change based on the derived transfer function, $\Delta P_t/\Delta P_a$. Assume a level bias of 15 in. of water.

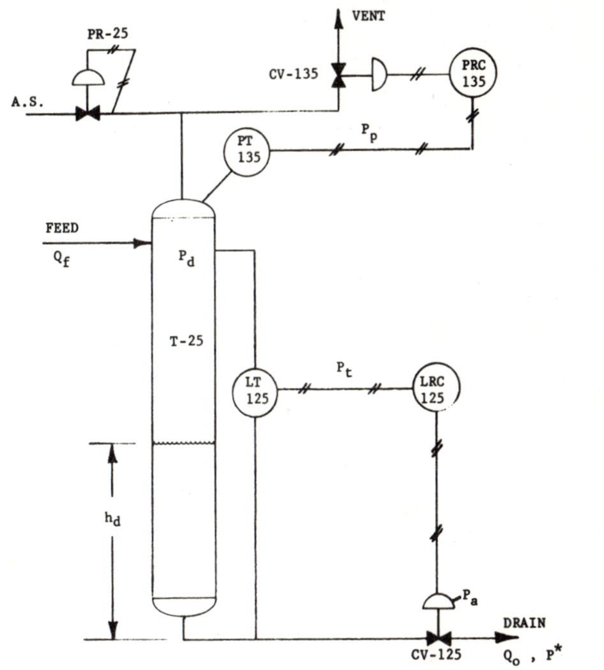

Figure 12-P2. Level control problem, piping and instrument drawing.

Level Control Problem—Design Data

T-25: Tank, 4 in. wrought iron pipe (I.D. = 4.026 in.) Feed, Q_f, design flow 0.5 gpm, maximum 1.0 gpm

LT-125: Level transmitter, 0 to 30 in. water input range, h
3 to 15 psig output range, P_t
design level 15 in. water, h_d
From manufacturer's literature: assume exponential lag with time constant of 0.05 min.

CV-125: Control valve, 3 to 15 psig input range, P_a
Flow characteristics for design pressure drop

P_a, psig	Q_o, gpm
3	0.0
6	0.12
9	0.5
12	0.82
15	1.00

design output flow Q_o = 0.5 gpm; downstream pressure, P^* = atmospheric

LRC-125: Level recorder controller, proportional; 3 to 15 psig input and output range; chart, 0 to 100 units for 0 to 30 in. water (scale factor 3.33 units/in. water)

Tank Pressure Control System:

PR-25, self-acting pressure regulator
PT-135, pressure transmitter
PRC-135, pressure recorder controller
CV-135, pressure vent control valve, design pressure $P_d = 15$ psig, maximum $P = 30$ psig

3. In Section 12.2 a "process" transfer function was derived for the heat exchanger loop of Figure 12-5. The transfer function is given in Equation (12-60) and the frequency response is shown in Figure 12-12. Controller settings were calculated in Equations (12-63) and (12-64) to be 22% P.B. and 0.3-min reset time. An experimental time response record for a 10°F set point change is shown in Figure 12-13b. Using these data and the frequency-to-step response program presented in Section 12.3, obtain the computer-calculated time response for a 10°F set point change and compare the result with Figure 12-13b.

References

1. Zoss, L. M., Gollin, N. W., and Edelman, R. I., "Dynamic Response Analysis of Air Heater Temperature Control System," *Ind. and Engr. Chem,* Vol. 48, pp. 1069-1073, 1956.
2. Hoyt, P. R., Stanton, B. D., and Union, D. C., "Dynamic Field Tests of a Steam Turbine," A.S.M.E. Paper No. 57-IRD-15, 1957.
3. Young, A. J., ed., *Plant and Process Dynamic Characteristics,* Proceedings of a Conference held at Cambridge, April 1956, Society of Instrument Technology, New York: Academic Press, 1957.
4. *Practical Approaches to System Engineering,* Proceedings of 14th Annual Symposium, New Jersey Section, I.S.A., April 1962, Newark, NJ.
5. Wilson, H. S., and Zoss, L. M., *Control Theory Notebook,* reprinted from *I.S.A. Journal,* Pittsburgh, PA: I.S.A., 1962.
6. Considine, D. M., *Process Instruments and Controls Handbook,* New York: McGraw-Hill, 1957.
7. Caldwell, W. I., Coon, G. A., and Zoss, L. M., *Frequency Response for Process Control,* New York: McGraw-Hill, 1958.
8. Gibson, J. E., *Nonlinear Automatic Control,* New York: McGraw-Hill, 1963.
9. Dransfield P., *Engineering Systems and Automatic Control,* Englewood Cliffs, NJ: Prentice-Hall, 1968.
10. Wolfgang, G. H., Wagner, C. A., and Zoss, L. M., "Computer Techniques for Frequency Response Analysis," Paper No. 65.1.63, 18th Annual I.S.A. Conference and Exhibit, Chicago, 1963.
11. Churchill, R. V., *Fourier Series and Boundry Value Problems,* New York: McGraw-Hill, 1941.

13 Specialized Systems

In the field of process control there are many systems designed to perform special functions. Some of these systems are very common and have achieved almost universal acceptance. In fact, some are so universally accepted that they are applied without question, even when not needed! Others are well documented in the literature, but have found only limited acceptance in process control applications. A number of these systems will be discussed in this chapter with regard to objectives and design theory.

13.1. Ratio Control

There are many applications in industry where two or more fluids must be blended according to specific ratios. While more than two fluids may be involved, the discussion at this time will be limited to two fluids. In most applications there will be a "wild" or primary flow that will vary as a function of process conditions. The objective is to blend a controlled or secondary flow according to some desired ratio. A typical system is shown in Figure 13-1a. In this system, primary flow (FT-105A) and secondary flow (FT-105B) measurements are transmitted to the flow ratio recorder controller (FFRC-105). The controller adjusts control valve (FFV-105) in the secondary flow line to achieve the required ratio. This same objective may be achieved using a ratio relay as shown in Figure 13-1b. To help the operator in this application, the flow ratio relay (FFY-106) often has indicators for both primary and secondary flow. This is shown in Figure 13-lb. Signal output from the flow ratio relay adjusts the flow controller set point to maintain the proper ratio.

There are many variations of the basic ratio control loop that may be used, and these are noted in the literature.[1,2] This text treats only the basic system as shown in Figure 13-1. It is desirable in designing a ratio control system to have a ratio setting as close to 100% as possible. For example, assume, that for every 3 liters of primary flow, 1 liter of secondary flow is required and

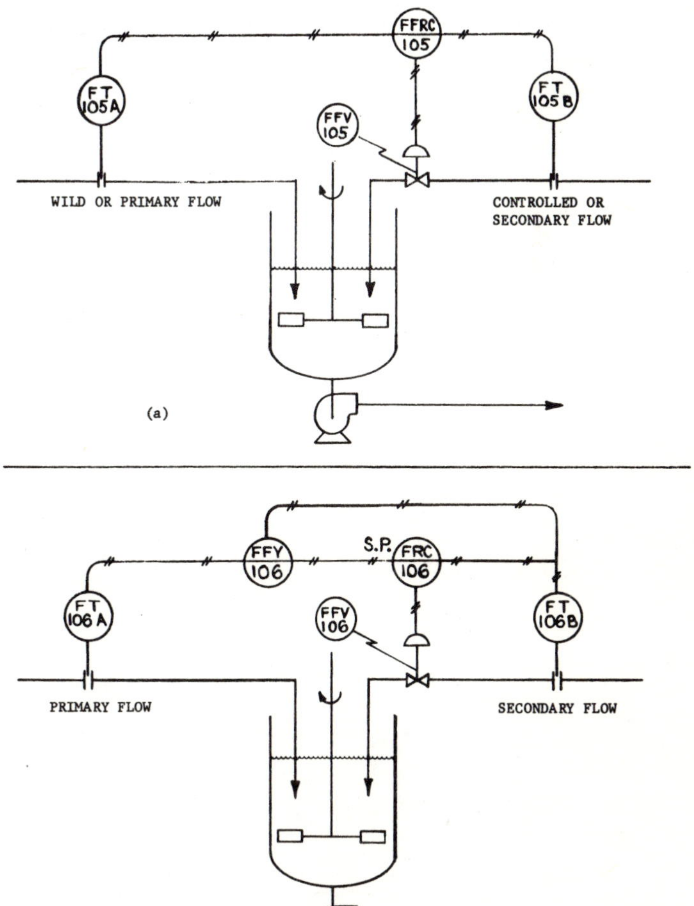

Figure 13-1. Flow ratio control systems.

that normal primary flow is 45 liters/min. If the primary meter range span is specified as 60 liters/min and the secondary range span as 20 liters/min, the result is

actual ratio = (secondary meter span/primary meter span) × ratio dial setting, (13-1)

or substituting values gives

actual ratio = (20 liters/min/60 liters/min) × 100% = 33.3% (13-2)

As shown in Equation (13-2) the actual ratio of 1:3 or 33.3% is achieved using a ratio dial setting of 100% and selecting primary and secondary meter spans to produce the design ratio. Selecting the range spans in this way is in agreement with good flow metering practice, which calls for normal flow at about 70% of range span. In this case both the 45-lpm primary and 15-lpm secondary flow occur at 75% of range span. For primary flow elements with square-root characteristics, these flow rates produce flow transmitter outputs of 9.75 psig. This is close to the middle of the standard 3 to 15 psig range, which is desirable from both linearity of mechanisms and response to changes around the operating point. For the 4 to 20 ma range of electronic components the output would be 13 ma, which is close to the 12 ma midrange point.

Note that not all flow elements have square-root characteristics. Many of the more modern devices, such as magnetic, turbine, and vortex flow elements, have linear outputs. This does not provide any problem to the ratio mechanism, which will work on either square-root or linear signals. It is required, however, that primary and secondary signals be compatible, namely, both square root or both linear. Where one element has a square-root output and the other is linear, a square-root extractor should be used to produce common linear signals. This result is shown in Figure 13-2.

Flow ratio controllers or ratio relays may be purchased with either square-root or linear dial calibrations. The mechanisms or components are identical with exception of ratio dial calibration. The basis for this statement can be seen by referring back to Figure 13-1a and b. Both FFRC-105 and FFY-106 work on 3 to 15 psig signals coming from primary and secondary flow transmitters. If linear elements are assumed with 60 and 20 liter/min range spans as noted earlier, a 50% ratio dial setting would give an actual ratio of

actual ratio = (20 lpm/60 lpm) × 50% = 16.7%, (13-3)

or 1 liter of secondary flow for every 6 liters of primary flow. Thus, at the normal primary flow of 45 lpm, the secondary flow would be 7.5 lpm, not 15 lpm as previously desired. Using the 3 to 15 psig range, the 45-lpm flow gives a 9-psi change in output from 3 to 12 psig, and 7.5 lpm gives a 4.5-psi change from 3 to 7.5 psig. The ratio is 4.5:9.0 or 50%. If the measuring elements are square-root, the 45-lpm flow changes the output pressure only 6.75 psi to 9.75 psig. Likewise, the 7.5-lpm flow changes the output pressure by only 1.69 psi to 4.69 psig. Thus, to achieve the same result using square-root elements, the ratio should be 1.69:6.75 or 25%. By offering precalibrated linear and square-root ratio dials, the manufacturer eliminates necessity for all the above calculations. The user must only check to see that flow element characteristics agree with ratio dial characteristics and use Equation (13-1) to calculate the ratio dial setting.

Calling the primary flow "wild" can often be misleading. As shown in Figures 13-1 and 13-2, primary flow is "wild" or uncontrolled. Wild or primary flow may also be controlled as shown in Figure 13-3, and this is a very common type of application. For either uncontrolled or controlled primary flow, ratio agreement between primary and secondary flow is dependent on system dynamics.[3] A typical block diagram showing control for both primary and secondary flow is given in Figure 13-4a. Response of secondary flow Q_s as a function of variations in Q_p and U_2 may be derived as follows from Figure 13-4b:

$$Q_s = \left[h_1 k_R g_3 g_4/(1 + g_3 g_4 h_2)\right] Q_p + \left[1/(1 + g_3 g_4 h_2)\right] U_2 \qquad (13\text{-}4)$$

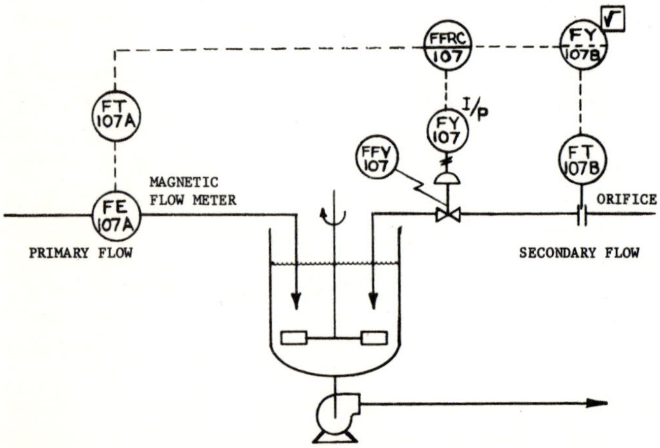

Figure 13-2. Flow ratio control system with linear and square-root elements.

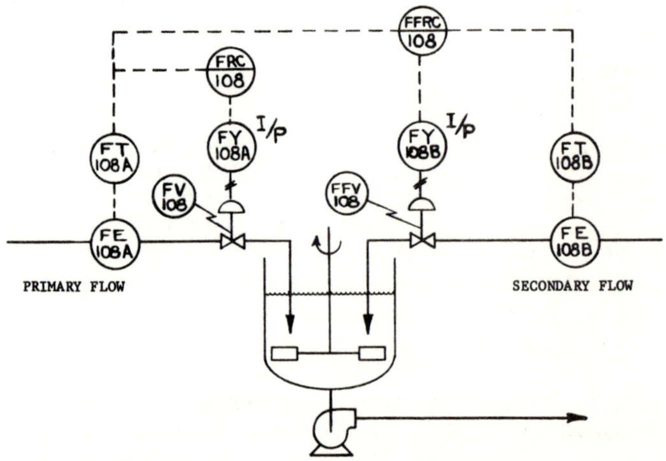

Figure 13-3. Flow ratio system with controlled primary flow.

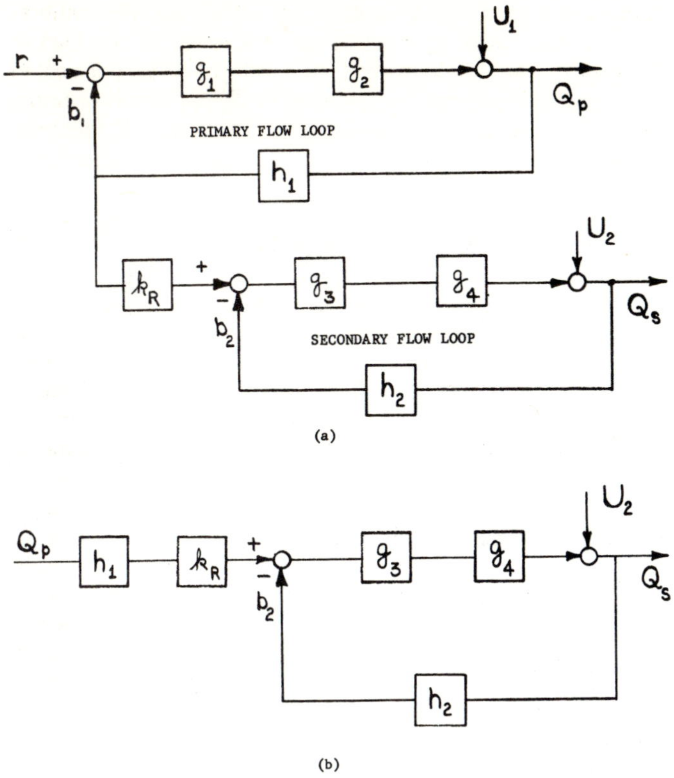

Figure 13-4. Block diagrams for a flow ratio control system.

If load change U_2 is assumed zero for purpose of analysis, it is possible to determine the ability of Q_s to follow Q_p, where

Q_s = secondary flow
Q_p = primary flow
h_1 = $k_1(\text{psi/lpm})/(1 + T_1 s)$ (primary measuring element and transmitter)
k_R = ratio dial setting
g_3 = $k_c(1 + [1/(T_r s)])$ (PI controller)
g_4 = $k_v(\text{lpm/psi})/(1 + T_v s)$ (secondary flow control valve)
h_2 = $k_2(\text{psi/lpm})/(1 + T_2 s)$ (secondary measuring element and transmitter)

From Equation (13-4),

$$Q_s/Q_p = h_1 k_R g_1 g_2/(1 + g_1 g_2 h_2) \tag{13-5}$$

Substituting in Equation (13-5) and reducing to conventional form gives

$$Q_s/Q_p = [(1 + T_2 s)/(1 + T_1 s)][k_v k_1/k_2]\{(1 + T_r s)/[(T_2 T_v T_r/(k_c k_v k_2))s^3 + (T_r(T_2 + T_v)/(k_c k_v k_2))s^2 + (T_r/(k_c k_v k_2))s + 1]\} \tag{13-6}$$

If T_1, T_2, T_r, and T_v are very small time contants, the dynamics of Equation (13-6) will be negligible and can be reduced to

$$Q_s/Q_p = k_R[k_1/k_2] \tag{13-7}$$

If the primary and secondary ranges of the earlier example of 60 and 20 liters/min are assumed, then

$$k_1 = 12 \text{ psi}/60 \text{ lpm and } k_2 = 12 \text{ psi}/20 \text{ lpm} \tag{13-8}$$

Substituting in Equation (13-7) and taking k_R to be 100% as defined earlier gives

$$Q_s/Q_p = 100\% [(12 \text{ psi}/60 \text{ lpm}) \times (20 \text{ lpm}/12 \text{ psi})] = 33.3\% \tag{13-9}$$

Thus, the steady-state or zero frequency term of Equation (13-6) is the actual flow ratio, as shown earlier in Equation (13-1).

In systems where two flows are ratioed, as has been discussed in this section, the assumption that T_1, T_2, T_r, and T_v are very small is valid. Integral time settings of 0.02 min, for example, are not uncommon, and a valve lag of 0.1 to 0.15 min will often be the longest time constant in the loop. Thus, flow ratio control is quite common and usually does not present problems. When other types of ratio control are suggested, a dynamic analysis such as shown in Figure 13-4 and Equation (13-6) should be made to determine ability of the secondary variable to follow the primary variable.

Unless the time constants of Equation (13-6) are all equal to zero, there will be an error in the ratio. This will be particularly true during start-up and for major disturbances in the primary flow. Controlling primary flow, as shown in Figure 13-3, will help to minimize the disturbance problem. The start-up problem is a function of run time. In general, ratio control is considered a continuous process, and under these conditions, start-up error can be negligible. For batch mixing or small quantity blending, the designer must use discretion. This is particularly true where short run times are involved and accuracy is truly necessary. In these cases it is common practice to use positive displacement pumps or meters, or to predetermine ingredients on a weight or volume basis.[1,2]

13.2. Cascade Control

Depending on the source, a variety of definitions may be found for cascade control. For process control applications, cascade control normally implies a multiple loop system involving interacting primary and secondary control loops. A typical cascade control system for a distillation column is shown in Figure 13-5. In this system the primary variable is column temperature, which is measured and transmitted using TT-125. Output of the primary controller TRC-125 provides the set point for the secondary controller FC-125. The secondary control loop regulates steam flow to the reboiler, which influences the primary variable, namely, column temperature. In evaluating performance of a cascade control loop, results are usually compared to those obtained using single loop control. For single loop control the output of TRC-125 in Figure 13-5 would be directly connected to FV-125.

A very common system for cascade control is the reactor vessel or jacketed kettle. A typical system is shown in Figure 13-6. In a system like this, where both heating and cooling are

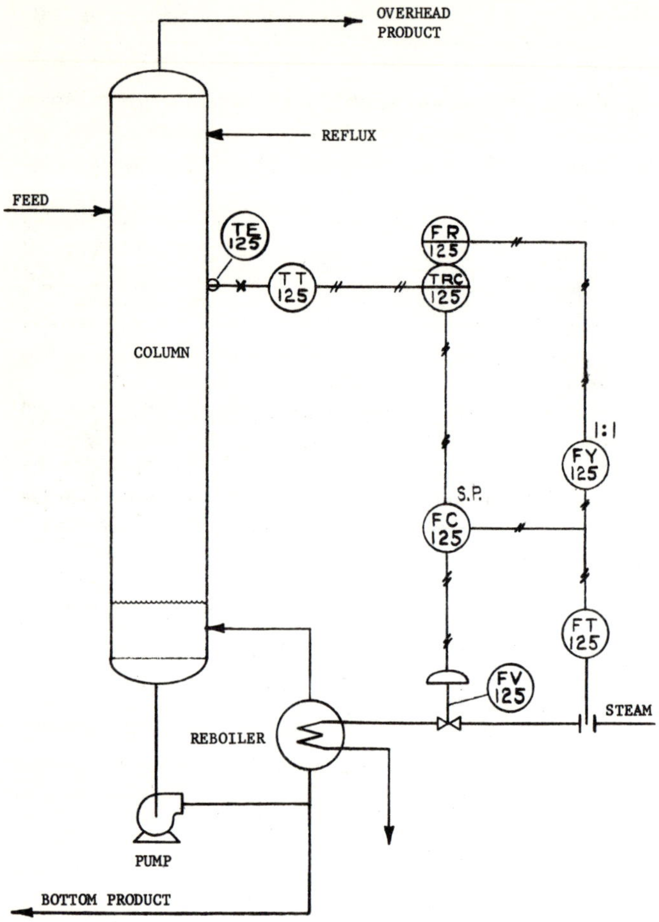

Figure 13-5. Cascade control system for a distillation column temperature control scheme.

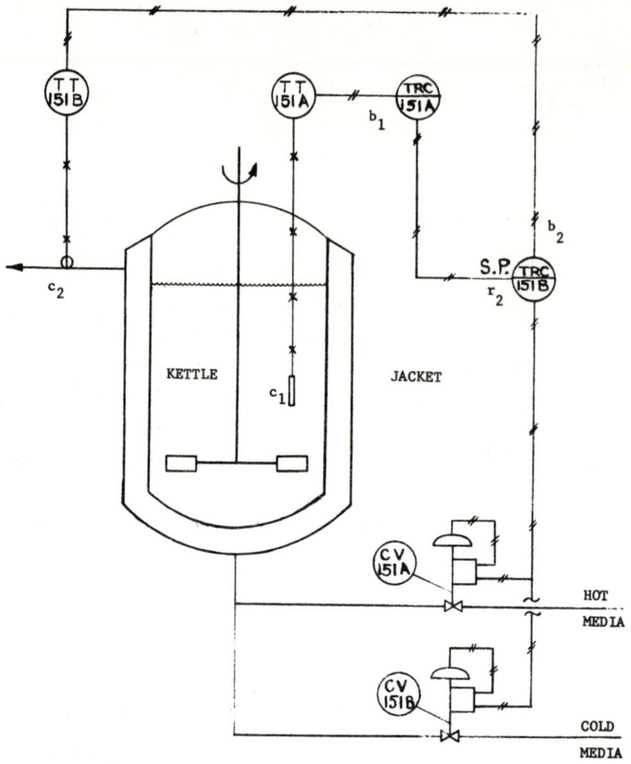

Figure 13-6. Cascade control system for a reactor vessel or jacketed kettle.

required, it is quite common to use split range valves. For pneumatically operated valves this may be accomplished by having one valve operate from 3 to 9.2 psig, and the other from 8.8 to 15 psig. Overlapping is usually employed to avoid any dead spot. For the kettle of Figure 13-6, cold media flow could be open at 3 psig and closed at 9.2 psig, while hot media flow would start to open at 8.8 psig and be fully open at 15 psig. This arrangement is usually accomplished using valve positioners with a 6-psi spring in place of the normal 12-psi spring (see Figure 10-16).

Block diagrams for both single loop and cascade control of the jacketed kettle of Figure 13-6 are shown in Figure 13-7. Associations between actual components and blocks are noted in both figures. Load change U_1 represents some change in hot or cold media flow, which would be caused by temperature or pressure changes in the supplies. Load change U_2 would normally result from heat generated by a reaction process in the kettle. To demonstrate and analyze the cascade control loop, the following values were assigned to transfer functions of Figure. 13-7 and modeled on a Control System Analog.[4]

g_1 = PI controller
g_2 = P controller
g_3 = valve transfer function = $1/(1 + 0.05s)$
g_4 = jacket transfer function = $1/(1 + 0.5s)$
g_5 = kettle transfer function = $1/(1 + 0.75s)$
h_1 = sensing element and transmitter = $1/(1 + 0.05s)$
h_2 = sensing element and transmitter = $1/(1 + 0.05s)$

While time-constant and gain values may vary in an actual system, the model is adequate to demonstrate the problem and solution.

From the discussion of load changes in Chapter 8, the reader should recognize a major problem with single loop control as shown in Figure 13-7a. It was found in Chapter 8 that for best results in response to load changes, time constants in the forward loop or downstream of a load change should be large. While this condition may hold for U_1 on Figure 13-7, it is fairly obvious that U_2 will cause a problem. Actually, in this case, U_1 also creates a problem. As noted in Chapter 9, the ideal "process" is one, big time constant located just upstream of the controlled variable. With jacket and kettle time constants of 0.5 and 0.75 min, respectively, the "process" is not ideal. There is no problem with stability. The problem is with a long, closed loop period, which results from the low-frequency crossover caused by two, large time constants.

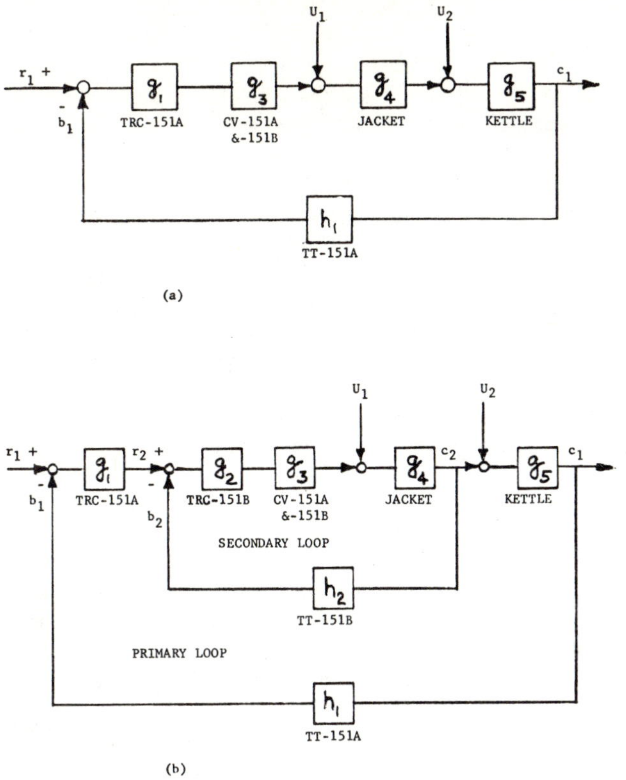

Figure 13-7. Block diagrams for jacketed kettle temperature control: (a) single loop control; (b) cascade control.

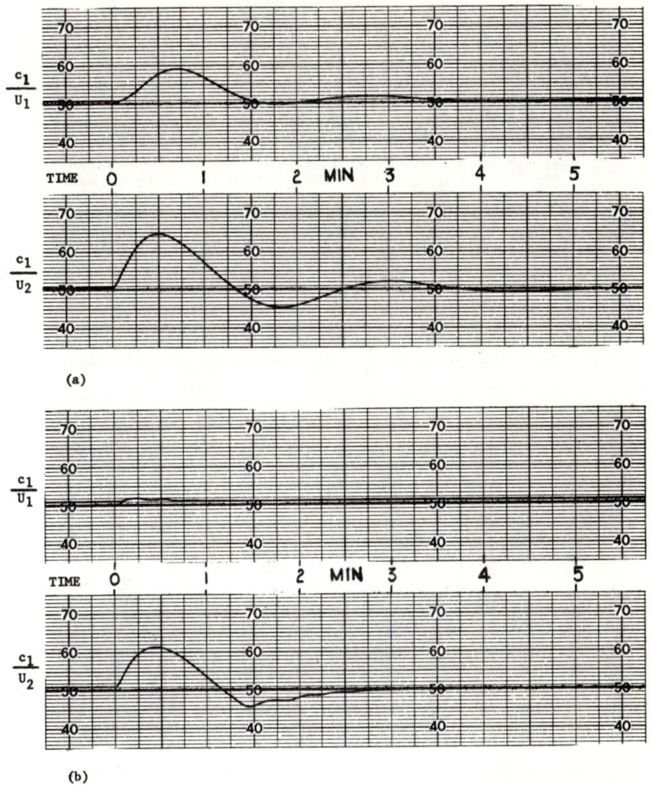

Figure 13-8. Control response to load changes: (a) single loop control; (b) cascade control. Chart range, 0–100%. Chart speed, 1 in./min. Load change magnitude at U_1 and U_2 equals 40% of chart range. Note: Set point and measurement have been offset approximately 1% for clarity.

Physically the problems can be explained as follows. To overcome a change in U_2 the control system must change jacket conditions. Since the jacket has a large time constant, this is a slow process, and kettle temperature will have a large overshoot and long period. Any change in U_1 must be recognized at c_1 before corrective action can be taken by the controller. Unfortunately, by the time the sensing element in the kettle recognizes a jacket change, the jacket is well on its way to assuming a new condition imposed by U_1. A very common example of this condition can be observed when boiling water on an electric stove. Even if power to the heating coil is turned off as soon as the water starts to boil, it is still too late. Thermal storage in the coil will continue to boil the water beyond that required for tea. This same phenomena exists in the jacketed kettle and will cause large overshoots and long periods. Typical results for load changes at U_1 and U_2, using single loop control, are shown in Figure 13-8. Having noted and observed the problem, it is now necessary to obtain a solution.

The solution of course is the cascade control system as shown in Figures 13-6 and 13-7b. Addition of a second sensing element and transmitter, namely, TT-151B (or h_2) and a second controller TRC-151B (or g_2), gives rise to a control loop within a control loop. The inner and outer loops of Figure 13-7b are defined as the secondary and primary loops, respectively. The prime objective of the secondary loop is to isolate U_1 from the remainder of the system. This is done by quickly correcting for any changes in the secondary loop and reducing their influence on the primary loop. Thus, response time of the secondary loop determines effectiveness of a cascade system. A prime example of this is the temperature-to-flow system of Figure 13-5. The secondary flow loop of this system can quickly correct steam flow changes caused by upstream pressure variations and thus prevent these changes from influencing column temperature.

While control improvement using the cascade technique depends on the ratio of major time constants in the primary and secondary loops,[5] the resultant improvement can be dramatic even when the ratio approaches unity. The kettle-to-jacket time-constant ratio in the present example is 0.75:0.5, or 1.5. Load change results for U_1 and U_2 using cascade control are also shown in Figure 13-8. As noted, the control improvement is dramatic and is typical of what can be accomplished with cascade control. Another way of interpreting this improved control is by looking at what was done when the cascade system was added. The two major time constants of the jacket and kettle were isolated. As shown in Figure 13-7a, the secondary loop is now an "ideal" process, namely, one big time constant located upstream of the controlled variable. For example, T_4 is much bigger than T_3 and T_{h2}. The result is a very simple control loop with fast response. If the response of c_2/r_2 is replaced by its closed loop box as shown in Figure 13-9, the same type of "ideal" process exists in the primary loop. In this case the kettle time constant T_5 is the dominant lag. While the improvement for c_1/U_2 is not as dramatic as for c_1/U_1, there is reduction in initial overpeak and a change in period following the initial disturbance. The point to remember is that the cascade system has done its job. In a system where the basic theory of Chapter 8 has indicated a major load change problem, cascade control has accomplished a definite improvement in control. The only other solutions to the problem are to redesign the process and/or isolate the system from the load changes.

A cascade control loop requires a second transmitter and controller, which increases initial cost, maintenance, and complexity of controller adjustment. Thus, even though a marked improvement in control can usually be obtained, further consideration should be given when specifying cascade control. Some of the questions to be answered are:

1. Do load changes U_1 and U_2 exist, and are they of appreciable magnitude?
2. Is the secondary control loop considerably faster than the primary loop?
3. Is the ratio of the major primary loop time constant to the major secondary loop time constant equal to or greater than one?

If the answers to above questions are "yes," then a cascade system should be specified. One additional point to consider is controller action for the inner loop. For the example of this chapter, g_2 was a simple proportional controller. It should be remembered that there is no requirement on jacket temperature c_2 other than it produce a desired kettle temperature c_1. Thus, there is no need to worry about offset, since c_2 will be driven by the primary controller g_1 to keep kettle temperature at the desired condition. For this reason reset and rate action, if required, are usually in the primary controller. When the secondary loop involves flow control as shown in Figure 13-5, reset action is also used in the secondary loop. The decision concerning reset action for the secondary loop is based on the concept of tight control, or high loop gain. When one large time constant exists, as in the text example, proportional action with high gain will cause c_2 to quickly follow any changes in r_2. If the secondary loop is a flow system, however, controller proportional gain will be very low, and reset is required to have c_2 follow r_2.

When used correctly, cascade control can make a major contribution to isolating a system from load changes. If there is any doubt on the part of the reader, a second inspection of Figure 13-8 is recommended.

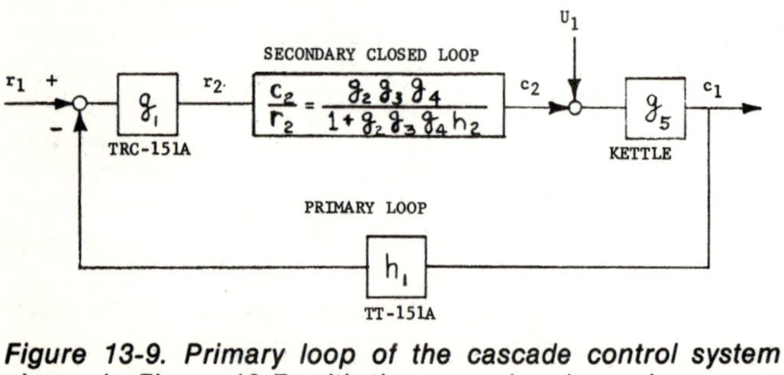

Figure 13-9. Primary loop of the cascade control system shown in Figure 13-7, with the secondary loop shown as a closed loop.

13.3 Feedforward Control

The concept of feedforward control is almost self-explanatory. Controller action based on load change information is "forwarded" to the control loop to minimize load change effect. Objective is similar to cascade control, namely, to provide good control when basic theory indicates load changes could be a problem. To help explain feedforward control it is helpful to look at cascade control results in more detail. As can be seen from Figure 13-8 cascade control, if properly applied, can almost eliminate the influence of secondary loop load changes U_1. The effect of primary loop changes U_2, however, is another story. While some help is obtained, it is evident from Figure 13-8 that load changes at U_2 could still present a problem. Primary loop load changes U_2 can result from variation in feed flow, feed temperature, feed composition, and//or reflux flow, as shown in Figure 13-5. In Figure 13-6, U_2 is a function of reaction rate variations. The purpose of this section will be to analyze the feedforward concept, see how it works, and determine where and when it can be applied.

To gain basic concepts a feedforward control loop will be applied to a simple heat exchanger. Flow and block diagrams for the system are shown in Figure 13-10. A close examination of Figure 13-10 reveals a conventional feedback control loop, where product temperature is controlled by regulating steam flow to the heat exchanger. Feedforward control is provided by

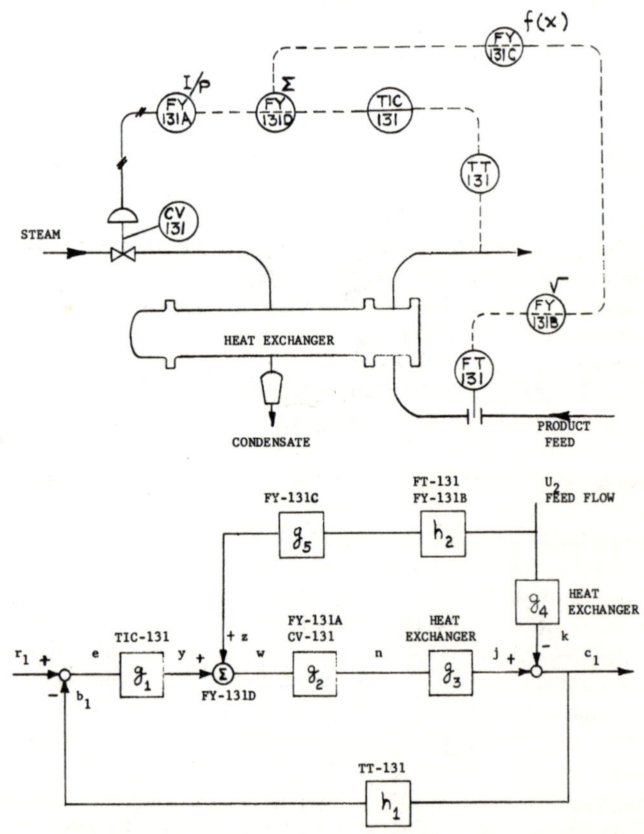

Figure 13-10. Flow and block diagrams for a feedforward control loop.

Table 13-1
Transfer Function Data from Figure 13-10

Flow Plan Symbol	Description	Transfer Function
TIC-131	Temperature controller	$(r_1 - b_1)\, g_1 = y$
FY-131D	Summing relay	$y + z = w$
FY-131A and CV-131	I/P transducer and control valve	$n = g_2 w$
Heat exchanger	Heat exchanger transfer function, steam flow to temperature	$j = g_3 n$
Heat exchanger	Heat exchanger transfer function, feed flow to temperature	$k = g_4 U_2$
Heat exchanger	Heat exchanger summation	$c_1 = j - k$
TT-131	Sensing element and transmitter	$b_1 = c_1 h_1$
FT-131 and FY-131B	Flow element, transmitter, and square-root relay	$b_2 = U_2 h_2$
FY-131C	Feedforward controller	$z = g_5 b_2$

changing steam flow in response to changes in product feed rate. *The existence of the feedback loop is a must in any feedforward scheme. By itself, feedforward control is an open loop system.* For a given product feed rate, a fixed steam flow would be set, with no regard to product temperature leaving the heat exchanger. For this kind of open loop system to work is a "pipe dream" demanding that all variables but product flow remain constant. To see how the combination of feedforward and feedback works, it is necessary to define transfer function values and solve for c_1/U_2 using the methods of Chapter 8.

The transfer functions will be defined according to prior work and logic as shown in Table 13-1.

The problem is to find c_1/U_2 and determine g_5, the feedforward controller transfer function. Starting with the controlled variable gives

$$c_1 = j - k \tag{13-10}$$

Substituting for j and k in Equation (13-10) gives

$$c_1 = g_3 n - g_4 U_2 \tag{13-11}$$

Replacing n in Equation (13-11) gives

$$c_1 = g_2 g_3 w - g_4 U_2 \tag{13-12}$$

Trading w for $y + z$ in Equation (13-12) gives

$$c_1 = g_2 g_3 (y + z) - g_4 U_2 \tag{13-13}$$

If the controller set point is constant, $r_1 = 0$. Under this normal condition, $y = -b_1 g_1$, and since $z = b_2 g_5$, Equation (13-13) reduces to

$$c_1 = -g_1 g_2 g_3 b_1 + g_2 g_3 g_5 b_2 - g_4 U_2 \tag{13-14}$$

Substituting for b_1 and b_2 gives

$$c_1 = -g_1 g_2 g_3 h_1 c_1 + g_2 g_3 g_5 h_2 U_2 - g_4 U_2 \tag{13-15}$$

Equation (13-15) may now be solved for c_1/U_2:

$$c_1/U_2 = (g_2 g_3 g_5 h_2 - g_4)/(1 + g_1 g_2 g_3 h_1) \tag{13-16}$$

Note that in Equation (13-16) the feedforward controller g_5 exists only in the numerator. Likewise, it is worth noting that the denominator is the familiar (1 + open loop transfer function), which has not been influenced by feedforward hardware. For the feedback controller to be 100% effective, there should be no change in c_1 following any change in U_2. Or as noted in Chapter 8, $c_1/U_2 = 0$. The reader is cautioned to remember that both c_1 and U_2 are delta or change values, and not absolute values. The only way that c_1/U_2 can be zero is to have the numerator of Equation (13-16) equal zero.

$$g_2 g_3 g_5 h_2 - g_4 = 0 \tag{13-17}$$

To determine the feedforward controller transfer function, it is necessary to solve and evaluate Equation (13-17) for g_5. The solution for g_5 is

$$g_5 = g_4/(g_2 g_3 h_2) \tag{13-18}$$

It is necessary at this point to check the transfer functions of Equation (13-18) from Figure 13-10 and Table 13-1. Values are

g_4 = heat exchanger transfer function, feed flow to temperature,
g_3 = heat exchanger transfer function, steam flow to temperature
g_2 = I/P transducer and control valve,
h_2 = flow element, transmitter, and square-root relay.

There are two assumptions at this point:

—Each of the noted transfer functions (g_4, g_3, g_2, and h_2) may be approximated by a single exponential lag; and
—Time-constant values for g_2 and h_2 are small in comparison to g_4 and g_3 and may be neglected.

If these two assumptions are accepted, then Equation (13-18) may be revised as follows:

$$g_5 = [k_4/(1 + T_4 s)]/\{[k_2/(1 + T_2 s)]\,[k_3/(1 + T_3 s)]\,[k_{h2}/(1 + T_{h2} s)]\} \tag{13-19}$$

Eliminating T_2 and T_{h2}, which is a valid assumption for many systems, and grouping terms gives

$$g_5 = K(1 + T_3 s)/(1 + T_4 s), \tag{13-20}$$

where $K = k_4/(k_2 k_3 k_{h2})$

Equation (13-20) shows a lead-lag transfer function with variable gain. To interpret this result it is necessary to examine

Figure 13-10. When a change occurs at U_2, the feedforward controller g_5 will proportion the corrective action by K, delay it by the time constant T_4, and advance it by T_3. Thus, there should be no change in c_1 following a change in U_2. This assumes that K, T_4, and T_3 in the feedforward controller can be adjusted to match process values.

The physical analysis shows that any change in feed flow is multiplied by k_4 and delayed by T_4. Therefore, the feedforward controller must do the same as shown by Equation (13-20). The only way to do this, however, is to measure feed flow and change steam flow. Feed flow is multiplied by k_{h2} and steam flow, in turn, by k_2 and k_3. Likewise, steam flow is delayed in producing its temperature change by T_3. To overcome these effects, g_5 must divide by $k_2k_3k_{h2}$ and advance steam flow by T_3. If the feedforward controller does its job, the disturbance k in Figure 13-10 will be matched by corrective action at j. The result will be no change in c_1. This of course is an ideal condition. Figure 13-11 shows results from tests taken on a laboratory control system. While this test is very much ideal as shown by test values on Figure 13-11, it is typical of the degree of improvement that can be made using feedforward control. One reference, for example, shows a 4-to-1 improvement in overhead composition control on a superfractionator.[6] Because of improved control, is was possible to increase tower capacity.

(a)

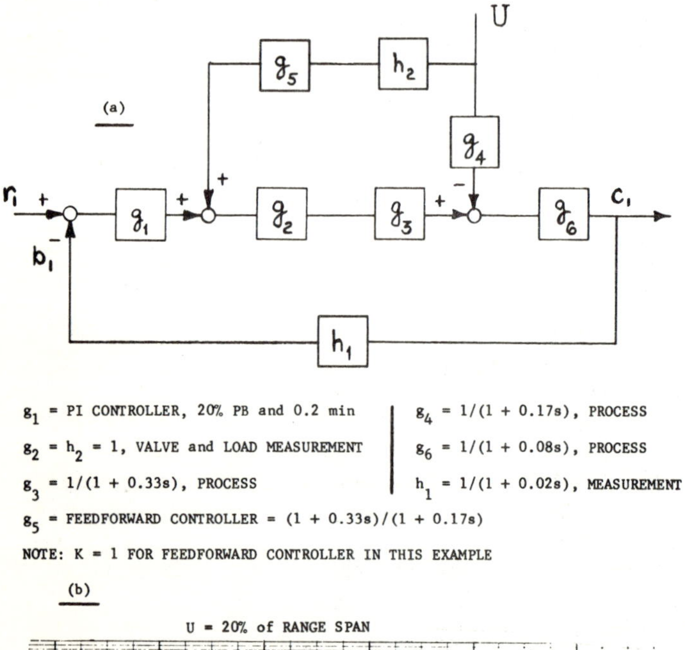

g_1 = PI CONTROLLER, 20% PB and 0.2 min

g_2 = h_2 = 1, VALVE and LOAD MEASUREMENT

g_3 = 1/(1 + 0.33s), PROCESS

g_4 = 1/(1 + 0.17s), PROCESS

g_6 = 1/(1 + 0.08s), PROCESS

h_1 = 1/(1 + 0.02s), MEASUREMENT

g_5 = FEEDFORWARD CONTROLLER = (1 + 0.33s)/(1 + 0.17s)

NOTE: K = 1 FOR FEEDFORWARD CONTROLLER IN THIS EXAMPLE

(b)

U = 20% of RANGE SPAN

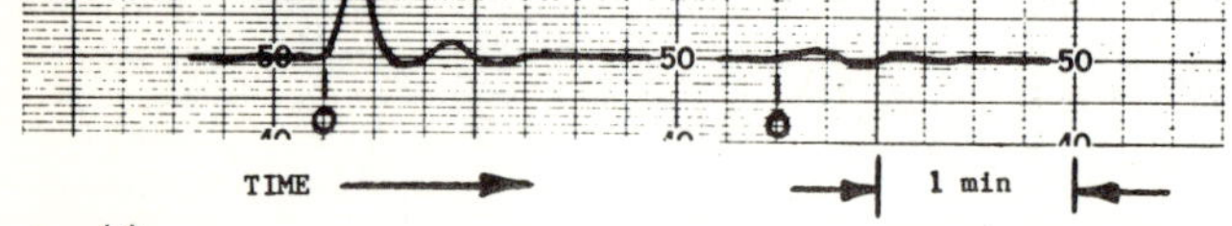

(c)

Figure 13-11. Feedforward control loop: (a) block diagram, (b) transfer function values, (c) feedback only and feedforward with feedback responses to a 20% of range load change.

While the heat exchanger example of this section is relatively simple it does point out basics of feedforward control. The column control system in Figure 13-5, as previously noted, is a typical example for feedforward control. Variations in feed flow, feed temperature, and feed composition could all be used as feedforward variables to modify steam flow. The jacketed kettle of Figure 13-6, however, is not a good example for conventional feedforward control. As shown in Figure 13-6, ingredients are charged, and the process goes through a specific cycle. Load change U_2 is a function of reaction rate and shows up as a temperature change. This variable is the primary variable of the cascade system and is already being used. Thus, there is nothing to be gained from a feedforward system. If, however, the system is run with periodic or continuous feeds and/or additives, then a feedforward system could be applied. Any change in feed rate or additives could be used to give advance information concerning hot or cold media flow. Many such reactor systems exist in industry and could benefit from feedforward control.

The variable gain lag-lead transfer function g_5 may be constructed from either pneumatic or electronic control components.[6,10] These are standard commercial items and may be purchased from most suppliers. The transfer function may also be easily programmed for systems using computer control. As a general rule it is safe to note that feedforward control will increase as more systems are computer controlled. The reason is simple: As long as measurements are available, computation can be handled in the existing computer. This avoids "extra equipment" necessary in an analog system and puts maintenance and adjustment responsibility at the computer level. As Figure 13-11 shows, major improvement can result from feedforward control. It is a powerful control scheme and should be considered where load change problems can occur.

13.4. Adaptive Control

As noted in Chapter 2, control theory is generally based on linear systems. In practice, however, few systems are really linear. The saving characteristic is that most systems operate continuously at a fixed reference or set point and do not experience large load changes. Thus, any small section of a nonlinear curve may be assumed linear. This sort of technique has been applied in Chapters 2 and 12. As with the English language, however, there are always exceptions to rules. Several common exceptions are process start-up, batching operations, and pH control. Start-up and batch operations experience large changes and variable set points, and they cannot be assumed to operate continuously. Thus, linear assumptions cannot be applied, and control problems are quite common. The pH system, on the other hand, can offer some rather abrupt nonlinearities in response to small changes. This is true even when operating in the continuous mode at a fixed set point. The reason of course is the nonlinear pH titration curve as shown in Figure 13-12.

In all these examples the basic problem relates to changing process characteristics. A simple example would be a change in product flow rate to a heat exchanger. A flow rate variation will change process dead time. It will also change the heat transfer coefficient which influences the time constant. For a fixed heat input it can also cause a gain change. For a small change around

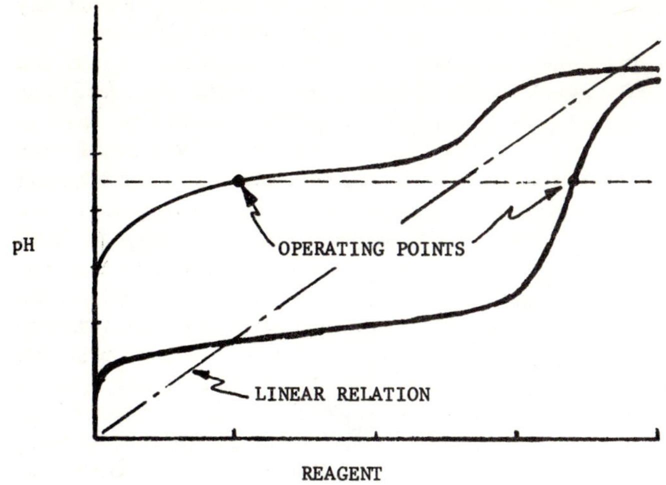

Figure 13-12. Typical pH titration curves.

a given set point these changes are negligible, and linearity may be assumed. With a large change, as in a start-up, variations in process characteristics can be appreciable. The result is that system stability varies with operating conditions. This type of problem can be handled in three basic ways. The first is to operate in the manual mode. While this may be acceptable for start-up of a continuous process, it is not a solution for batch or pH control. A second solution is to adjust the controller to give stability at the most critical process conditions. Although this technique may work, the system can be highly overdamped for other operating conditions. This will result in large overshoots and long response times. The third technique is adaptive control. In this solution, controller adjustments are varied or adapted as a function of some process variable to maintain loop stability for any process condition. Some examples will be reviewed to detail this concept.

Early adaptive control schemes were involved with characterized valves[7,8] The rule is to select a valve characteristic such that valve gain changes are opposite to process gain changes. A typical example is a flow control loop[8] using an orifice plate and Δp transmitter as shown in Figure 13-13. In a well-designed flow loop the "flow process" transfer function is 1. This is particularly true for liquid flow, since compressibility is negligible and no storage term exists. Stated another way, when valve flow changes, flow is immediately changed at the orifice. In gas flow, where compressible flow is possible, this assumption is still true if the orifice drop is much less than line pressure. As an example assume 100 in. of water drop at the orifice for a line pressure of 100 psig. The 100 in. of water is equal to approximately 3.6 psi. This is a small change when compared to line pressure and will have negligible storage. The reader should refer to Chapter 2, Section 2.1, for a more detailed explanation if necessary.

The flow control loop is, therefore, nothing more than a combination of transmitter, controller, and valve as shown in Figure 13-13. If the controller is assumed to be linear, the selected valve characteristic should be the reciprocal of the transmitter characteristic. The product would then produce a constant gain at any flow. To show this in detail assume the following orifice plate and transmitter equation:

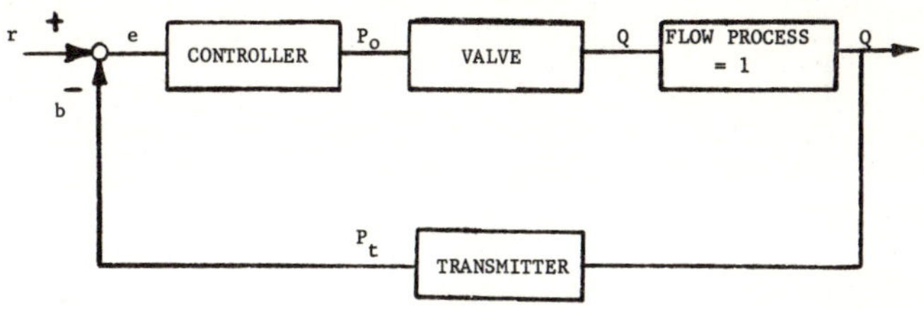

Figure 13-13. Flow control loop.

$$P_t = K_1 Q^2 + K_0, \tag{13-21}$$

where P_t = transmitter output pressure, 3 to 15 psig,
Q = product flow, 0 to 100%,
K_1 = proportional constant, psi/%^{2},
K_0 = zero constant, psig.

When product flow is zero, P_t = 3 psig. Thus, K_0 must equal 3 psig. When Q equals 100%, P_t must equal 15 psig. Thus, K_1 must equal 0.0012 psi/%^{2}. The steady-state gain at any product flow rate may be found by taking the slope of Equation (13-21). This is similar to what was done in Chapter 12. The result is

$$dP_t/dQ = 2K_1Q \tag{13-22}$$

To provide a constant gain product for transmitter and valve, the valve characteristic should be

$$Q = \left[K_2(P_0 - 3)\right]^{1/2} \tag{13-23}$$

where P_0 = valve input pressure, 3 to 15 psig,
K_2 = proportional constant, %^{2}/psi.

In this case when P_0 = 3 psig, product flow Q equals zero. When P_0 = 15 psig, Q = 100%, and K_2 = 833%^{2}/psi. The steady-state gain for any product flow may again be found by taking the slope of Equation (13-23):

$$dQ/dP_0 = K_2/(2Q) \tag{13-24}$$

The gain product for any flow rate may be found by multiplying Equations (13-22) and (13-24):

$$dP_t/dP_0 = (dP_t/dQ)(dQ/dP_0) \tag{13-25}$$
$$= 2K_1Q\,K_2/(2Q) = K_1K_2$$

$$dP_t/dP_0 = K_1K_2 = \left[0.0012\,(\text{psi}/\%^2)\right]\left[833\,(\%^2/\text{psi})\right] = 1$$

The result of Equation (13-25) is obviously an ideal case and involves two very important assumptions that were not stated.

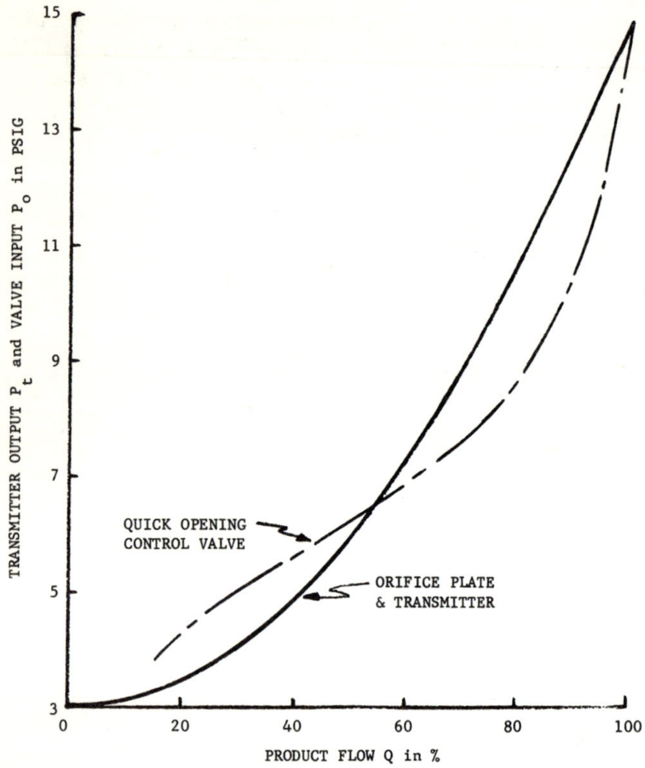

Figure 13-14. Transmitter and valve characteristics.

The first assumption is that the control valve has a constant pressure drop. If this is true then the installed characteristic for the valve will be the same as the inherent characteristic of Equation (13-23). If the valve pressure drop is not constant, the installed characteristic will exist and gain product will not stay at 1. The second assumption concerns dynamic characteristics of transmitter and valve. Unless valve and transmitter time constants are equal, gain product will vary during any disturbance. While a system with both constant pressure drop and equal time constants is possible, its existence is highly improbable. However, use of a quick-opening valve on flow control is still desirable, since it approximates the inherent characteristics of Equation (13-23). This is particularly true if constant pressure drop exists. Typical transmitter and valve plots are shown in Figure 13-14 for constant pressure drop. If the pressure drop is not constant it may be advisable to use a linear valve. Under conditions of variable pressure drop, the linear valve, as shown in Figure 10-12, approximates the quick-opening characteristic. As the reader should suspect by this point, it is almost impossible to get perfect compensation. However, the proper choice of valve characteristic can act as a self-adaptive system and provide considerable improvement in control.

A sequel to the flow control example is what the author likes to call "operator" adaptive control. In this example controller settings are written out on a card attached to the control panel. These controller settings are noted as a function of controlled variable operating point. This concept may be shown by using Equation (13-22) and calculating slope or gain at 20, 50, and 80% flow rates. The results are shown in Table 13-2.

If a valve with a linear-installed characteristic was used on this process, it would be necessary to change controller gain at each operating point to provide uniform stability. Such a case can exist for a flow loop using an equal percentage valve with variable pressure drop. As noted in Chapter 10, an equal percentage characteristic approaches the linear characteristic when operating with variable pressure drop. When this happens, operators will find controller gains that work and post them on the panel. Values given in Table 13-2 assume a constant valve gain of 8.33%/psi and a steady-state loop gain of 1. Results may be calculated as follows:

$$\text{controller gain} = 1/(\text{valve gain} \times \text{transmitter gain}) \qquad (13\text{-}26)$$

It is fairly easy to see from past stability studies that a gain change from 0.63 to 2.5 could cause trouble. This is equivalent to a 4:1 controller gain variation for product flow changes from 20 to 80% and is sure to cause stability problems.

If a constant set point is maintained, this condition of variable gain will only occur during large load changes and is not apt to cause any problem. The quick-opening valve of Figure 13-14 is another solution. If slopes are calculated at 20, 50, and 80% of product flow, using Figure 13-14, the results are shown in Table 13-3. By using the quick-opening valve characteristic, the controller gain variation has been reduced from 4:1 to approximately 3:1. Note that these calculations are based on Figure 13-14, which shows the inherent valve characteristic based on constant pressure drop. Under variable pressure drop the installed characteristic will more closely approximate the ideal square-root function of Equation (13-23). This will provide

Table 13-2
Controller Gain Adjustment Required for Uniform Stability*

Product Flow Q (%)	Transmitter Slope or Gain (psi/%)	Controller Gain
20	0.048	2.5
50	0.120	1.0
80	0.192	0.63

*Based on Equation (13-22) and $K_1 = 0.0012$ psi/%^{2}.

Table 13-3
Controller Gain Adjustment Required for Uniform Stability*

Product Flow Q (%)	Transmitter Slope or Gain (psi/%)	Valve Slope or Gain (%/psi)	Controller Gain
20	0.048	12.3	0.6
50	0.120	16.7	2.0
80	0.192	7.1	1.4

*Based on Figure 13-14.

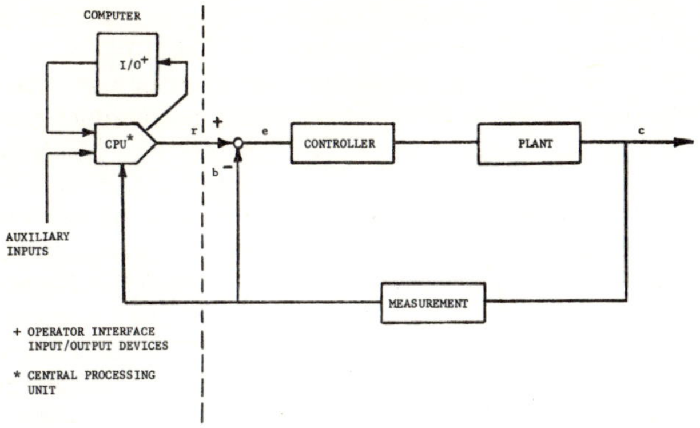

Figure 13-15. Supervisory computer control with only one control loop shown.

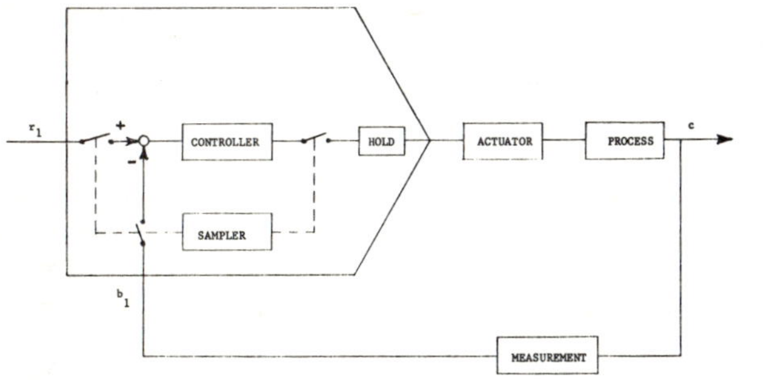

Figure 13-16. Direct digital control with only one control loop shown.

additional help such that controller gain variation will approach the ideal 1:1 ratio for all product flow rates. The result will not be perfect as previously noted. A major improvement, however, will be achieved.

With the advent of computer control it is possible to allow the computer to handle the "operator adaptive" process. While there is no intention to cover computer control in this text, it does appear necessary to note how computer control can be used on certain specialized control systems. To do this requires defining two main forms of computer control. Probably the most common form of computer control is "supervisory control." In this system the computer acts as a clearing house for information from numerous sources, processes this information according to programmed instructions, and initiates output action. A typical system is shown in Figure 13-15. The normal output action for supervisory control is set point adjustment of a conventional analog controller. Output messages may also be provided for operator information.

The second form of computer control is "direct digital control," or DDC. In this system the computer replaces the conventional analog controller in each control loop. Each loop is sampled in sequence or on demand, a comparison of measurement and set point is performed, and output action is initiated. Figure 13-16 presents a possible system. Output action may be position or rate sensitive, but in either case is maintained by a "holding"device until the next sample time. Output action is based on error signal, as it would be in a conventional analog control loop. The major difference occurs in how the computer is instructed to function. This instruction is called an algorithm and may be nothing more than a computerized expression for PI or PID controller action. A typical comparison to Equation (9-7) for a PI controller is shown in Equation (13-27).

$$\Delta P_0 = Ke + (K/T_r) \int e\, dt \tag{9-7}$$

$$\Delta P_0 = Ke_n + (KT_s/T_r) \sum_{k=1}^{n} e_k \tag{13-27}$$

In a sampled system the continuous error function e is not available to the computer.[9] What is available is a discreet error e_n for each sample. If T_s is the sample period, then an effective time-error product may be obtained as follows:

$$\int e\, dt = e_1 T_{s1} + e_2 T_{s2} + e_3 T_{s3} + \cdots + e_n T_{sn} \tag{13-28}$$

However,

$$T_{s1} = T_{s2} = T_{s3} = \cdots = T_{sn} = T_s.\text{Thus,}$$

$$\int e\, dt = T_s \sum_{k=1}^{n} e_k \tag{13-29}$$

Substituting Equation (13-29) into Equation (9-7) gives Equation (13-27). In this type of algorithm it is necessary to select the sample time T_s in addition to proportional gain K and reset time T_r. If T_s is very small, the computer algorithm will perform very much like the analog or continuous controller. As T_s is increased, however, it will have a direct influence on stability. Sampled data control, however, is beyond the scope of this book. Many references to this subject can be found, if the reader is interested.

Additional algorithms are available for DDC. These include error-squared functions, dead band allowance in which no action is taken, so-called "bang-bang" (on-off), and modified bang-bang algorithms. For additional information concerning computer control in general, the reader is referred to specific company literature and other texts.[11,12,13]

A very simple adaptive application is shown in Figure 13-17. It is a combination of the field elements of Figure 13-13 along with DDC control as shown in Figure 13-16. In this case it is assumed that loop stability is a function of product flow rate and that the computer must function to replace the previously noted "operator adaptive" procedure. The PI algorithm

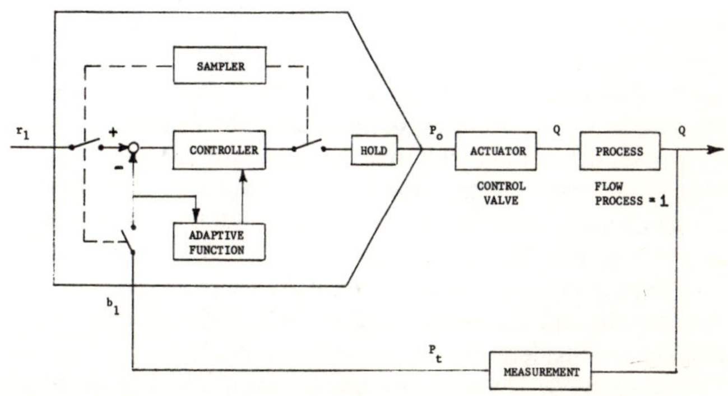

Figure 13-17. The DDC adaptive flow control system.

of Equation (13-27) will be selected as the control function shown in Figure 13-17. In addition, an adaptive function must also be programmed as shown in Figure 13-17. The adaptive function will change gain K and reset time T_r, as a function of measured flow P_t, to maintain loop stability. To some extent the adaptive function works as an input device to the computer memory. It changes gain and reset time settings just as an operator could do. The difference is that it is done on an automatic basis according to some prescribed instruction. While this is a very simple example, it does demonstrate what can be done. For the interested reader there are many examples where the adaptive procedure includes both static and dynamic considerations.[14] The objective at this point is merely to provide basic concepts so that the reader can follow more detailed reports.

Up to this point only systems with large upsets or major operating point variations have been examined. Control of pH, as previously noted, is one exception where abrupt nonlinearities can arise even for small changes. This is clearly shown by the nonlinear titration curves of Figure 13-12. Stability problems can exist if the set point or reference is at the "knee" of a particular titration curve, or if the titration curve is altered by additives or buffering action within the process. If process gain decreases, the loop can become sluggish, with large overshoots and wide variations in pH. When control shifts to a steep portion of the titration curve, process gain increases, and unstable control is the usual result. The ideal solution of course would be an automatic titration analyzer. A process gain slope or pH change per reagent addition could be calculated at the desired set point. By using an adaptive function similar to that shown in Figure 13-17, controller gain could then be adjusted to match process gain and maintain loop stability.

Lack of an on-line automatic titration has led to several alternate techniques, most of which make use of a nonlinear controller. While there may be numerous variations, the basic nonlinear controller is as shown in Figure 13-18.[15,16] There are two adjustable parameters, namely, deadband width and gain within the deadband. For a fixed titration curve, deadband width and gain can be field adjusted by trial and error to provide a constant loop gain. This technqiue is similar to the characterized valve example, where controller gain becomes the reciprocal of process gain. When the titration curve is variable as it is in many applications, it is necessary to take additional adaptive action to maintain constant loop gain. One such technique is based on process damping.[16] A discriminator or adaptive controller senses whether the process is sluggish (overdamped, $\zeta > 1$) or oscillatory (underdamped, $\zeta < 1$) and adjusts the deadband width accordingly. For a sluggish response, deadband width can be reduced to give increased controller gain. For an oscillatory response, deadband width must be increased. The author notes that after several tests, superb control was achieved and that "load changes were handled smoothly, with prolonged periods of cycling and offset eliminated."[16]

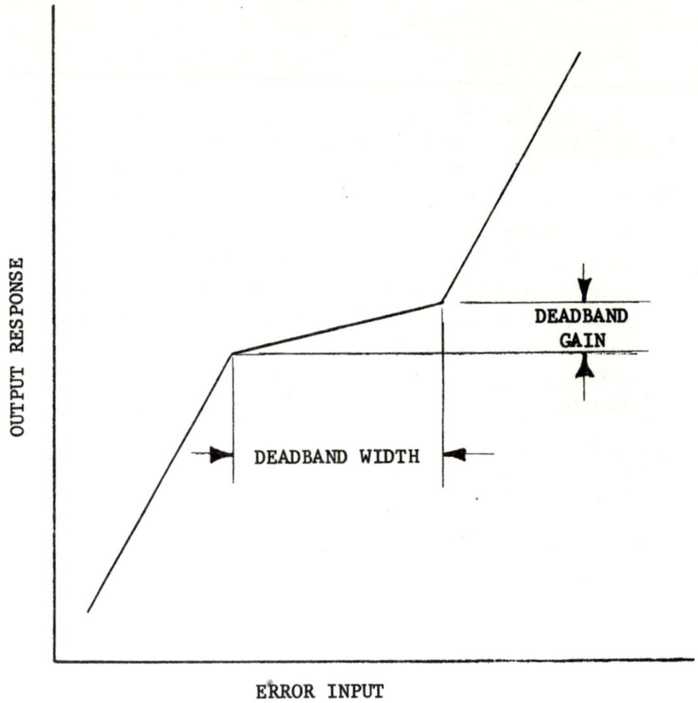

Figure 13-18. Nonlinear controller action.

The treatment of adaptive control in this text is of an elementary nature and has been given very limited coverage. While dynamic consideration was noted, no detailed coverage has been presented. There are two reasons for this action. Although much has been written on the subject, there are very few standard techniques that may be applied at this time. Typical examples include dynamic compensation,[17] variable sampling,[18] statistical filtering, pattern recognition, and model identification.[19,20] To cover these methods in detail would require a second book. The second reason involves actual application of adaptive techniques. While much has been written, industrial acceptance has been very limited. Reasons range from a lack of understanding of method and "process" to the complexity which often results when numerous calculations must be carried out with analog components. It is the author's opinion that these roadblocks will gradually be overcome at a rate proportional to application of integrated circuit elements and computer control. Hopefully, this brief introduction will supply the necessary background for reading the literature on adaptive control with some understanding.

The problems of start-up and batch control will also be covered in more detail in Chapter 14.

Chapter 13—Study Questions and Problems

(Answers appear on pages 170-172)

1. Two flows are to be ratioed such that for every 5 lpm of primary flow, 1 lpm of secondary flow is required. Primary flow will be controlled using a magnetic flow meter as the measuring element. An orifice plate will be used to measure secondary flow. Although the design ratio is 20%, it is expected that variations in operations could require ratios

from 15 to 25%. Assume all measurement and control elements to be electronic, with I/P conversion at the final control elements. Sketch a P&ID (piping and instrument drawing), specify flow element ranges for a design primary flow of 60 lpm, and specify scale type and initial setting.

2. Sketch a cascade control block diagram (similar to Figure 13-7) for the distillation column temperature control system of Figure 13-5. Assume load changes in steam supply pressure U_1 and feed flow rate U_2. Define transfer functions in general terms such as g_4, reboiler, g_5, column from bottom to TE-125 and so forth. Apply basic concepts of this section to explain why cascade control should be used for this system.

3. Based on Problem 2 and its solution, select controller actions for TRC-125 and FC-125. Justify your selection.

4. The prime objective of feedforward control is to provide good control even though __________ changes exist, which could cause problems in a conventional __________ control system.

5. Feedforward is an __________ system and must therefore always include a __________ control loop.

6. It was noted in the text that feed composition to the distillation column of Figure 13-5 could cause a primary load change U_2. Assuming existence of this problem, design a feedforward control system to help overcome this load change. Assume a continuous composition transmitter. If a sampling device is used, a "holding" device must be added to hold signals between samples. Use a composition transmitter on feed flow and incorporate feedforward control with the existing cascade system. Use whatever additional control components are necessary and show your results on the feedforward/cascade control system below left.

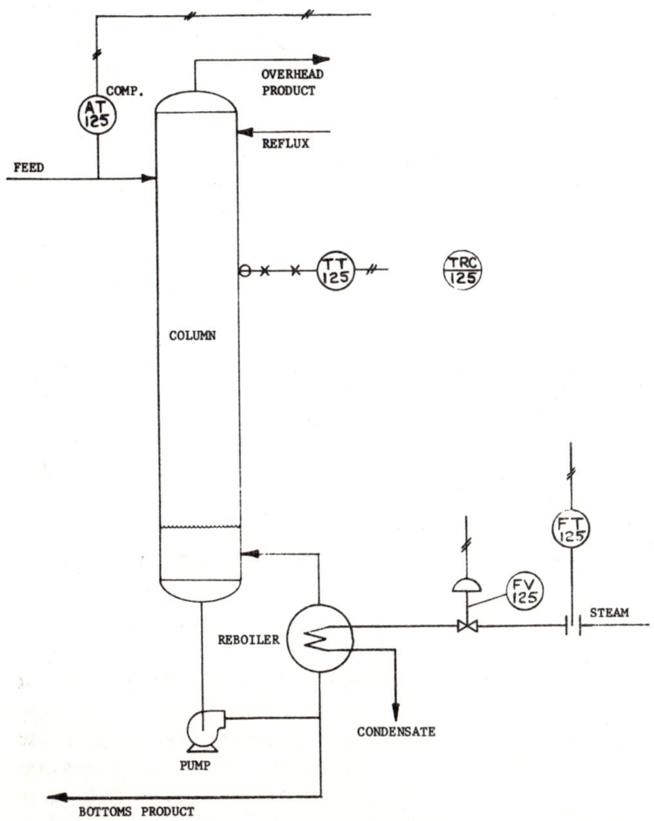

Figure 13-P6. Complete the feedforward/cascade system.

7. Sketch a block diagram for the feedforward control system designed in Problem 6, assuming the following transfer functions:

g_1 = TRC-125
g_2 = FC-125
g_3 = CV-125
g_4 = reboiler and column up to the sensing element for TT-125
g_5 = column from feed tray down to the sensing element for TT-125
g_6 = FY-125B
h_1 = TT-125
h_2 = FT-125
h_3 = AT-125

Again assume a continuous composition transmitter. If a sampling device is used, a "holding" device must be added to hold signals between samples.

8. Is control system theory usually based on linear or nonlinear system analysis?

9. System stability will vary with the operating point for any __________ system.

10. In an adaptive control system, __________ are varied or adapted as a function of some process variable.

11. A simple example of an adaptive system is a __________ valve.

12. The idea of a characterized valve is to introduce a nonlinearity into the control loop, which has the __________ characteristic of an existing nonlinear component.

13. In an "operator adaptive" system, controller adjustments are manually changed as a function of __________.

14. In a computer control system the instruction for control action is called an __________.

15. Assume a heat exchanger control system as shown on the next page, in which product feed rate variations cause stability problems. A possible explanation would involve variations in transfer functions due to time constant and

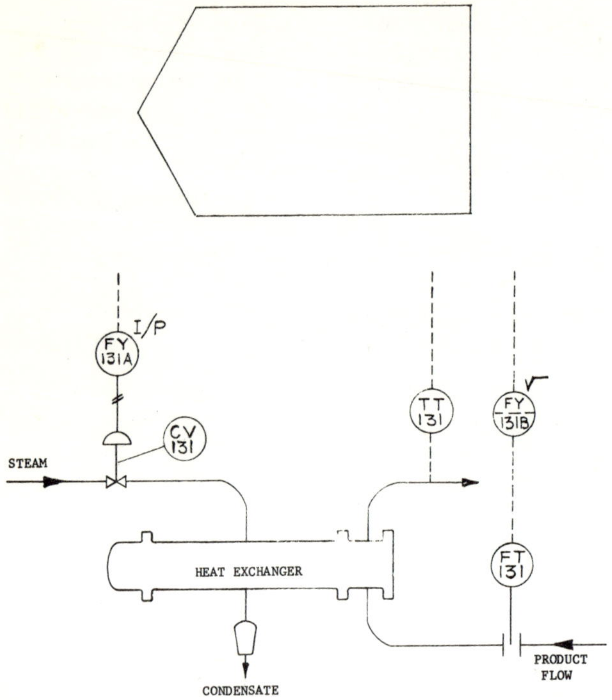

Figure 13-P15. Heat exchanger. Complete the adaptive DDC system.

dead time lag changes. Using DDC and adaptive control, complete the adaptive DDC system and explain your adaptive logic.

References

1. Considine, D. M., *Process Instruments and Controls Handbook,* New York: McGraw-Hill, 1957.
2. Buckley, P. S., *Techniques of Process Control,* New York: John Wiley, 1964.
3. Eckman, D. P., *Automatic Process Control,* New York: John Wiley, 1958.
4. Zoss, L. M., and Wagner, C. A., "Two Process-Control Simulators, Pneumatically Simulating Control," *ISA Journal,* Vol. 11, No. 2, 1964, pp. 61-68.
5. Gollin, N. W., "Cascade Control Systems," *Control Engineering,* Vol. 3, No. 7, 1956.
6. MacMullan, E. G., and Shinskey, F. G., "Feedforward Analog Computer Control of a Super-fractionator," *Control Eng.,* Vol. 11, No. 3, 1964, pp. 69-74.
7. Ziegler, J. G., and Nichols, N. B., "Valve Characteristics and Process Control," *Instruments,* Vol. 22, No. 1, January 1949.
8. Ramsey, J. R., and Fournier, C. D., "Matching Valves with Process Requirements," *Instruments and Control Systems,* Vol. 46, No. 7, 1973, pp. 65-67.
9. Hartwigsen, C. C., Mortimer, K., Ruopp, D. E., and Zoss, L. M., "Analysis of Sampling Rates and Control Settings for Direct Digital Control," I.S.A. Preprint 30. 1-4-65, Presented at 20th Annual I.S.A. Conference and Exhibit, 1965.
10. Wright, R. M., and Johncock, A. W., "A Better Approach to Distillation Control," *Instruments and Control Systems,* Vol. 49, No. 6, June 1976, pp. 49-53.
11. Bartee, T. C., *Digital Computer Fundamentals,* New York: McGraw-Hill, 1972.
12. Rhyne, V. T., *Fundamentals of Digital Systems Design,* Englewood Cliffs, NJ: Prentice-Hall, 1973.
13. Luyben, W. L., *Process Modeling, Simulation, and Control for Chemical Engineers,* New York: McGraw-Hill, 1973.
14. Unbehaven, H., and Schmid, Chr., "Status and Industrial Application of Adaptive Control Systems," *Auto Cont. Theory and Application,* Vol. 3, No. 1, January 1975, pp. 1-12.
15. Shunta, J. P., and Fehervari, W., "Nonlinear Control of Liquid Level," *Instrumentation Technology,* Vol. 23, No. 1, 1976, pp. 43-48.
16. Shinskey, F. G., "Adaptive pH Controller Monitors Nonlinear Process," *Control Engineering,* Vol. 21, No. 2, 1974, pp. 57-59.
17. Bakke, R. M., "Analysis of the Dynamics of an Adaptive Direct Digital Control System," A.S.M.E. Paper no. 67-WA/AUT-14, Pittsburgh, PA, November 1967.
18. Sutherland, A. A., "Variable Sampling Frequency—A New Adaptive Technique," *Control Engineering,* Vol. 20, No. 4, 1973, pp. 33-35.
19. Bristol, E. H., "Adaptive Process Control: Versatile On-Line Tool," *Control Engineering,* Vol. 20, No. 4, 1973, pp. 41-44.
20. Bristol, E. H., "Adaptive Process Control: Versatile On-Line Tool," *Control Engineering,* Vol. 20, No. 6, 1973, pp. 41-43.

14 Batch Control Systems

In most control texts, continuous system operation is assumed. With the exception of Chapter 13, where nonlinear problems of start-up were noted, the same assumption has been made so far in this text. There is, however, a large group of industrial processes that operate in the batch mode. For the purpose of this text, a batch process will be defined as one that behaves in the following manner:

1. Ingredients are charged according to some recipe or sequence.
2. Action is undertaken, based on a procedure involving one or more measured variables, to achieve some end result.
3. Product is withdrawn upon cycle completion and forwarded in a specified manner.

While a batch process is in the procedure phase, it may very well resemble a continuous process. The batch reactor or jacketed kettle of Chapter 13 (used to demonstrate cascade control; see Figure 13-6) is a typical example. There are, however, several specific problems related to batch control that merit consideration. In steps 1 and 3, for example, charging and withdrawal of ingredients and product may present sequential control problems. In step 2 (the procedure phase) problems of start-up and programming will normally exist. The purpose of this chapter will be to analyze these problems of start-up, procedure, and sequential control. The order of study has been selected to match normal process development and should become apparent as the reader advances through the text.

14.1. Batch Control Start-up

Many batch processes operate primarily in the manual mode. In some cases combinations of manual and automatic modes may be used with operators starting and stopping various functions. When a decision is made to control all or part of each

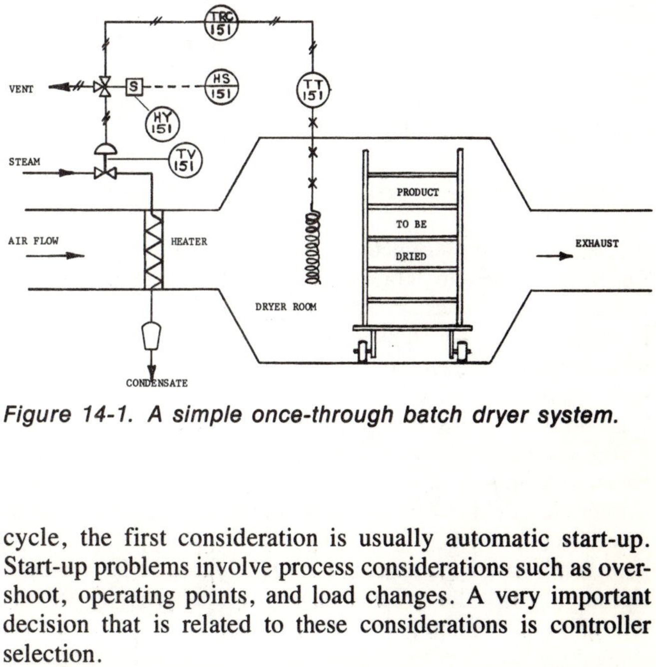

Figure 14-1. A simple once-through batch dryer system.

cycle, the first consideration is usually automatic start-up. Start-up problems involve process considerations such as overshoot, operating points, and load changes. A very important decision that is related to these considerations is controller selection.

To demonstrate this in more detail, assume a batch dryer as shown in Figure 14-1. To start, assume that the dryer has been charged or loaded with product at room temperature. The procedure calls for increasing temperature to 80°C and holding at this temperature for 8 hr. At the end of this time it is assumed that product is dry enough to be removed for further processing. The system is a simple once-through-air heater, using steam as heating media. The control system uses a capillary-type bulb recommended for air systems, along with a conventional pneumatic temperature transmitter, TT-151. To avoid overheating of product during a loss of air supply, an air-to-open valve is used as the final regulating element (TV-151). A proportional plus reset controller (TRC-151) has been specified to compensate for any load changes due to variations in air flow rate or temperature, and steam supply pressure. A hand switch (HS-151) activates a solenoid valve (HY-151) in the air line between the controller and valve. When HS-151 is open, HY-151 vents air from the valve motor closing TV-151. When HS-151 is closed, HY-151 is switched to close the vent and connect controller output directly to the valve. Thus, once product is loaded in the dryer, automatic start-up is initiated when the operator closes HS-151 on the control panel. At the end of 8 hr, HS-151 is opened to complete the procedure and allow for removal of product. In addition to the specified 80°C drying temperature, process requirements list a maximum allowable temperature of 85°C. To state this in another way, the process must be started and controlled with little or no overshoot.

The stage has now been set to demonstrate the classical problem of batch control start-up. Prime ingredients are process requirements of little or no overshoot, and selection of a PI controller. Assume, for example, that a batch has been completed and that the operator has opened HS-151. This in turn shuts off steam flow to the heater. Dry product is removed, and a new charge of product is loaded into the dryer room. These operations take time and may vary from a few minutes to an hour or more. During this time, dryer room temperature decreases and may even approach ambient conditions. A typical record is shown in Figure 14-2a. The actual level to which temperature falls is not as important as the product of error times unloading and loading time. To show why this is true, it is necessary to reexamine Equation (9-7) for a PI controller:

$$\Delta P_o = Ke + (K/T_r)\int e\,dt \qquad (9\text{-}7)$$

As can be seen from Equation (9-7), integral or reset action is a function of $e\,dt$. Integration of $e\,dt$ over the unloading and loading time will cause a large change in output ΔP_o. The effect is similar to an error step change, as shown in Figure 9-6, and will cause output pressure P_o to drive to saturation (supply pressure). Thus, as soon as HS-151 is closed to start the new cycle, TV-151 will be exposed to approximately supply pressure and be driven wide open. This will provide maximum steam flow to the heater and, in turn, a maximum rate of increase in dryer room temperature. This of course is desirable and might appear as shown in Figure 14-2b.

As temperature increases as shown in Figure 14-2b, the error signal decreases. To get some idea of what happens, the following values and conditions will be assumed:

P_o = 20 psig at $t = 0$ (start of new cycle)

K = 100/PB = 100/50 = 2 (50% Prop. Band)

T_r = 5 min

Input range: 3 to 15 psig (100°C = 12 psi)

Output range: 3 to 15 psig

From Figure 14-2b it is now possible to obtain error values shown in Table 14-1.

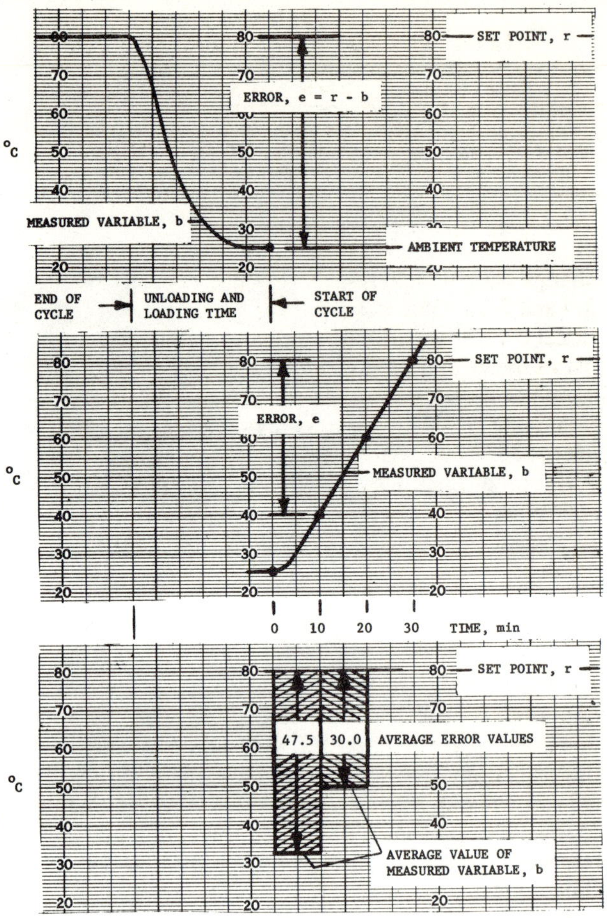

Figure 14-2. (a) Shutdown, (b) start-up, and (c) average error curves for a batch control system.

From the temperature error, scale factor, and controller gain, change or reduction in output pressure due to proportional action in the first 20 min may be calculated as shown in Equation (14-1):

$$\Delta P_o(0) = Ke = 2(55°\text{C})(12\text{ psi}/100°\text{C}) = 13.2\text{ psi}$$

$$\Delta P_o(10) = Ke = 2(40°\text{C})(12\text{ psi}/100°\text{C}) = 9.6\text{ psi}$$

$$\Delta P_o(20) = Ke = 2(20°\text{C})(12\text{ psi}/100°\text{C}) = 4.8\text{ psi}$$

or

$$\Delta P_o(\text{total}) = \Delta P_o(20) - \Delta P_o(0) \qquad (14\text{-}1)$$

$$= 4.8\text{ psi} - 13.2\text{ psi}$$

$$= -8.4\text{ psi}$$

It is worth noting again at this time that calculations to arrive at Equation (14-1) are not absolute values. They merely show the

Table 14-1
Error Values as Measured from Figure 14-2b

Time t (min)	$r - b = e$, °C	Scale Factor
0	80 − 25 = 55	(12 psi/100 °C)
10	80 − 40 = 40	(12 psi/100 °C)
20	80 − 60 = 20	(12 psi/100 °C)

change from 0 to 20 min. The reader should refer to Sections 3.4 and 7.1 if review is needed. Or stated another way, proportional controller action has reduced output pressure P_o by 8.4 psi as the error signal decreased from 55 to 20°C. The result would be a decrease in P_o from 20 to 11.6 psig if only proportional action were present. This, however, is not the case. To get the complete picture it is necessary to also include the integral term of Equation (9-7). To show effect of integral or reset action, an average error value will be used in each 10-min time increment. Using values from Table 14-1 for 0 and 10 min, the result is

$$e_{avg} = (e_0 + e_{10})/2 = (55 + 40)/2 = 47.5\,°C \qquad (14\text{-}2)$$

For the 10- to 20-min time increment e_{avg} equals 30°C. The effective result is as shown in Figure 14-2c. It is important at this time, however, to reexamine two fundamentals of reset or integral control. First, integral action as shown by Equations (13-28) and (13-29) is a summation process. Thus, unlike the change due to proportional action which is equal to the difference at the end of 20 min, integral action will be equal to the sum of pressure changes for each increment. Second, whereas proportional action in response to a decreasing error reduced output pressure, reset or integral action is still trying to drive output pressure to a value that will provide zero offset or error. Therefore, the change due to integral action will be positive. By using noted values for K and T_r, a time increment of 10 min, and the scale factor of Table 14-1, values may be calculated for each increment as follows:

$$\Delta P_o = (K/T_r)\int e\,dt$$

$$\Delta P_o(\text{1st}) = (2/5\text{ min})(47.5°\text{C})(12\text{ psi}/100°\text{C})(10\text{ min}) = 22.8\text{ psi}$$

$$\Delta P_o(\text{2nd}) = (2/5\text{ min})(30°\text{C})(12\text{ psi}/100°\text{C})(10\text{ min}) = 14.4\text{ psi}$$

Total change as shown in Equation (14-3) is the sum of both the first and second incremental changes.

$$\Delta P_o(\text{total}) = \Delta P_o(\text{1st}) + \Delta P_o(\text{2nd}) = 22.8 + 14.4 = 37.2\text{ psi} \qquad (14\text{-}3)$$

If combined changes of proportional and reset are added to P_o at time zero, the result is

$$P_o(20) = P_o(0) + P_o(\text{prop}) + P_o(\text{reset}) \qquad (14\text{-}4)$$

$$P_o(20) = 20\text{ psig} - 8.4\text{ psi} + 37.2\text{ psi} = 48.8\text{ psig}$$

The result is absurd when an air supply pressure of approximately 20 psig is used. In actual practice, output pressure saturates at 20 psig and stays there until the error is zero. This is the absurd part and is defined as reset windup. When the error is zero as shown in Figure 14-2b, the measured variable is at the set point. At this time, however, the steam valve is still wide open, and there is no way to avoid a very large overshoot. A typical record is shown in Figure 14-3a. Reset windup is a very common problem. Because of it, a conventional PI controller cannot provide start-up without overshoot.

Fortunately, there are ways to overcome reset windup and the problem of overshoot on start-up. The simplest solution of course is to use only proportional control. If a fixed set point is used and load changes are negligible this can be a satisfactory solution. Controller gain must be adjusted to give an overdamped response. While this method can avoid overshoot, it will produce a slow rise time and response to any disturbances. If load or set point changes occur, control will be very poor, and large offsets will occur. This procedure is seldom used and is not recommended. Before specifying proportional control, the reader should apply the fundamental considerations of Chapters 7 and 8 to determine possible offset magnitudes.

A second solution to reset windup is some form of limiting or lockout system.[1] Many devices are available from manufacturers for both pneumatic and electronic controllers. These range from simple limiting devices that will not allow output signals to exceed valve range limits to very complicated two controller systems. In many cases they do not eliminate reset windup, but only work to minimize the effect. Before specifying such a system, the reader should check very closely to see if the proposed system really meets his needs.

A very practical solution for many applications is to use a PD controller. Figure 14-3b shows start-up results using a PD controller. Process conditions are identical for both the PI response of Figure 14-3a and the PD response of Figure 14-3b. As noted in Section 9.8, rate action will reduce overshoot and offset. In the case of offset, the key word is reduce. Rate action allows higher proportional gain to be used, and offset is reduced; it is not eliminated. Thus, in considering the application of a PD controller to a batch system, load and set point changes should be examined. Here again, the theory of Chapters 7 and 8 will help to estimate offsets. A comparison of 10% load and set point changes for the process of Figure 14-3, using both PD and PI controllers, is shown in Figure 14-4. Here again, the advantages of a PD controller are shown, if small offsets can be tolerated.

When load changes and variable set points are expected, and overshoot and offset cannot be tolerated, other techniques must be used. If a computer is involved, there are many ways to overcome reset windup. It is also possible to use different algorithms that avoid potential for reset windup.[2,3,4] There are, however, many processes that will for one reason or another not be computer controlled and the question becomes, "What then?" Here again, most manufacturers offer what is called a series, non-interacting or cascade controller (other names may also be used). The author prefers the series term as the most descriptive, and this will be used. To expand on this idea it is necessary to review the ideal PID controller:

$$\Delta P_o/e = K([1/(T_r s)] + 1 + T_d s) \qquad (9\text{-}47)$$

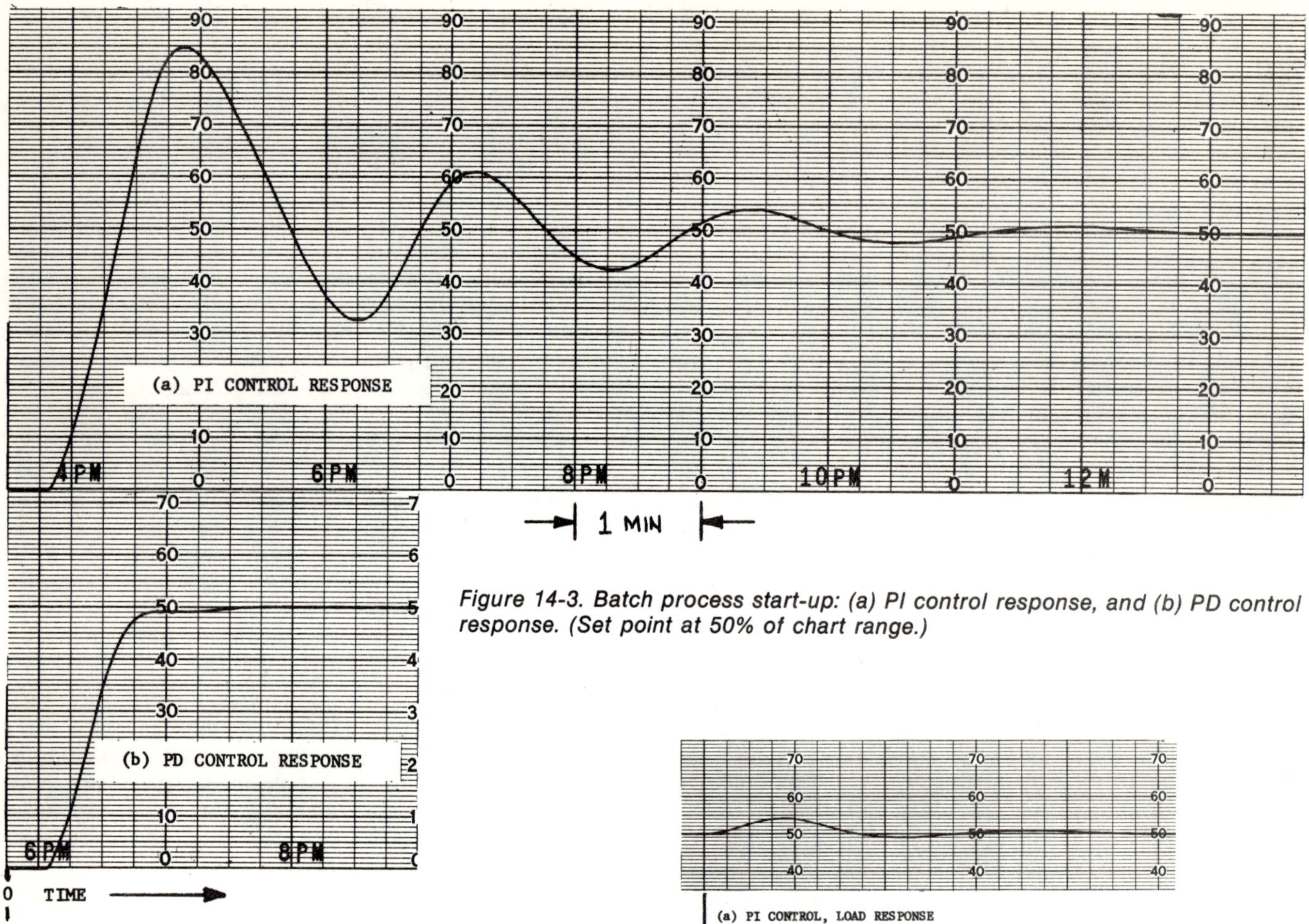

Figure 14-3. Batch process start-up: (a) PI control response, and (b) PD control response. (Set point at 50% of chart range.)

Equation (9-47) is ideal in that it does not show the high- and low-frequency cutoffs of a real controller, and it does not show the interaction problem of "parallel feedback" controllers noted in Section 9.9. If high and low cutoffs are omitted, a real "parallel feedback" controller could be defined as shown in Equation (14-5). Note that prime signs are used in Equation (14-5) to denote dial setting values for a real "parallel feedback" controller. This concept will be used when comparing effective settings of real and ideal controllers.

$$\Delta P_o/e = K'\left\{\left[1 + (T_d'/T_r')\right]/\left[1 - (T_d'/T_r')\right]\right\} \left\{1/\left[T_r'(1 + (T_d'/T_r'))s\right] + 1 + T_d's/(1 + (T_d'/T_r'))\right\} \quad (14\text{-}5)$$

The denominator of the gain term of Equation (14-5) clearly shows why the ratio of T_d'/T_r' is limited to one quarter (see Equation (9-50). If $T_d'/T_r' = 1$, the denominator is zero. This gives infinite gain or on-off control. A similar description for a series controller is shown in Equation (14-6). As in Equation (14-5) where prime signs were used to distinguish dial settings, degree signs will be used in Equation (14-6):

$$\Delta P_o/e = K^\circ(1 + (T_d^\circ/T_r^\circ))\left\{1/\left[T_r^\circ(1 + (T_d^\circ/T_r^\circ))s\right] + 1 + T_d^\circ s/(1 + (T_d^\circ/T_r^\circ))\right\} \quad (14\text{-}6)$$

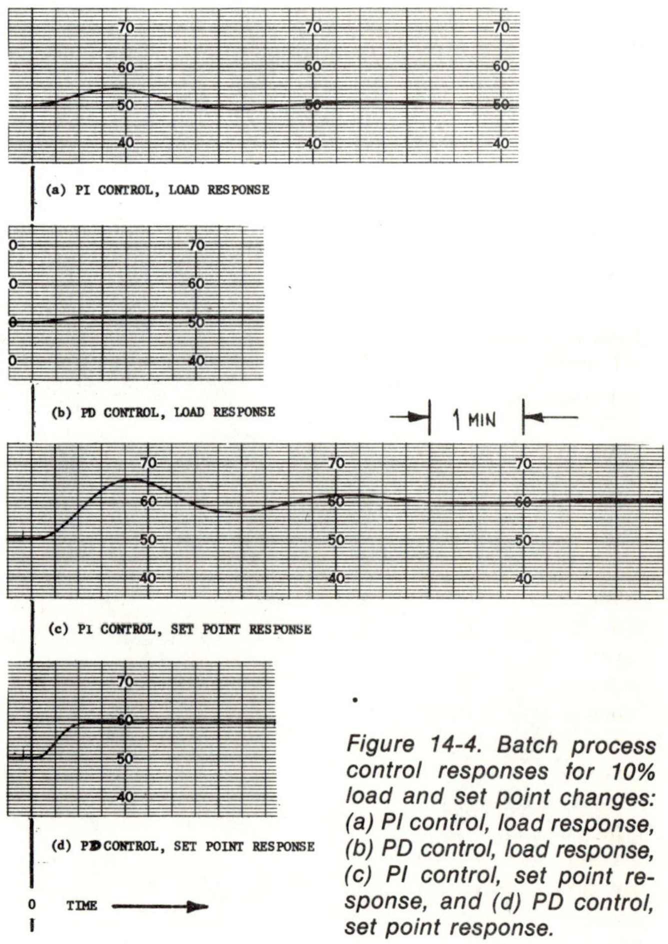

Figure 14-4. Batch process control responses for 10% load and set point changes: (a) PI control, load response, (b) PD control, load response, (c) PI control, set point response, and (d) PD control, set point response.

Note for later use that the controller does not have any limit on T_d°/T_r°. It can become a high-gain or on-off controller only when K° is very large, or when $T_d^\circ >> T_r^\circ$. In practice neither of these two conditions is apt to occur without being evident to anyone with even minimal control experience.

To see why a series controller will help on start-up, it is first necessary to see how one is constructed. A simple flow plan for a series controller is shown in Figure 14-5a. Actually there are three controllers, namely, PD, P, and PI. If proportional gain product of the PD and PI stages is made equal to 1, then each stage has only one adjustment. To explain how the controller works, all three proportional gain terms will be set at 1. Assume that a batch process has been shut down, giving stage 3 of the series controller time to "windup." When the start button is pushed, the initial response as shown in Figures 14-2b and 14-3 is almost a ramp function. To help describe the series controller action, a pure ramp will be assumed as shown in Figure 14-5b. The error signal e, however, is only seen by stage 1 or the PD section of the series controller. Output of stage 1 will lead the measured variable b by the amount of derivative or rate time T_d°. A typical curve is shown in Figure 14-5b. If the reader has any question about this procedure, Section 9.6 should be reviewed.

Through internal controller circuitry, stage 2 (the P section) sees an input signal of e'. Since this stage has no particular importance to the present explanation, it was assumed, as noted earlier in the text, to have a gain of 1. Thus, $e' = e''$, which is the input to stage 3 (the reset section). At some time t_x, as shown in Figure 14-5b, e'' is driven to zero. The effect is to make the reset stage believe that the measured variable is at the set point. When this occurs, reset unwinds, with an associated change in P_o. By this technique the measured variable b can be driven to the set point without any appreciable overshoot. Some cycling will occur as indicated on Figure 14-5b. This results from "fooling Mother Nature," and as everyone knows, there is a price to pay.

The process of Figure 14-3 was started under identical conditions using a series-type PID controller. Start-up, load, and set point results are shown in Figure 14-6. When compared to the PD start-up of Figure 14-3, the results are not impressive. It is important at this time to remember the original objective, namely, to handle both start-up, and load and set point changes. Thus, to make a fair comparison, load and set point results must be compared to Figure 14-4. When this comparison is made, it becomes apparent that the series PID controller has done its job. The reader is cautioned in regard to comparing PD versus series-type PID using only Figures 14-4 and 14-6. While PD control is a very desirable choice, the user should check potential offsets closely. The process used for this analysis does not create large offsets, as can be seen from Figure 14-4. Another process could yield large offsets to load and set point changes when using PD control. The series PID controller would still work. The reader should check his system according to basic concepts of Chapters 7 and 8 before selecting the controller.

Although Figure 14-5 helps to explain the concept of a series-type PID controller and why start-up without overshoot can be accomplished, questions still remain. Two common questions are: how to adjust controller settings, and how stabiity can be maintained with the large derivative or rate time required to prevent overshoot. To answer these questions requires a com-

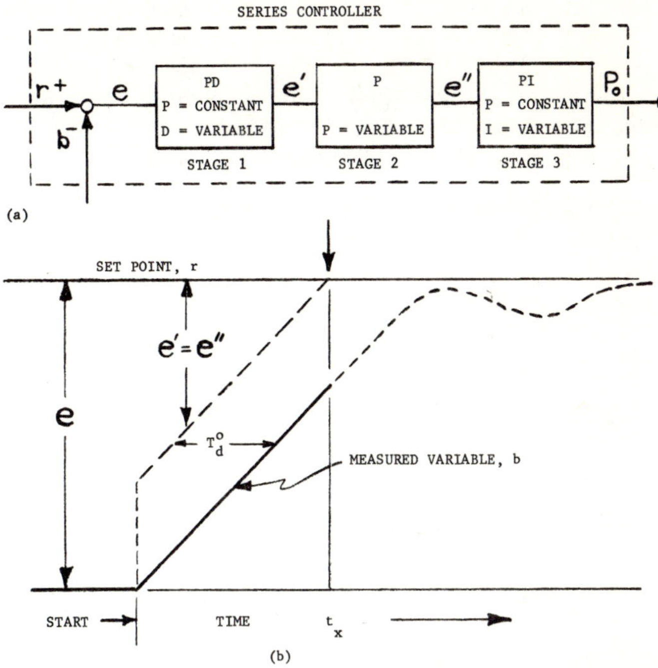

Figure 14-5. Series-type controller: (a) flow plan, and (b) start-up response.

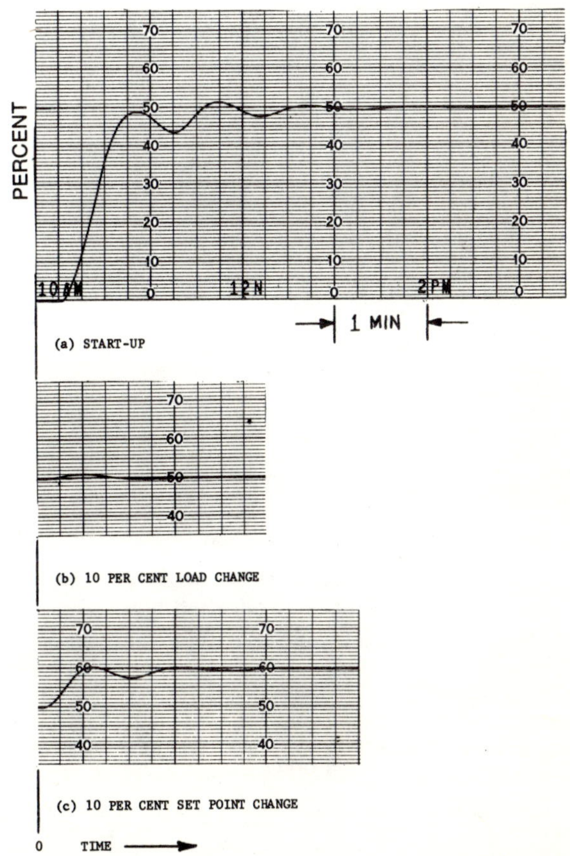

Figure 14-6. Batch process with series-type PID control: (a) start-up, (b) 10% load change, and (c) 10% set point change.

parison of controller settings for an ideal, parallel, and series controller. Table 14-2 shows equivalent terms from Equations (9-47), (14-5), and (14-6).

The reader must now be fully aware of the difference between dial settings and effective values. For example, K', T_r', and T_d' are dial settings. The effective value of K for a parallel controller is the combined result shown in Table 14-2. Assume, for example, that Ziegler-Nichols settings have been calculated for a given process with the following values:

$$K' = 3,\ T_r' = 8 \text{ min, and } T_d' = 2 \text{ min}$$

thus,

$$T_d'/T_r' = 1/4 \text{ (according to Equation (9-51))}$$

The equivalent ideal or effective settings may now be calculated using values from Table 14-2:

$$K = K'\{[1 + (T_d'/T_r')]/[1 - (T_d'/T_r')]\} = 3\{[1 + (1/4)]/[1 - (1/4)]\} = 5 \quad (14\text{-}7)$$

$$T_r = T_r'\{1 + (T_d'/T_r')\} = 8\{1 + (1/4)\} = 10 \text{ min} \quad (14\text{-}8)$$

$$T_d = T_d'/\{1 + (T_d'/T_r')\} = 2/\{1 + (1/4)\} = 8/5 \text{ min} \quad (14\text{-}9)$$

Now use the series controller equations from Table 14-2 with the following dial settings:

$$K^\circ = 1,\ T_r^\circ = 2 \text{ min, and } T_d^\circ = 8 \text{ min, or } T_d^\circ/T_r^\circ = 4$$

Remember that this ratio is acceptable in the series controller.

Equivalent ideal or effective values are

$$K = K^\circ\{1 + (T_d^\circ/T_r^\circ)\} = 1\{1 + 4\} = 5 \quad (14\text{-}10)$$

$$T_r = T_r^\circ\{1 + (T_d^\circ/T_r^\circ)\} = 2\{1 + 4\} = 10 \text{ min} \quad (14\text{-}11)$$

$$T_d = T_d^\circ/[1 + (T_d^\circ/T_r^\circ)] = 8/[1 + 4] = 8/5 \text{ min} \quad (14\text{-}12)$$

Note that effective settings for both the parallel and series controllers are identical. This, however, is true only for small set point and load changes. For start-up, T_d° is four times greater than T_d' and provides the increased derivative time required to prevent overshoot. This means that the series-type controller can provide control equal to the parallel controller on set point and load changes, and it will prevent overshoot which the parallel controller cannot do.

Table 14-2
Equivalent Controller Settings for Ideal, Parallel, and Series-type Controllers

	Ideal	Parallel	Series
Proportional	K	$K'\left\{\dfrac{1 + (T_d'/T_r')}{1 - (T_d'/T_r')}\right\}$	$K^\circ\{1 + (T_d^\circ/T_r^\circ)\}$
Reset	T_r	$T_r'\{1 + (T_d'/T_r')\}$	$T_r^\circ\{1 + (T_d^\circ/T_r^\circ)\}$
Rate	T_d	$\dfrac{T_d'}{1 + (T_d'/T_r')}$	$\dfrac{T_d^\circ}{1 + (T_d^\circ/T_r^\circ)}$

To adjust settings on a specific series-type controller, the reader should follow the manufacturer's recommended procedure. With the controller used for examples of this chapter, a T_d°/T_r° ratio of 4 will usually result in control responses that are much too close to cycling to be acceptable. This is true for many series-type controllers. Unless other information is available, the following settings are recommended as a starting point:

$$G_O^\circ = 0.3G_U \quad (14\text{-}13)$$

$$T_r^\circ = T_d^\circ = P_U/4 \quad (14\text{-}14)$$

This gives a T_d°/T_r° ratio equal to 1. This still gives an increase in T_d° for start-up and helps to provide a more stable result for all operating conditions. As with all methods, some trimming is usually necessary. The process used for this chapter had an ultimate proportional band of 10% and an ultimate period of 1.0 min. Using Equations (14-13) and (14-14) gives initial values of

$$K^\circ = 3 \text{ (PB} = 33\%)$$

$$T_r^\circ = T_d^\circ = 1/4 = 0.25 \text{ min}$$

To provide the responses shown in Figure 14-6, these values were trimmed to

$$K^\circ = 4 \text{ (PB} = 25\%)$$

$$T_r^\circ = 0.25 \text{ min}$$

$$T_d^\circ = 0.35 \text{ min}$$

The increased derivative was needed to prevent overshoot. Lowering the proportional band setting or increasing proportional gain is normally a step in the right direction and was possible for this process.

Using Equations (9-47), (9-48), and (9-49), parallel controller settings were calculated to be

$$K' = 6 \text{ (PB} = 17\%)$$

$$T_r' = 1/2 = 0.5 \text{ min}$$

$$T_d' = 1/8 = 0.13 \text{ min}$$

Start-up, load, and set point responses similar to Figure 14-6 are shown for a parallel controller using these settings in Figure 14-7. Although some trimming was attempted, these values proved to be as good as any. A comparison of effective values for both controllers using Table 14-2 gives results shown in Table 14-3.

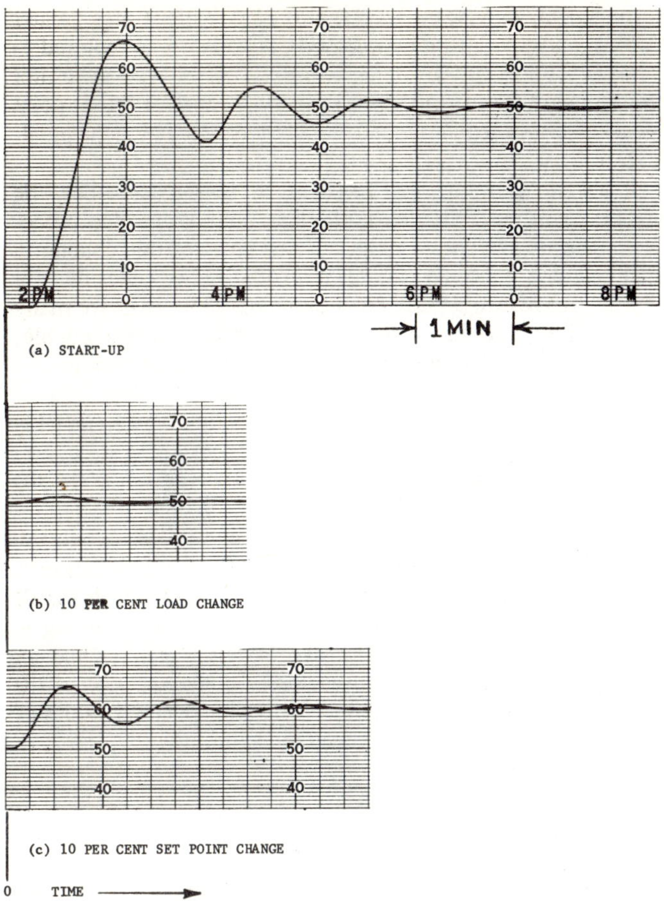

Figure 14-7. Batch process with parallel-type PID control: (a) start-up, (b) 10% load change, and (c) 10% set point change.

Table 14-3 again shows the equivalency of effective values for normal operation. The load changes of Figures 14-6 and 14-7 clearly confirm this fact. While small set point changes will probably have similar responses, a change of 10% already points out the series controller's advantage. Start-up and set point results of Figures 14-6 and 14-7 clearly show how a series-type controller can eliminate overshoot. Another application of the series controller will be noted in the next section.

14.2. Batch Process, Procedure Control

Often in a batch system the procedure phase may have a specific program involving both on/off instructions and changes in controlled variables.[5] A very common variable for batch systems is temperature. Control of a specific temperature profile will be examined in this section, as a batch system advances through the procedure phase of its cycle. Since start/stop and on/off operations are more directly associated with sequential control, they will be examined in the next section.

As an example of procedure control, assume a chemical reactor charged with products at ambient temperature. The procedure for temperature control might be as shown in Figure 14-8.

Table 14-3
Calculated Settings for Parallel and Series Controllers

Action	Parallel	Series
K	10 (10% PB)	9.6 (10.5% PB)
T_r	0.63 min	0.6 min
T_d	0.1 min	0.15 min

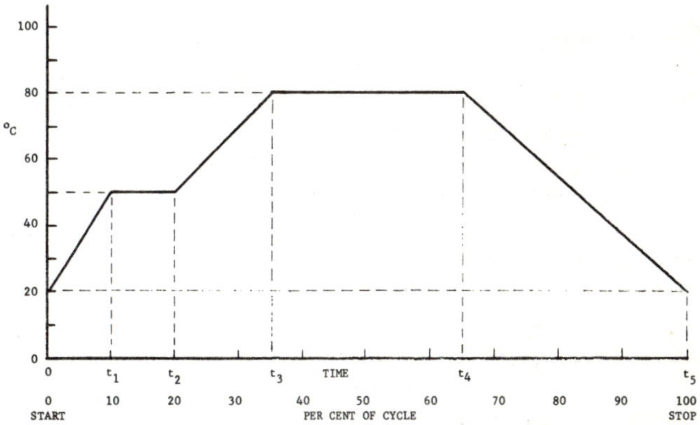

Figure 14-8. A typical batch process temperature control profile.

When the cycle is started, the temperature set point is increased from 20 to 50°C over t_1 min. A holding period exists from t_1 to t_2, at which time the set point is increased to 80°C at t_3. Product is again held at constant temperature until t_4. From t_4 until the end of the cycle, product is cooled. Figure 14-8 establishes the procedure; it does not explain how this can be done or what type of control response will be obtained.

There are several methods that may be used to obtain the temperature profile of Figure 14-8. Early control techniques involved a cam with a mechanical follower to adjust the set point.[6] There are many of these devices still being used. A primary limitation of this type of device is with procedural changes involving either the controlled variable or time. When either variable is changed, a new cam must be cut. On large case circular chart instruments, time can be an additional problem in that one or more cycles must be completed in one chart revolution. For example, if t_5 of Figure 14-8 was equal to 8 hr, the problem could be solved in either of two ways. A chart clock having one revolution in 8 hr could be specified, or a conventional 24-hr clock (1 rev/day) could be used. If the 24-hr clock is used, the procedure could be cut three times on one cam. This works fine with even multiples of time such as 4, 6, or 8 hr. However, when the procedure calls for 7 hr and 21 min, the problem is not so simple.

For odd time increments there are also two solutions. The first is to use an interrupt during any hold period. In the example of Figure 14-8 the drive motor turning the cam could be stopped during either or both hold periods for a total of 1 hr and 21 min. The cycle is thus reduced to 6 hr, and a standard chart clock could be used. This would involve using a start/stop sequence and one or more timing devices. A second solution is to use a homing device. In this solution a chart drive is selected to a scale that exceeds maximum process time requirements. When the cycle is completed, a fast-drive homing motor is activated to return the cam to the start position. When the start position is reached, the homing motor stops, and the cycle may be repeated. This solution also involves start/stop operations and timing devices. In some cases a manual reset is used in which the operator positions the cam at the start of each cycle.

While procedural changes tend to emphasize limitations of this type of device, they also tend to point out advantages. For a simple, well-established repetitive process, for which no changes can be anticipated, a cam device can provide a reliable control mechanism at a very reasonable cost. In more recent equipment designs, mechanical followers have been replaced by electronic and pneumatic followers.[7]

When complexity of procedure, numbers of units, and procedural changes increase, application of individual control devices can become cumbersome and unwieldy. In those circumstances some type of modern computer circuitry usually can provide a solution. The equipment used in these applications includes process control computers, special purpose mini- and microcomputers, and programmable logic controllers. At any point in time, variety and capacity of available computer hardware present a problem for even an experienced instrument engineer. For this reason no particular piece of equipment will be analyzed, and the reader is referred to company literature for details. What will be emphasized are process requirements. The basic functions are start/stop, control, time, and procedure programming, as noted in Figure 14-8. As previously noted, start/stop operations will be considered in the next section on sequential control.

Controlling, timing, and procedure programming can be done within a computer system or on an external basis. When these functions are performed on an external basis, required auxiliary equipment is normally sequenced by the main controller unit. The choice of how a particular operation is handled will depend on capability of the main controller unit, user preferences, availability of existing hardware, and safety considerations. If a process computer were applied to the system of Figure 14-8, for example, all three functions (controlling, timing, and set point programming) could be carried out by either DDC or supervisory control. The opposite extreme would be a simple programmable controller limited to on/off logic. In this type of system, individual controllers, timers, and set point function generators would be switched on and off as required. Figure 14-9 shows the jacketed kettle of Figure 13-6 with a programmable controller added. This is a typical piping and instrument drawing (P&ID). The interlocking logic, which is generally presented as a ladder diagram, is normally shown on an electrical drawing and will be presented in detail in the next section. A verbal description will be given at this point.

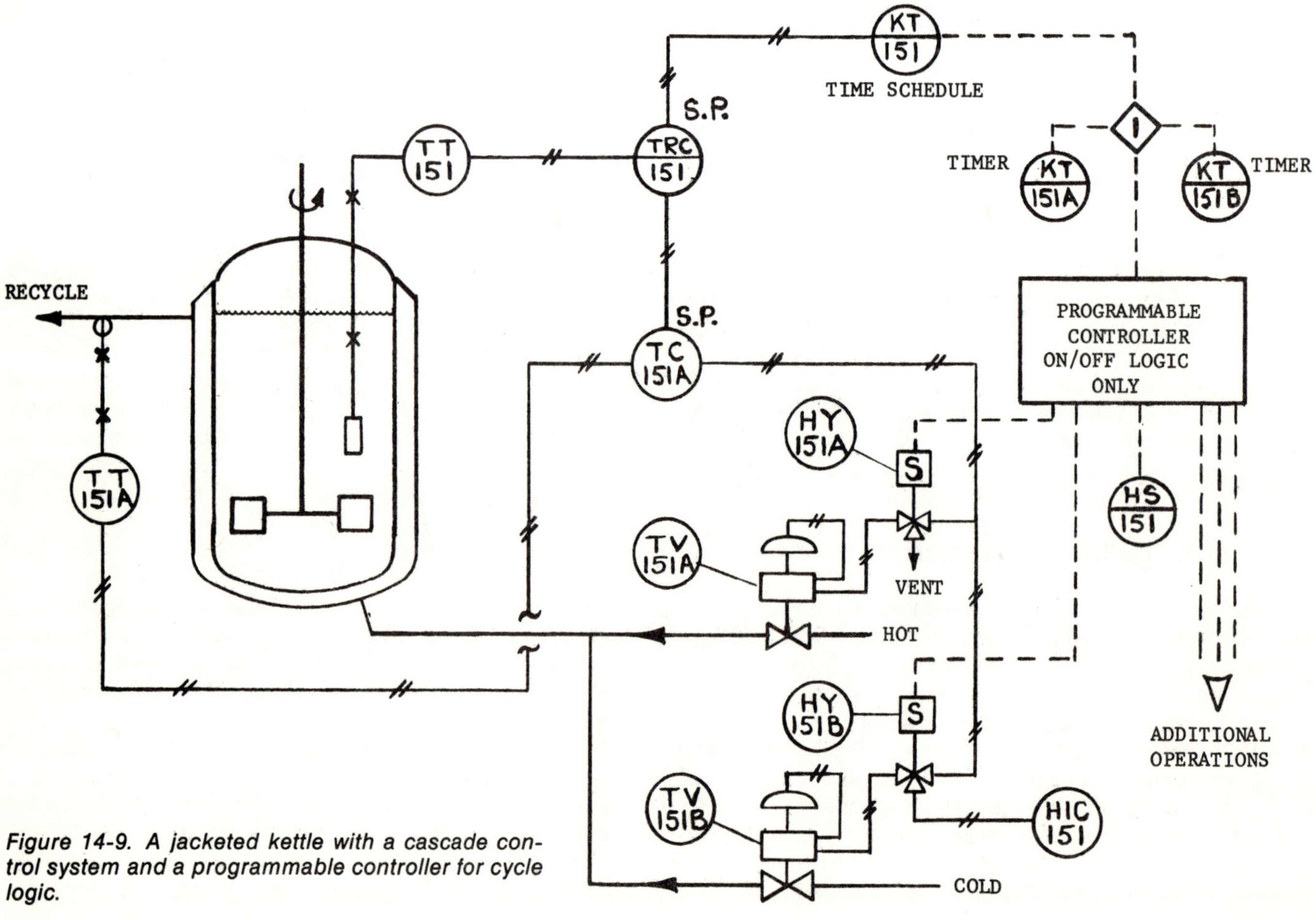

Figure 14-9. A jacketed kettle with a cascade control system and a programmable controller for cycle logic.

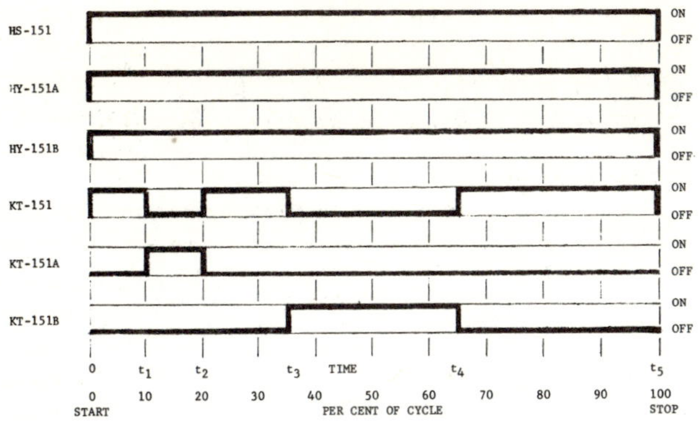

Figure 14-10. Time-cycle diagram for the jacketed kettle of Figure 14-9.

The reader should refer to Figures 14-8, 14-9, and the time-cycle functional diagram shown in Figure 14-10 as procedure is detailed.

STEP 1—Push start button HS-151

Action: Placing HS-151 in the closed or on position starts three additional operations. In the shut down or load condition, control valves TV-151A and TV-151B should be closed. When the cycle is started these valves should be switched such that they are regulated by TC-151A. This switching is performed by turning solenoid valves HY-151A and HY-151B to the on position. Action of the solenoid valves is a function of control valve action. As noted in Chapter 13 and shown in Figure 14-9, split range valves with positioners are used. For this example assume that TV-151B is open at 3 psig and closed at 9.2 psig, and that TV-151A starts to open at 8.8 psig and is fully open at 15 psig. When HY-151A is in the off position, the diaphragm on TV-151A is vented and the valve is closed. When HY-151B is in the off position, a manual loading station HIC-151 is connected to the actuator on TV-151B. If the output of HIC-151 is greater than 9.2 psig, then TV-151B is closed. A less expensive alternate may be achieved by replacing HIC-151 by a direct connection to a 20-psig air supply. Solenoid valves HY-151A and HY-151B go to the on position when HS-151 is turned on, as shown in Figure 14-10.

In addition to the solenoid valves, a time-schedule controller KT-151 is turned on by HS-151. As shown in Figure 14-10, KT-151 is some type of function generator that will produce a pneumatic output for the set point signal to the primary controller TRC-151. This may be any one of the previously described or referenced devices using mechanical, pneumatic, or electrical function generating techniques. Output of KT-151 will drive the set point of TRC-151 from 20 to 50°C during the time increment 0 to t_1, or 10% of cycle time as shown in Figure 14-8. When KT-151 reaches 50°C, it turns off, activating STEP 2.

STEP 2—Start timer KT-151A

Action: When the time-schedule controller KT-151 reaches a 50°C set point, it is turned off and held at this position. Turning off KT-151 may be accomplished by a mechanical, pneumatic, or electric limit switch. In a very simple system such as shown here, turning off KT-151 could be used as a signal to start timer KT-151A. As shown in Figure 14-9, however, this action is relayed back to the PC, which then turns on timer KT-151A. The only justification is that this information is used by the PC to initiate additional operations not shown in Figure 14-9. In truth, this is still a weak justification. The real reason is the need to demonstrate a PC on a simple system. At time t_2 or a cycle time of 20%, time-schedule controller KT-151 is restarted as KT-151A times out.

STEP 3—Restart time-schedule controller KT-151

Action: Time-schedule controller KT-151 is restarted and drives TRC-151 set point from 50 to 80°C. This change uses 15% of cycle time and requires a different rate of change than previously used. The mechanism for rate change within KT-151 will depend on the device selected. When the 80°C set point is reached, KT-151 stops, and timer KT-151B is started.

STEP 4—Start timer KT-151B

Action: Procedure is similar to STEP 2. When KT-151B times out at t_4 or 65% of the cycle, KT-151 is restarted.

STEP 5—Restart time-schedule controller KT-151

Action: Set point for TRC-151 is driven from 80 to 20°C to provide controlled cooling of product. When 20°C set point is reached, KT-151 stops. This final action also drives HS-151, HY-151A, and HY-151B to the off position. Where required, reset devices should also be activated to return all components to the cycle start conditions. When this action is completed the system is ready for start of the next cycle.

Having followed through one example, it should be evident to the reader the wide variety of processes that require this type of control. Choice of equipment can vary from relays, timers, and time-schedule devices to full-scale computer installations using DDC control. The method chosen for any given application will depend on system complexity, needs for program changes, operator and maintenance capability, management directives, and last but not least, economic considerations. Whatever method is chosen, the start-up problem discussed in the first section of this chapter should still be considered. Although this problem may be eliminated or reduced by programming the set point, it is not automatic and should be checked for each installation. Product of error magnitude and shut down time will still be the major factor in determining reset windup. In many applications a series controller or reset windup lookout device may still be required to avoid start-up problems. A wiring diagram for this system will be shown in the next section.

14.3. Batch Process, Sequential Control

In all batch control systems there are a number of start/stop or on/off operations. These operations follow a given sequence and may be related by time or specific conditions of selected variables. While the last section was primarily related to procedural or set point programming, it was obvious that many on/off operations were involved. It was also assumed that the jacketed kettle was charged, without asking how. In this section an attempt will be made to see how a complete system could be automated and controlled. To accomplish this it will be necessary to introduce some basic logic concepts and then apply these concepts to a complete system. The author's objective in this section is to introduce sequential control and provide basic tools for further study.

On/off operations require switches. These switches may take a variety of forms such as disconnect, push button, toggle, cam-operated limit switches, or switches activated by process variables such as temperature, pressure, flow, or level. A switch is a binary device having an open and closed position. In the application of Boolean algebra to switching logic, the closed position is represented by a 1 and the open position by a 0.[8,9] Some typical switch symbols are shown in Figure 14-11. Note that N.O. and N.C. refer to "normally open" and "normally closed." The term "normally" refers to the unactuated position.

The control relay is another very common device used in sequential control circuits. A typical relay circuit in both pictorial and schematic form is shown in Figure 14-12. When the coil is energized, the core moves downward, opening the N.C. contacts and closing the N.O. contacts. When the coil is de-energized, a spring returns core and contacts to the normal position. Here again is a binary device having either an open or closed position, which can be represented by a 0 or 1, respectively. The control relay is not limited to one set of contacts as shown in Figure 14-12. Relays may be obtained with multiple contacts and with a variety of special features. For example, the relay of Figure 14-12 is a break-before-make type of circuit. It is also possible to get make-before-break and time-delay relays. In the make-before-break circuit the N.O. contacts will close before the N.C. contact opens. In time-delay relays, contacts may take time to activate after the coil is energized or time out to the normal position after the coil is de-energized. Symbols for a time-delay relay are also shown in Figure 14-12. These and other special features allow the relay to perform a multitude of operations within a sequential control circuit.

Another sequential control device is a timer. A common timer circuit is shown in Figure 14-13, along with operational details.[11] In a timer such as this, a specific time is set on a dial indicator. When the clutch coil is energized, the synchronous motor starts timing and, as noted in Figure 14-13, activates certain instantaneous contacts. When the motor times out, delayed contacts are activated. The timer and contacts are reset to the original position when the clutch coil is de-energized. A variety of time ranges and special features may be obtained. These include pulsed, percentage, and pneumatic timers, as well as those with digital setting to increase accuracy. Timer contacts are again binary devices, for which a 0 is an open or off position and 1 is closed or on.

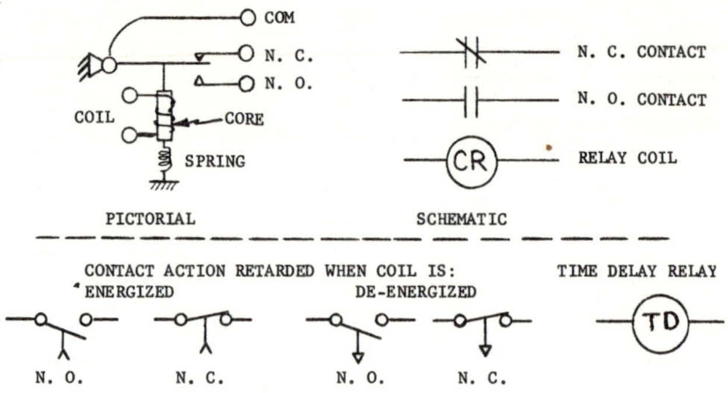

Figure 14-12. Control and time-delay relay symbols.

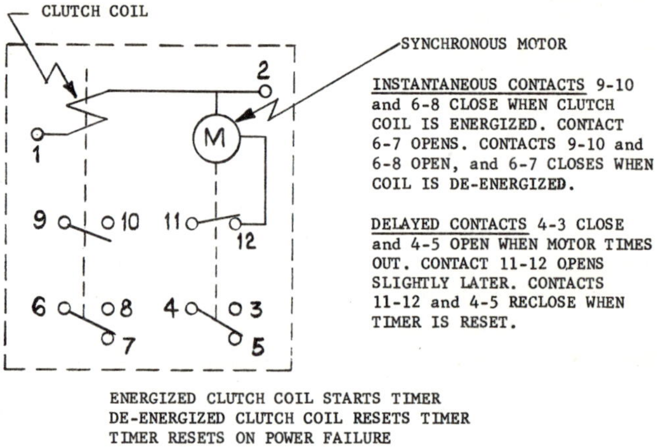

Figure 14-13. A common industrial timer circuit. Courtesy of Eagle Signal Division, Gulf & Western Manufacturing Co., Davenport, Iowa.

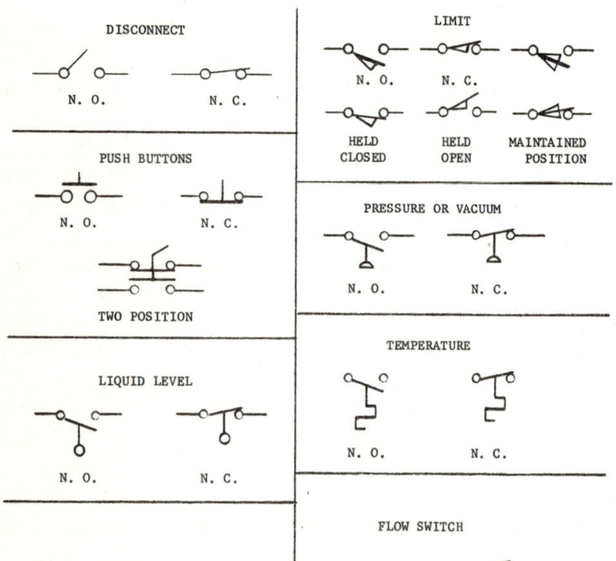

Figure 14-11. Examples of common switch symbols.

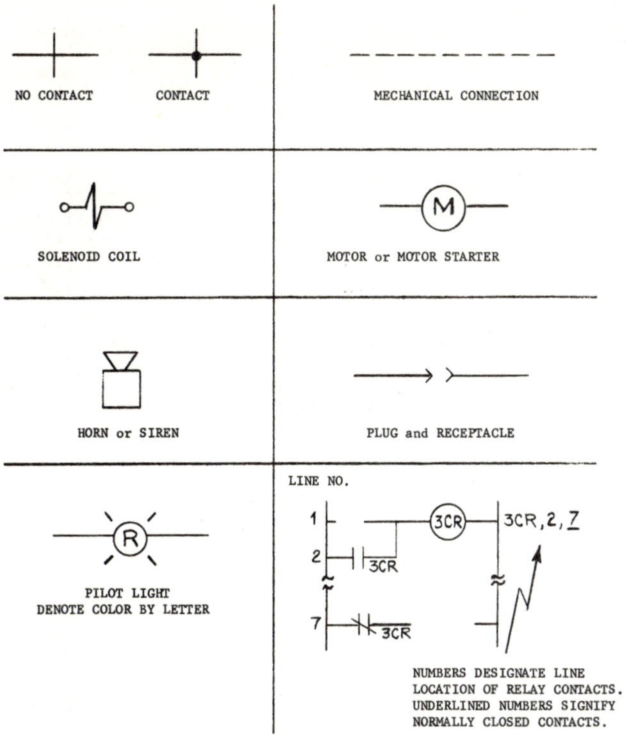

Figure 14-14. Electrical symbols for components that are common to process control applications.

There are many other symbols for devices such as solenoid coils, pilot lights, motor starters, transformers, alarm horns, and other electrical, hydraulic, and pneumatic equipment. Some of the more common process components are shown in Figure 14-14. For a more detailed list the reader should obtain both JIC (Joint Industry Conference) and NEMA (National Electrical Manufacturers Association) standards.[10,12,13]

With this brief introduction to sequential components and symbols it is now possible to design a complete control cycle for the jacketed kettle of Figure 14-9. In addition to the procedure control of the previous section, fill and drain operations will also be included. the complete P & ID (piping and instrument drawing) is shown in Figure 14-15. The logic will be as follows:

—Fill with x gallons of product A
—Fill to high level with product B
—Start agitator motor
—Start procedural control scheme
—At end of procedural cycle stop agitator
—Drain kettle to low level
—Reposition all components for start of next cycle

In addition to the above, it is necessary to include safety and interlocking check circuits. For example, filling operations should not start unless the kettle is empty and the discharge valve is closed. Conditions such as these will now be integrated with hardware to meet process requirements.

Before starting any system, an emergency stop circuit is usually a basic consideration. Reasons for inclusion will range

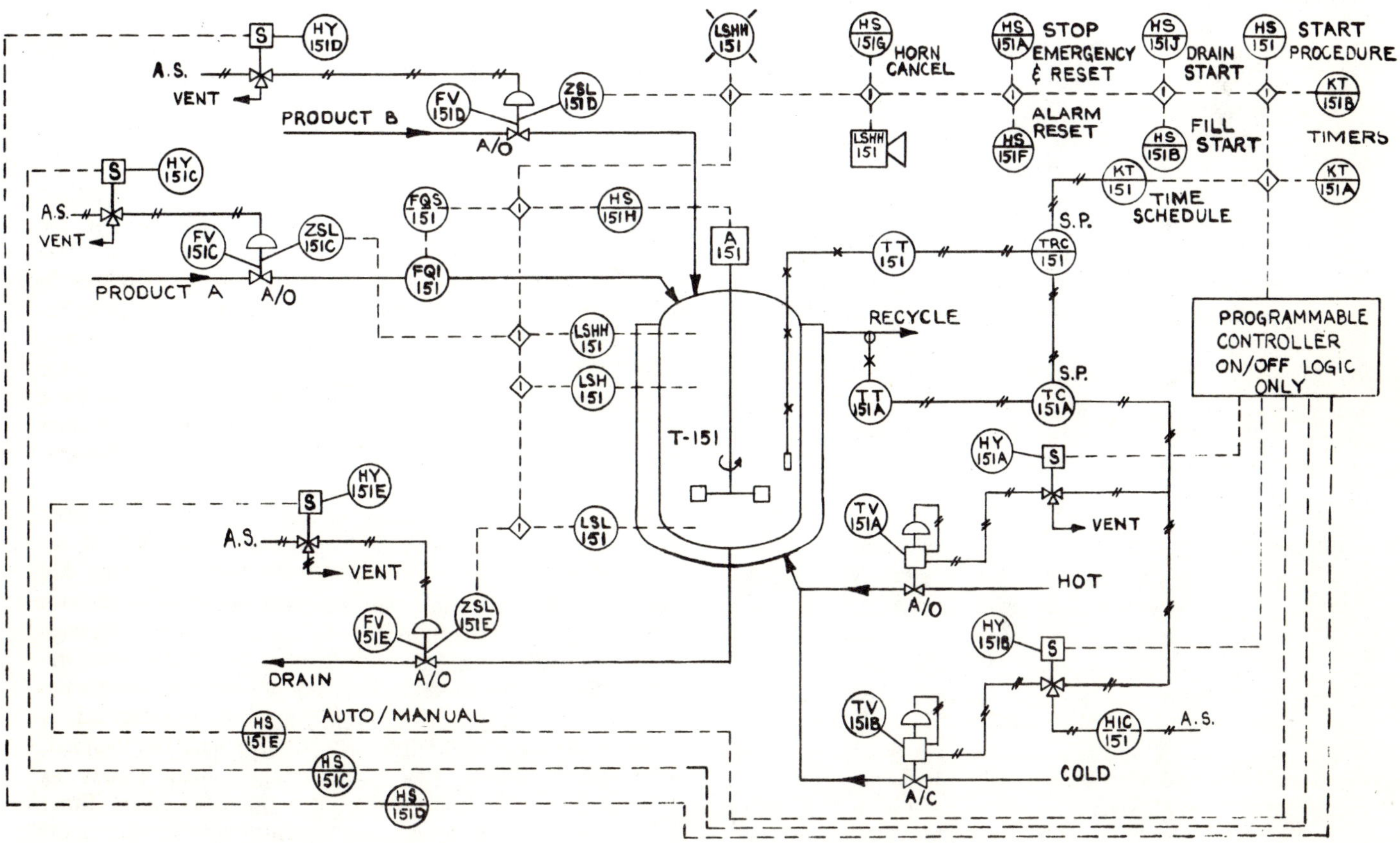

Figure 14-15. A jacketed kettle control scheme: fill, procedure, and drain sequence.

from safety to process considerations. Many standard circuits are available and are referenced in literature and manufacturer's catalogs.[14] One such circuit is shown on lines 1 and 2 of Figure 14-16. The format of Figure 14-16 is called a ladder diagram, represented by the "hot" and ground risers L1 and L2, respectively. Ladder rungs are assigned line numbers so that interlocking circuits may be checked. To show how this works, the emergency stop circuit will be described. Pushing the emergency stop button HS-151A will de-energize relay 1CR. As shown to the right of 1CR in Figure 14-16, relay 1CR has two sets of contacts. Both of these contacts are located on line (or rung) 2. When the relay is de-energized, these contacts go to their normally open position as shown in Figure 14-16. Two functions are performed:

1. When contacts in L1 open, all power is removed from the system;
2. When contacts around the reset switch HS-151A open, they cause relay 1CR to remain de-energized even though the emergency stop switch is returned to the closed position.

Thus, to restore power to the system it is necessary for an operator or supervisor to close the reset switch HS-151A. When this is done, 1CR is energized and the contacts are closed. Power is restored to the system, and 1CR is held in the energized state by its own contacts. With power available and provisions available for an emergency stop, the system is ready for filling.

The first step in the cycle is to fill the kettle with x gallons of product A. As previously noted, however, several conditions should be checked before this action is undertaken. These conditions in itemized form are

1. Is the tank empty? LSL-151 closed.
2. Is drain valve closed? ZSL-151E closed.
3. Is flow total switch reset to zero? FQS-151 closed.
4. Is product B valve closed? ZSL-151D closed.

The first three conditions are mandatory. If level switch LSL-151 is closed, the tank is empty and ready for filling. Before filling, drain valve FV-151E must be closed as shown by position switch ZSL-151E on the valve stem. To measure the proper fill of product A, the flow integrator or totalizer must be reset to zero or closed. Condition 4 is optional and would depend on relative feed rates of products A and B. For example, if B feeds much faster than A, the high-level trip could be reached before all of A was in the tank. By making B wait until A is complete, this possibility is removed. All these conditions are shown on line 3 of Figure 14-16. When all switches are closed and start button HS-151B is closed, relay 2CR will be energized to start the fill cycle.

As shown in Figure 14-16, relay 2CR has contacts on lines 4 and 7. Line 7 shows that when auto-to-manual switch (A/M) HS-151C is in the automatic position, contacts on 2CR are energized or closed, and contacts on 5CR are not energized, solenoid HY-151C will be energized. When HY-151C is energized, FV-151C is opened and flow of product A starts. This can be seen in Figure 14-15. The additional contacts of 2CR on line 4 provide hold-in action or memory. These contacts parallel the start switch, HS-151B and level switch LSL-151. Without this set of contacts, filling would stop when the operator took his hand off HS-151B, or as soon as LSL-151 opened due to increasing level.

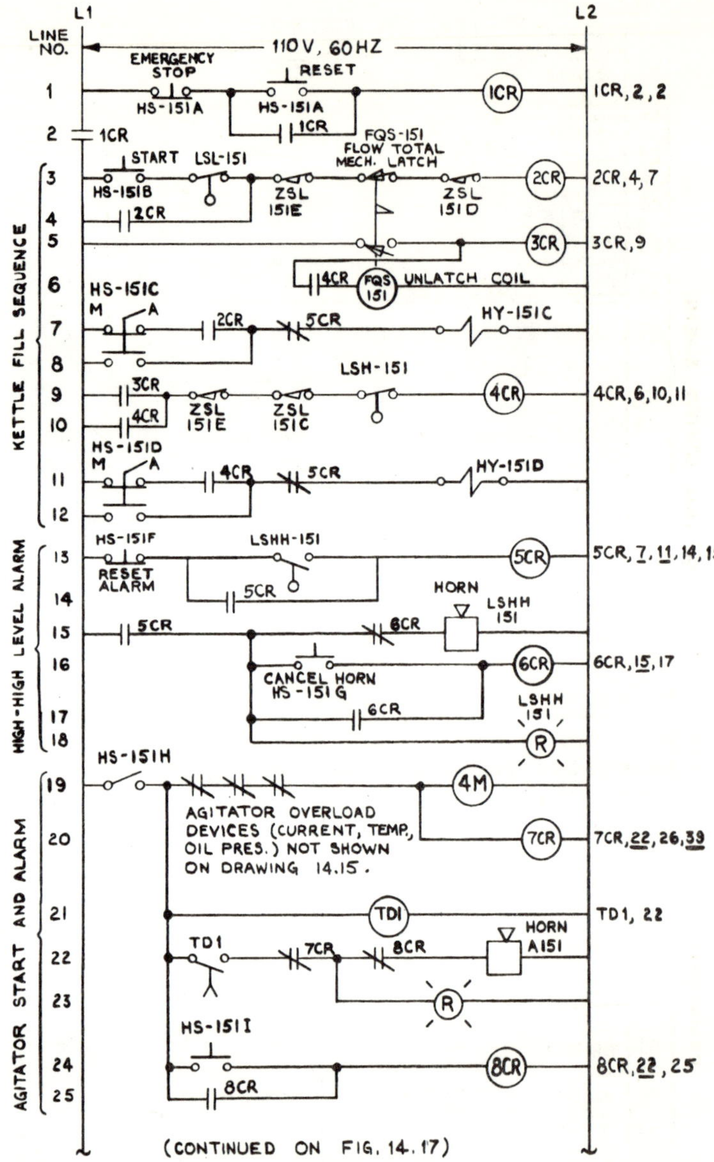

(CONTINUED ON FIG. 14.17)

Figure 14-16. Ladder diagram showing emergency stop, kettle fill cycle, level alarm, and agitator start and alarm circuits for the jacketed kettle of Figure 14-15.

Neither of these operations is a valid reason to stop the filling operation.

The normally closed contacts of 5CR on line 7 provide a fail-safe redundancy. The relay 5CR is on line 13 and is activated by high-high level switch LSHH-151. If for some reason a false signal is obtained from any of the components on line 3 and a high-high level exists in tank T-151, contacts of 5CR on line 7 will open and prevent start of the filling operation. This safety interlock will also prevent an overflow in the event of a flow totalizer (FQS-151) failure after the filling operation has started.

The A/M switch HS-151C can be used to manually load product A. Turning this switch to manual opens contacts on line 7 and closes contacts on line 8. Under this condition, solenoid HY-151C will be energized to open valve FV-151C. This manual

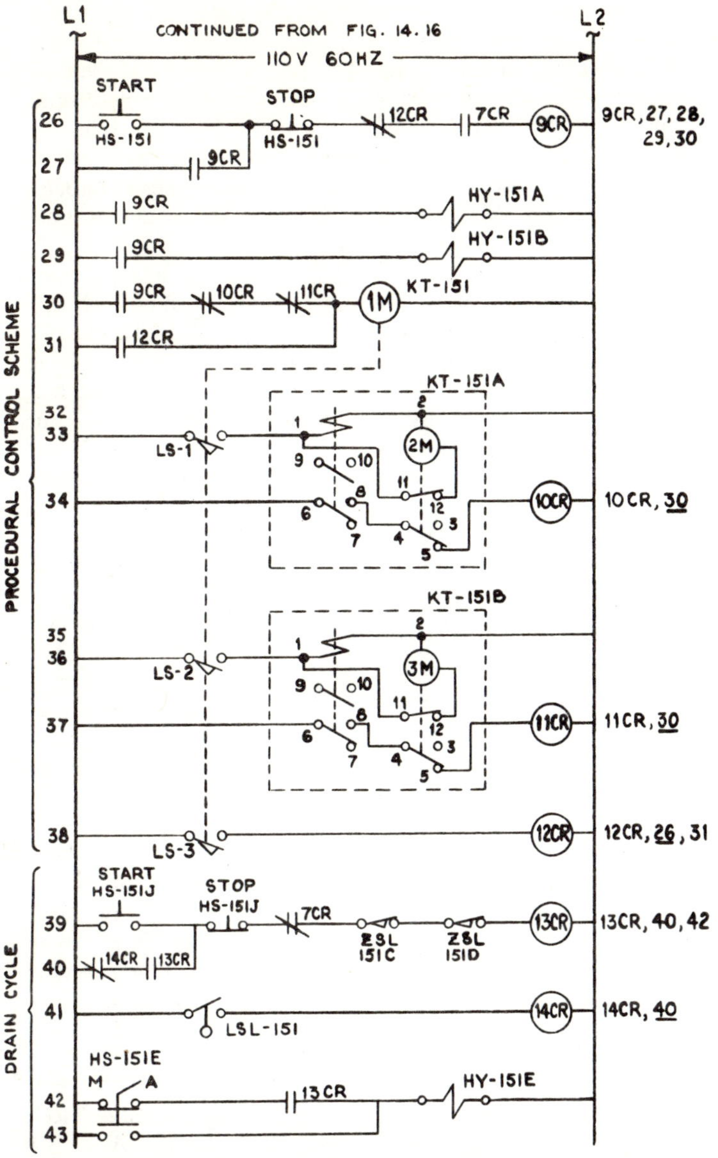

Figure 14-17. Ladder diagram showing procedural control scheme and drain cycle for the jacketed kettle of Figure 14-15. This figure is a continuation of Figure 14-16.

operation allows for overriding all conditions of line 3, but will not override a high-high level condition.

As product A flows into tank T-151, integrated or total flow is measured by FQ1-151. When this value is equal to set value x on FQS-151, the switch opens and de-energizes 2CR. This is a mechanical latch-type switch which will remain in the open position until reset. A second pair of contacts on this switch close and energize relay 3CR on line 5. The objective is to stop flow of product A and start B. The action is as follows. When 2CR is de-energized, contacts on line 7 open to de-energize HY-151C. This vents air from FV-151C and stops flow of product A. Relay 3CR has contacts on line 9 which will energize relay 4CR to start product B flowing.

Before starting product B, however, it is necessary to again check process conditions as follows:

1. Is drain valve closed? ZSL-151E closed.
2. Is product A valve closed? ZSL-151C closed.
3. Is there space in T-151? LSH-151 closed.

Item 1 is a routine check. In item 2, a check is made to be sure flow of product A has in fact stopped. Product B is to fill any vacant tank volume until the high-level switch LSH-151 opens. Item 3 checks to see that space is available and also provides the method for stopping product B flow. If all these conditions are met, 4CR is energized.

Relay 4CR has 3 sets of contacts on lines 6, 10, and 11. Contacts on line 11 will energize solenoid HY-151D if A/M switch HS-151D is in the automatic position and if relay contacts 5CR are closed. This is identical to line 7 and will open valve FV-151D as long as LSHH-151 is not closed. Here again, manual operation may be achieved using HS-151D with the same limitations as previously noted.

The two additional sets of contacts on 4CR are used to reset FQS-151 and maintain flow of product B. When 4CR is energized, contacts on line 6 close to energize the unlatch coil of FQS-151. This resets the flow total switch for the next cycle and at the same time de-energizes 3CR. The additional 4CR contacts on line 10 maintain 4CR after the 3CR contacts on line 9 open. Thus, flow of product is maintained until the high level is reached and LSH-151 opens.

The early reset of FQS-151 and associated need for 4CR hold-in contacts on line 10 appears to be an unnecessary step. There is, however, a good reason for this step, which involves noise of level measurement. Flow of product B into T-151 can cause LSH-151 to go on-and-off around the high-level limit. If 3CR was maintained in the energized state, this would cause current surges through lines 9, 10, and 11. It would also cause solenoid HY-151D and valve FV-151D to pump open and closed. By resetting FQS-151, which must be done at some point, and de-energizing 3CR, this problem is avoided. The first time LSH-151 opens, 4CR is de-energized and stays de-energized due to loss of contacts on line 10.

At this point the kettle is charged and ready for start of the agitator motor. As previously noted, however, there are always some additonal safety and interlocking circuits to be considered. One such circuit is the high-high level circuit. As shown on line 13 (for superstitious designers line 13 may be omitted), relay 5CR will be energized if and when LSHH-151 is closed. As noted earlier, this will open contacts on lines 7 and 11 and de-energize solenoids HY-151C and HY-151D. This closes valves FV-151C and FV-151D and prevents any additional flow of either product to T-151. Additional contacts on line 14 hold in relay 5CR to avoid chatter or noise problems noted earlier. At the same time, contacts on line 15 close to energize an alarm horn and panel light indicating high-high level.

There is a variety of alarm circuits that may be used. The circuit on lines 15-18 functions as follows. The horn gains operator attention and may be cancelled as soon as possible. This step is accomplished by closing HS-151G on line 16 and energizing relay 6CR. By this time the reader should recognize that contacts on line 17 close to hold in relay 6CR and that contacts on line 15 open to de-energize the horn. The light on line 18, however, remains energized to define the problem. When the operator has taken corrective action to reduce the level, the light may be de-energized by opening reset switch HS-151F on line 13. If LSHH-151 is open, the light will go out. In the case where level has not been reduced, the horn will again be energized, and the light will stay on. Thus, opening the push button reset switch HS-151F will either reset the alarm or re-sound the horn to indicate continued existence of high-high level.

The agitator A-151 could be started on an automatic basis following the fill sequence or may be started by an operator. Manual operation has been used in this example and is represented by HS-151H. While details have been omitted in Figure 14-15, lines 19 through 25 have been included to show motor overload safety devices and another form of alarm circuit. When HS-151H is closed by an operator, motor starter 4M will be energized if no overload devices are energized. Typical motor overload devices include line current limits, motor or bearing temperatures, and oil pressure. Since most industrial motors operate on multiphase high-voltage circuits, instrument responsibility is normally limited to energizing the starter as shown on line 19. An additional control relay 7CR is used with contacts in the agitator alarm circuit, procedure circuit, and drain circuit. Procedure and drain circuits will be examined and explained later.

The alarm circuit for the A-151 is shown on lines 20 through 25 and works as follows. When HS-151H is closed, the motor starter on line 19, relay 7CR on line 20, and time-delay relay TD1 on line 21 are all energized. The result is that 7CR contacts on line 22 open and TD1 contacts time closed. This condition will persist until HS-151H is opened or an overload device opens. For example, high bearing temperature could cause an overload device to open. When this occurs, motor starter M4, agitator motor, and relay 7CR are all de-energized. The 7CR contacts on line 22 are now closed, which in turn energize the horn and panel light. As a rule, only one horn will be used though two are shown in Figure 14-16. Panel lights, however, which remain in the alarm state are usually individually assigned. As in the level alarm circuit, the horn may be acknowledged or cancelled by closing HS-151I on line 24. Relay 8CR is held in by contacts on line 25 and de-energizes the horn with contacts on line 22.

The alarm circuit for the agitator shows how a time-delay relay may be used. It also demonstrates some potential problems. The overload contacts on line 19 are shown as normally closed, opening on an overload to stop the motor. Two of the suggested overload variables, namely, line current and oil pressure, are capable of surge conditions. With plain contacts this could result in on/off cycling of the agitator motor. This type of action is undesirable for even small motors and can be responsible for all kinds of problems when large motors are involved. There are two steps that can be taken to overcome these problems. The first is to use a time-delay device on overload circuits. In this type of device an overload condition must exist for a given period of time before action is taken. For example, a line current spike would not trip the overload contact, but a sustained increase would. The second action is to use latch-type overload devices. In this type of device, contacts will be maintained in the desired action, once they are tripped. To restart the motor it is first necessary to remedy the problem causing the overload and then reset the overload device. This action is similar to the flow totalizer FQS-151 on lines 3, 5, and 6. In many applications time-delay and latch functions are combined. A device having these properties is a time-delay fuse. To reset in this case requires a replacement fuse. While the agitator circuit of lines 19 through 25 is of questionable practical value, it does help to demonstrate equipment and potential problems.

With the kettle charged and agitator started, the stage is set for the procedural control scheme described in Section 14.2. On/off operations for this sequence are shown in Figure 14-17, lines 26 through 38. The reader should recognize that Figure 14-17 is really a continuation of Figure 14-16 and that Figures 14-9 and 14-15 show piping and instrument connections. Figures 14-8 and 14-10 showing the temperature profile and time-cycle diagram, respectively will also prove helpful in understanding this program.

The start circuit for this sequence is shown on lines 26 and 27. When the start button HS-151 is closed, relay 9CR will be energized if 12CR is not energized and 7CR is energized. Relay 12CR is on line 38, and its function will be described at that point. Contacts 7CR are associated with relay 7CR on line 20 of Figure 14-16. Relay 7CR is energized when the agitator is running. The implied condition is that heating should not start without product agitation. When relay 9CR is energized, contacts on line 27 close and hold in the start circuit. This is another type of start/stop circuit which may be stopped at anytime by pushing stop button HS-151.

When relay 9CR is energized, it also closes contacts on lines 28, 29, and 30. Contacts on 28 and 29 energize solenoids HY-151A and HY-151B. These in turn activate temperature valves TV-151A and TV-151B. Temperature control loop 151 is now in service. The contacts on line 30 will energize the time-schedule controller KT-151 unless relays 10 and 11 are energized. Relays 10 and 11 are energized by timers KT-151A and KT-151B and provide the hold periods shown in Figure 14-8. These will be assumed at present to be de-energized. If this is true, KT-151 will run and program the set point to the primary controller TRC-151. Action will be as shown in Figure 14-8. A limit switch LS-1 on line 33 will be closed when the 50°C set point is reached. This action will start timer KT-151A and stop time-schedule controller KT-151.

When limit switch LS-1 closes on line 33, the clutch coil of timer KT-151A is energized. Instantaneous contacts 9-10 and 6-8 close as noted in Figure 14-13. In this application contacts 9-10 are not used. By using a jumper connection between timer terminals 4 and 8, contacts 6-8 can be used to energize relay 10CR. When 10CR is energized, its contacts on line 30 open to stop time-schedule controller KT-151. Timer motor 2M is also energized when LS-1 closes by using a jumper connection between timer terminals 1 and 11. When timer KT-151A times out, delayed contacts 4-5 and 11-12 open. When contacts 4-5 open, relay 10CR is de-energized. This causes 10CR contacts on line 30 to close and restart the time-schedule controller motor 1M. When contacts 11-12 open, timer motor 2M stops. As noted in Figure 14-13, contacts 11-12 open slightly later than 4-5. This insures that contacts 4-5 will open before timer motor 2M stops.

The limit switch LS-1 is held closed as noted in Figure 14-11. Thus, as time-schedule controller KT-151 advances through the next step, LS-1 will be released and opens. This de-energizes clutch coil of timer KT-151A, and as noted in Figure 14-13, resets the timer to its initial condition. Timer KT-151A is therefore ready for the next cycle.

The time-schedule controller KT-151 drives the set point of TRC-151 from 50 to 80°C during this portion of the cycle. At the 80°C set point, limit switch LS-2 is driven closed. The identical procedure of delaying time-schedule controller KT-151 is now repeated using timer KT-151B and relay 11CR.

When timer KT-151B times out, the time-schedule controller is restarted to program the cooling phase as shown in Figure 14-8. In the cooling phase the set point of TRC-151 is reduced from 80°C to 20°C. When the 20°C set point is reached, limit switch LS-3 closes to energize relay 12CR. Relay 12CR was noted earlier and has contacts on lines 26 and 31. Its function will now be described.

When 12CR is energized, contacts on line 31 close. Opening line 26 de-energizes relay 9CR and brakes the hold-in circuit on line 27. This action de-energizes solenoid valves HY-151A and HY-151B, closing control valves TV-151A and TV-151B. Although temperature control of the kettle has been stopped, there is still a problem: With 12CR contacts in the open position, this circuit could not be restarted. When relay 12CR was energized, however, contacts on line 31 closed. Thus, even with relay 9CR de-energized, the motor on time-schedule controller KT-151 will continue to run. This action will be quickly interrupted by limit switch LS-3, which must be held in to energize 12CR. As the switching mechanism drives by the closed position on LS-3, it will open and de-energize 12CR. This will stop KT-151 and close 12CR contacts on line 26. The procedural control scheme from lines 26 to 38 is now reset to the original condition and ready for the next batch. The system is now ready for the drain cycle.

When the start switch HS-151J on line 39 is pushed, one of two things will happen: The desired result will be to start the drain cycle by opening valve FV-151E; a probable result will be nothing. There are three conditions which could cause this problem. The first two are open feed valves, which would cause position switches ZSL-151C and/or ZSL-151D to be open. This result could also be caused by faulty position switches. The most probable cause will be failure to turn off the agitator. Agitator start and stop in this example is a manual operation. When the agitator is started, relay 7CR is energized and, as noted earlier, has contacts on lines 22, 26, and 39. Contacts on lines 22 and 26 have been explained and are associated with the agitator alarm circuit and procedural control start circuit, respectively. The contacts on line 39 prevent starting the drain cycle before stopping the agitator. The reason is to prevent an unloading of the agitator by draining the kettle. This may or may not be a problem depending on kettle and agitator size. In some applications it may be necessary to continue agitation to keep the product mixed. It should be remembered that parts of this example have been selected to demonstrate switching or sequence technique and may not always be good process technique.

Assuming all conditions are satisfied, relay 13CR will be energized closing contacts on lines 40 and 42. Contacts on line 40 provide a hold-in circuit for 13CR if relay 14CR is not energized. If A/M switch HS-151E is in the automatic position, contacts on line 42 will energize solenoid HY-151E and open drain valve FV-151E. When low-level switch LSL-151 on line 41 closes, relay 14CR is energized. Relay 14CR has contacts on line 40 which open to brake the hold-in circuit on relay 13CR. This closes drain valve FV-151E by de-energizing solenoid HY-151E. The complete cycle is now ready for the next batch.

There are two additional considerations that should be noted. On lines 28 and 29 of the procedural control circuit, no provision has been made for manual control. In general it would be good practice to supply an A/M switch, such as HS-151E on line 43, for both circuits. In the drain cycle no check was made on where the product was going. This step would be similar to checks used in the fill cycle, where the high-high level switch LSHH-151 could prevent filling. By leaving out this check, however, the author has a convenient way to end the example. If this makes the reader as relieved as it does the author, then everyone is happy.

There are, however, three items that must be noted before this section is ended. In the example just completed, a programmable controller was used with on/off logic only. As noted in Section 14.2, sequential control can be accomplished using equipment ranging from large-scale computers to individual control relays connected as shown in Figures 14-16 and 14-17. Choice of equipment will depend on process complexity, anticipated changes, number of program or procedural variations, and cost. For example, in the example of this chapter all relay devices and timers could have been included in a programmable controller. There is a wide variety of devices available.[15,16,17] The need for input measurement signals (such as flow, temperature, pressure, load and level) and for output action elements (such as solenoids, motor starters, alarms, and pilot lights) will be common to any piece of equipment. In addition, most devices may be programmed using the ladder diagram. Thus, as a general rule it is a good idea to decide what needs to be done, design the procedure and sequence, and then select the controlling device.

While the ladder diagram is a well-used technique in both process and electrical control applications, there are other methods for presenting the same information. One of these techniques which was noted earlier in this section is Boolean algebra. The other technique is the logic diagram. There are advantages and disadvantages of all techniques, as well as proponents for each method.[18] No attempt is intended to explain these techniques in detail. A simple example for one portion of Figure 14-16 will be given, however, to show the reader basic equalities of method. When need arises for this knowledge, the reader can find numerous references.[2,19]

The portion of Figure 14-16 to be examined includes lines 19, 24, and 25, and is shown in Figure 14-18a. In Boolean algebra or logic operations there are three basic operations: AND, OR, and NOT. The components such as switches and relay contacts are binary, which as previously noted in this section can be represented by a 1 when closed and a zero when open. For this reason and to simplify terminology, Figure 14-18a may be replaced by Figure 14-18b. Substitutions for terminology are shown in Figure 14-18b. In this particular circuit, the load may be energized in any one of three ways:

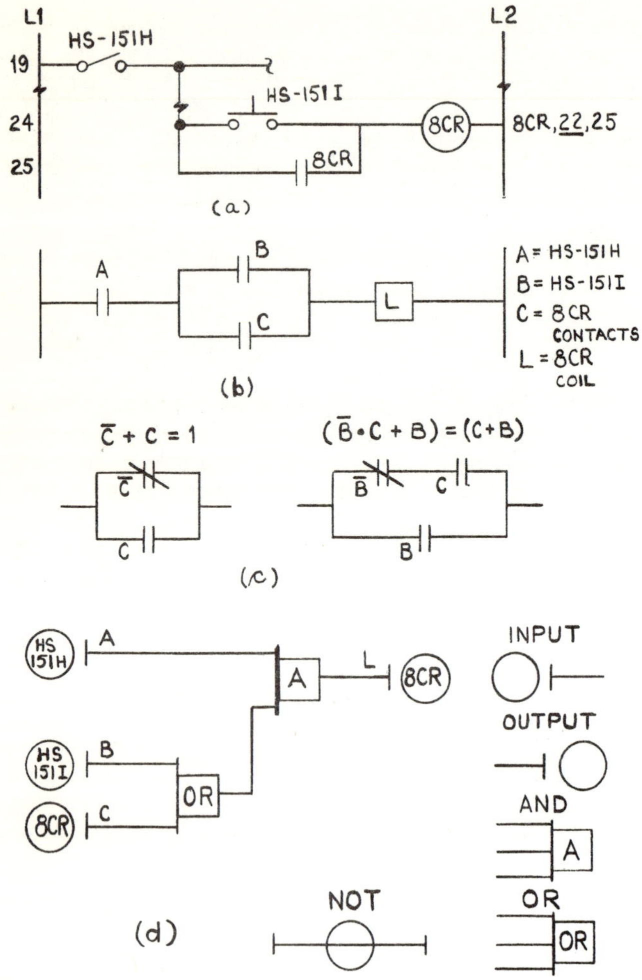

Figure 14-18. Ladder, Boolean, and logic diagrams.

1. If A AND B are energized, OR,
2. If A AND C are energized, OR,
3. If A AND B AND C are energized.

In Boolean algebra the AND operation is represented by multiplication and the OR by summation. Thus, Figure 14-18b may be represented in equation form as follows:

$$A \bullet B + A \bullet C + A \bullet B \bullet C = L \tag{14-15}$$

There is a number of Boolean identities, laws, and theorems[19] that may be used to reduce a Boolean equation to a particular solution. With some minor exceptions the process is very similar to algebra. Thus, Equation (14-15) may be reduced as follows:

$$A(B + C + B \bullet C) = L \tag{14-16}$$

and

$$A\left[B + C(1 + B)\right] = L \tag{14-17}$$

One of the identities in Boolean algebra is that $(1 + B) = 1$. The reason is simple when binary notation is considered. If 1 is a closed switch and is fixed, it is the same as a hardwired permanent connection. Thus, it makes no difference what B does in the above OR statement. Using this identity, Equation (14-17) may be reduced to

$$A(B + C) = L \tag{14-18}$$

Another way to arrive at the same result is to use a truth table. In the example at hand there are three binary inputs, namely, A, B, and C. A truth table lists in a systematic manner all possible combinations of these input variables. Since binary elements are used, there are 2^nth combinations, where n represents the number of variables. Table 14-4 shows the procedure for this particular example.

As shown in Table 14-4, combinations 6, 7, and 8 will energize the load. The Boolean algebra equation may then be written from Table 14-4 by combining combinations 6, 7, and 8.

$$A \bullet \bar{B} \bullet C + A \bullet B \bullet \bar{C} + A \bullet B \bullet C = L \tag{14-19}$$

The bar over the variable indicates the NOT function. Reducing Equation (14-19) gives

$$A(\bar{B} \bullet C + B \bullet \bar{C} + B \bullet C) = L \tag{14-20}$$

and

$$A\left[\bar{B} \bullet C + B(\bar{C} + C)\right] = L \tag{14-21}$$

In Boolean identities $(\bar{C} + C)$ is equal to 1. The basis for this is shown in Figure 14-18c. When NOT C and C are in a parallel path OR condition, the circuit is always closed. Thus Equation (14-21) reduces to

$$A(\bar{B} \bullet C + B) = L \tag{14-22}$$

The term $(\bar{B} \bullet C + B)$ is another identity equal to $(C + B)$. This is also shown in Figure 14-18c. Using N.O. and N.C. concepts for relay contacts, this concept can easily be shown. If B is not energized and C is energized, the circuit will be closed. If B is energized, the circuit will also be closed. Thus, if either B OR C

Table 14-4
Truth Table for Three Variables

Combination No.	Variables A	B	C	Load L	Action
1	0	0	0	0	L is off
2	0	0	1	0	L is off
3	0	1	0	0	L is off
4	0	1	1	0	L is off
5	1	0	0	0	L is off
6	1	0	1	1	L is on
7	1	1	0	1	L is on
8	1	1	1	1	L is on

is energized, the circuit will be closed. Therefore, Equation (14-22) reduces to

$$A(C+B)=L \qquad (14\text{-}23)$$

This is identical to Euation (14-18) and shows another method for obtaining the result.

The same circuit may be represented using logic diagrams, as shown in Figure 14-18d. Symbols follow as close as possible to recommended standards,[12,13,20] which are not totally consistent. If and when a question of meaning exists, a key should be given. The logic diagram of Figure 14-18d provides an identical story to Figure 14-18a, and Equations (14-18) and (14-23). Thus, for most applications, choice of method will depend on background of the designer or on the type of equipment used. The exceptions to this rule are large systems or a system that will be produced in quantity. In these applications Boolean algebra methods with qualifying conditions may be used to check for minimal component usage.

No matter what method is used, a statement of the problem, along with process and safety considerations, must be available. With procedure and restrictions defined, a sequential control scheme can be designed for batch operation. The techniques and examples presented in this chapter will obviously not provide all the answers. They should, however, give the reader enough background to understand existing systems and provide a starting point for further study if required.

Chapter 14—Study Questions and Problems

(Answers appear on pages 173-174)

1. When a batch process is shut down, what kind of large signal is generated?

2. During a batch process shut down, what will the error signal-time product cause the output of a PI controller to do?

3. When the output of a PI controller drives to high or low saturation because of a sustained error signal, this is called __________.

4. A PI controller cannot provide start-up without __________.

5. A proportional controller may be used on a batch system only if __________ point and __________ changes are not present.

6. Most reset lockout systems do not eliminate reset windup, but they do __________ the effect.

7. If small offsets can be tolerated, a __________ controller may be effectively used on batch systems.

8. A series-type controller uses __________ response ahead of __________ response to provide start-up without overshoot.

9. A series-type controller has been purchased to use on a batch control system. Ultimate gain and period in the procedure or run mode are 3.5 and 6 min, respectively. Calculate initial controller settings for start-up without overshoot.

10. A single loop control system for a jacketed kettle is shown below. In this system kettle temperature should be increased from 22 to 45°C in the first 15% of the cycle. The temperature should be held at 45°C for the next 20% of the cycle, and then driven to 60°C in 5% of cycle time. At this point, maximum cooling should be applied for the remainder of the cycle. The temperature set point device should return to the start position by the end of the cycle. Sketch the temperature control profile for this system and a time cycle diagram for each control element that must be turned on and off. Where necessary, provide verbal description.

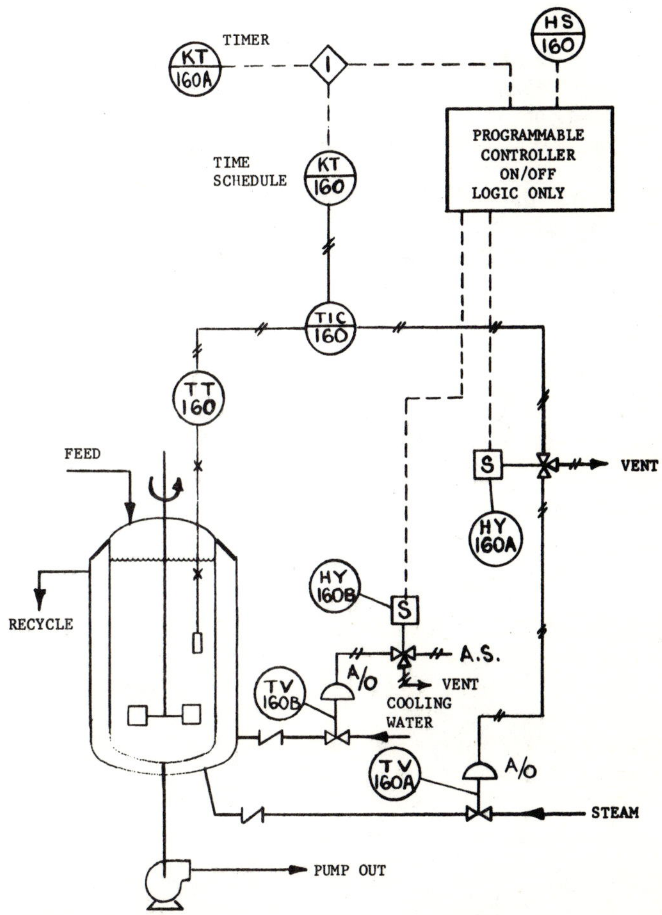

Figure 14-P10. Jacketed kettle control system.

11. When a switch is closed it may be represented by a __________, and a __________ when it is open.

12. The terms N.O. and N.C. mean __________ and __________, respectively.

13. The term normal refers to the ________ or ________ condition.

14. Assuming that fill, drain and agitator functions are manually operated, sketch a ladder diagram to sequence the jacketed kettle of problem 10 through the procedure state. Describe how the sequence differs from that given on Figure 14-17.

15. Write the truth table for lines 9 and 10 in Figure 14A-14. Show when 3CR is energized.

16. Write the Boolean algebra statement for lines 9 and 10 of Figure 14A-14.

17. Using logic symbols in Figure 14-18d, sketch a logic diagram for lines 9 and 10 of Figure 14A-14.

References

1. Hall, C. J., "Don't Live with Reset Windup," *Instrumentation Technology*, July 1974, p. 55.
2. Harrison, H. L., and Bollinger, J. G., *Introduction to Automatic Controls*, Scranton, PA: International Textbook Co., 1968.
3. "DDC Remolds Basic Process Control," *C and EN*, March 13, 1967.
4. *Advances in Instrumentation*, Vol. 29, Part 1, ISA Annual Conference, New York, 1974.
5. Zoss, L. M., and Delahooke, B. C., *Theory and Applications of Industrial Process Control*, Albany, NY: Delmar, 1961.
6. Considine, D. M., *Process Instruments and Controls Handbook*, New York: McGraw-Hill, 1957.
7. Sparks, A. C., and Falusy, R. F., "Colorful yarn: Sequencing Dye Cycles," *Instrumentation Technology*, August 1970.
8. Hohn, F. E., *Applied Boolean Algebra, An Elementary Introduction*, New York: Macmillan, 1960.
9. Scheid, F., *Introduction to Computer Science*, Schaum's Outline Series, New York: McGraw-Hill, 1970.
10. Electrical Standards for Industrial Equipment, Joint Industry Conference, 7901 W. Park Dr., McLean, VA 22101.
11. *Bulletin 125*, Eagle Signal, Division of Gulf & Western, Inc.
12. Hydraulic and Pneumatic Standards for Industrial Equipment, Joint Industry Conference, 7901 W. Park Dr., McLean, VA 22101.
13. National Electrical Manufacturers Association, "Industrial Control Equipment," Pub. No. 1C 1-1965, 155 East 44th St., New York, NY 10017.
14. Gollin, N. W., "Design of Automatic time and Sequence Control Systems for Processes," *ISA Journal*, April, May, and June 1955.
15. "Energy Management Systems, Programmable Type," Bulletin 1810P, Allen-Bradley Systems Division, Highland Heights, OH 44143.
16. "The Versatile Controller," Model 184 Programmable Controller, Modicon Corp., Andover, MA 01810.
17. Cycle Logic Systems," TI 45-201a, The Foxboro Co., Foxboro, MA 02035.
18. Coelho, W. A., "Putting Programmable Controls to Work on Material Handling" *Instrumentation Technology*, March 1974, pp. 39-43.
19. *Industrial Static Switching Handbook*, 1968, Jordon Controls, Inc., Milwaukee, WI 53218.
20. *Binary Logic Diagrams for Process Operations*, ISA Standard-S5.2, ISA, Pittsburgh, Pa 15222.

Appendix

Answers to Study Questions

Chapter 1

1. controlled

2. measured

3. subtracted

4. error, final control

5. measured

6. manipulated

7. *c*, *b*, *r*, *e*, *m*

8.

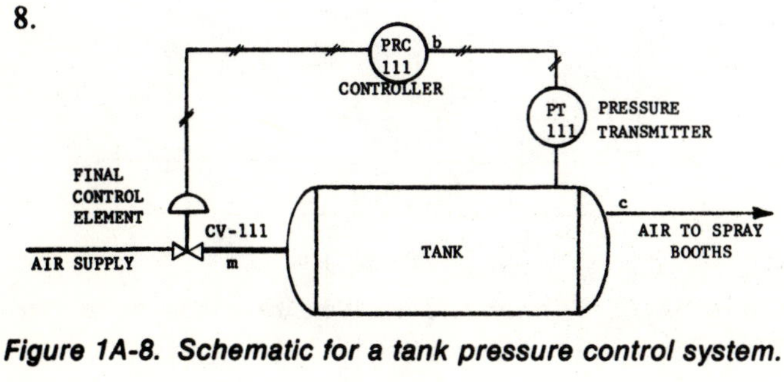

Figure 1A-8. Schematic for a tank pressure control system.

9.

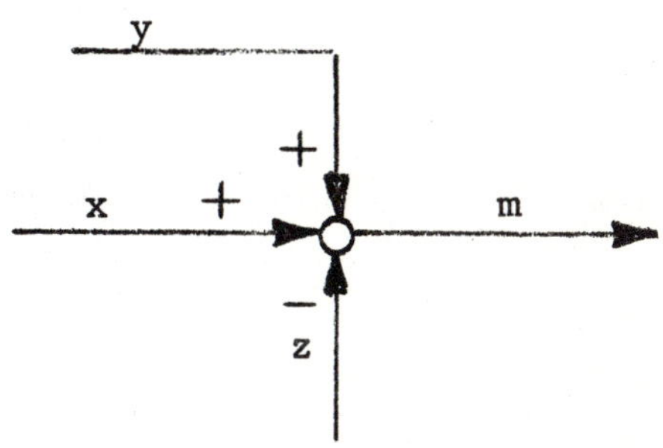

Figure 1A-9. Summing point block diagram for $m = x + y - z$.

10.

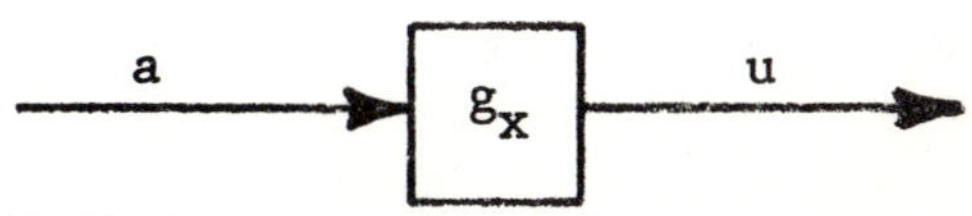

Figure 1A-10. Transfer function block diagram for $u = g_x a$.

11.

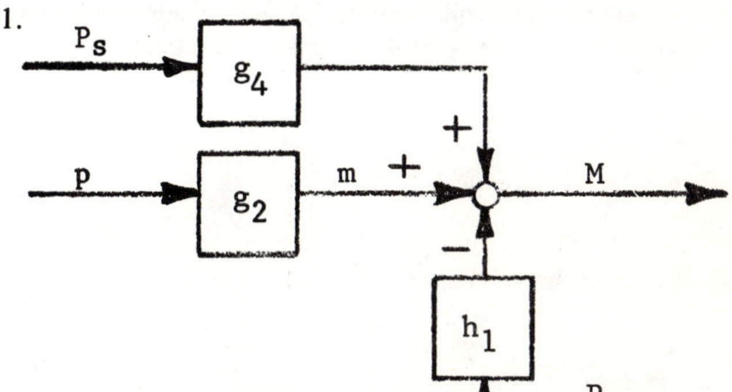

Figure 1A-11. Combined block diagram for $m = g_2 p$ *and* $M = m + g_4 P_s - h_1 P_t$.

12. block diagram

13. input, output

14. error, actuating

15. magnitude, time

16.

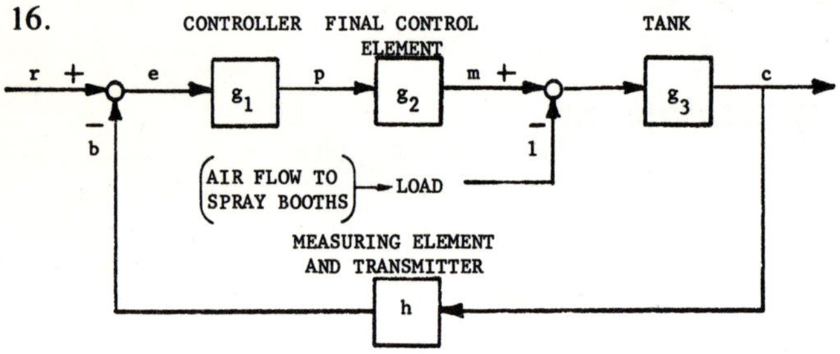

Figure 1A-16. Closed loop block diagram for the tank pressure control system of Question 8.

17.

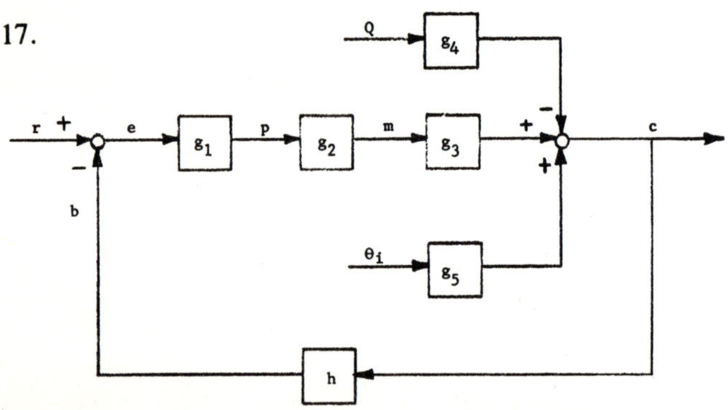

Figure 1A-17. Block diagram of Figure 1-10 showing external disturbances.

18. As product flow Q increases, the controlled variable c (outlet temperature) will decrease. As product inlet temperature Θ_i increases, the controlled variable c (outlet temperature) will increase.

19. external disturbances

20. physical characteristics

21. stability

22. one quarter

23. initial overpeak

24. external disturbances

Chapter 2

1. time

2. transfer function

3. straight line

4. start-up

5. small

6. energy, mass

7. first-order linear differential equation

8. resistance, storage

9. resistive, capacitive, time

10. time

11. slope = rise/run

12.

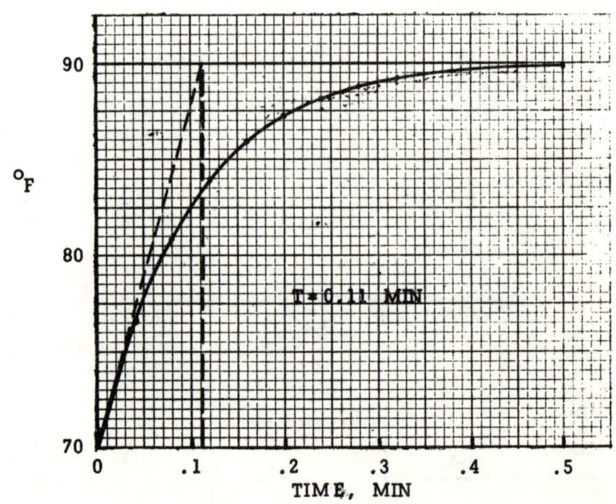

Figure 2A-12. Temperature vs. time response for a thermometer.

13. From Equation 2-15, $P_2(T) = A(0.632) \cong 0.63A$. From Problem 12, A = 20°F. Since $P_2(T) \cong \Theta_2(T)$ due to similarity of systems.

$$\Theta_2(T) = 0.63\,(20) = 12.6°F \text{ (or } 82.6°F)$$

Reading from Figure 2A-12, T = 0.102 min. Note: Discrepancy between values of T in Figure 2A-12 and the solution in this problem results from errors in graphical techniques and should be expected.

14. Incomplete response calculations are shown below. The response plot is shown in Figure 2A-14. Note: As time increases, it is necessary to subtract the difference of two large numbers to obtain ΔT. For example, at 0.5 min, ΔT = 90.0 - 89.9. When this condition exists, any small error in reading values from a graph or chart record will cause a large error. For this reason these data points are usually neglected, as shown in Figure 2A-14.

Incomplete Response Calculations

Time, min	Temperature, °F	ΔT, °F	Incomplete Response
0.00	70.0	20.0	20.0/20 = 1.0
0.04	76.6	13.4	13.4/20 = 0.67
0.08	80.9	9.1	9.1/20 = 0.455
0.12	84.0	6.0	6.0/20 = 0.30
0.16	85.9	4.1	4.1/20 = 0.205
0.20	87.3	2.7	2.7/20 = 0.135
0.24	88.2	1.8	1.8/20 = 0.09
0.28	88.8	1.2	1.2/20 = 0.06
0.32	89.2	0.8	0.8/20 = 0.04
0.50	89.9	0.1	0.1/20 = 0.005

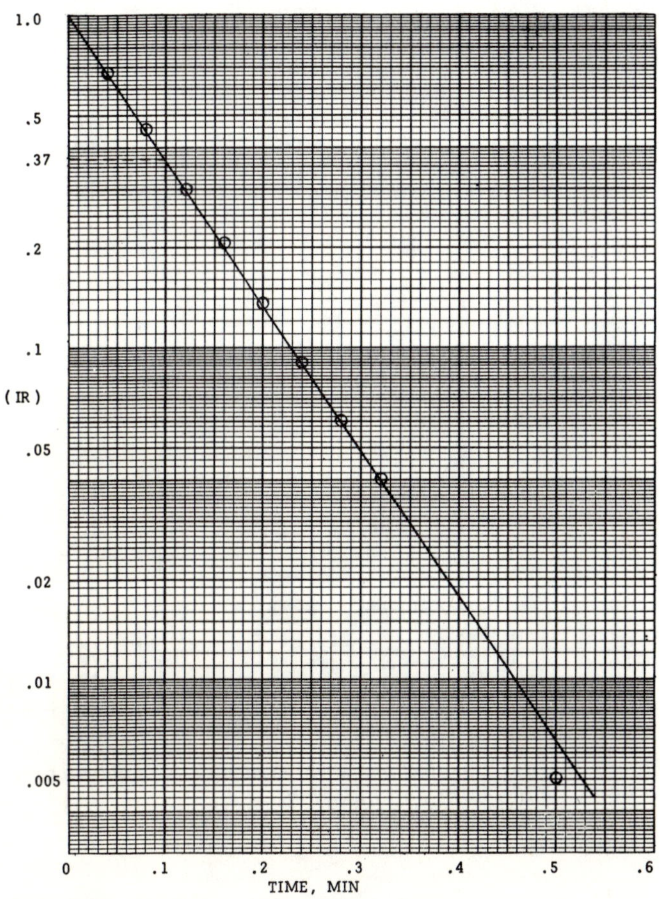

Figure 2A-14. Incomplete response plot.

15. Since the time constant is located at 0.63 in the complete response, its value in the incomplete response is at 0.37. Reading from Figure 2A-14 at 0.37, incomplete response gives a time value of 0.1 min; thus, T = 0.1 min. Here again, graphical technique is responsible for any discrepancy between this answer and those of Problems 12 and 13.

16. first-order, exponential

17. transportation or distance velocity lag

18. $L = (10 \text{ ft}/[3000 \text{ ft/min}])(60 \text{ sec/min}) = 0.2 \text{ sec}$

19. $L = (12 \text{ ft}/[4 \text{ ft/sec}])(1/[60 \text{ sec/min}]) = 0.05 \text{ min}$

20. plug flow and no heat loss to surroundings

21. A step input change to an exponential or first-order lag will result in an exponential output. Since the output of the first exponential lag is also the input to the second exponential lag, the "rise" term in Equation (2-11) will be reduced. The "run" term is equal to the time constant and will be a constant. Thus, as the "rise" term decreases, the "slope" or rate of change of output must also decrease. Because this process is repeated through each succeeding exponential or first-order lag, the initial "rise" term becomes very small after passing through two or three exponential lags. As a result, the output "slope" or rate of change is initially zero or very small. Thus, the initial output is equivalent to dead time lag.

22. pseudo

Chapter 3

1. direction

2. x and y coordinates, magnitude, angle; complex number

3. $3 + j4$

4. Magnitude $= \sqrt{[(3)^2 + (4)^2]} = \sqrt{25} = 5$
phase angle $= \tan^{-1}(4/3) \cong 53.1°$

5. magnitudes, phase angles

6. $15 + j6 + j20 + j^2 8$
$(15 + j^2 8) + j(6 + 20)$
$7 + j26$
magnitude $= \sqrt{[(7)^2 + (26)^2]} \cong 27.1$
phase angle $= \tan^{-1}(26/7) \cong 75°$

7. sine wave, yes

8. transient

9. steady-state sine wave

10. equal, zero

11. decrease, increase

12. output, input

13. lag, period

14. magnitude ratio, phase angle

15. root

16. two

17. By using the quadratic factor

18. $X = (-b \pm \sqrt{(b^2 - 4ac)})/(2a)$

19. $X_1 = -2 + j4$ and $X_2 = -2 - j4$

20. Heaviside operator p

21. $1 + 10p = 0, T = 10, p = 0.1$

22. $p = \pm\sigma \pm j\omega$

23. initial conditions

24. sP_2

25. time^{-1} or 1/time

26. frequency, sine wave

27.

Calculated Frequency Response for $T = 1/(2\pi)$

Frequency, cpm	Magnitude	Phase, degrees
0.1	1.00	− 6
0.2	0.98	−11
0.4	0.93	−22
0.7	0.82	−35
1.0	0.707	−45
2.0	0.45	−63
4.0	0.24	−76
7.0	0.14	−82
10.0	0.10	−84

28.

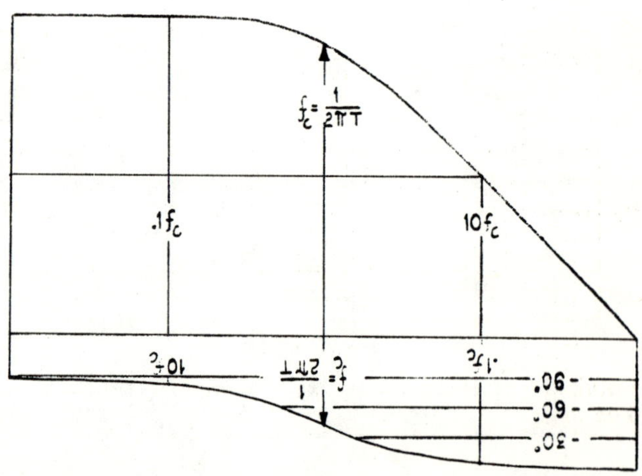

Figure 3A-28. The template is not to scale and is shown only to demonstrate form and indexing.

29. 0.32 cpm

30. plug flow and negligible heat loss

31. one

32. $-fL$ 360°

33. output/input = e^{-Ls}

34. 0.05 min

35.

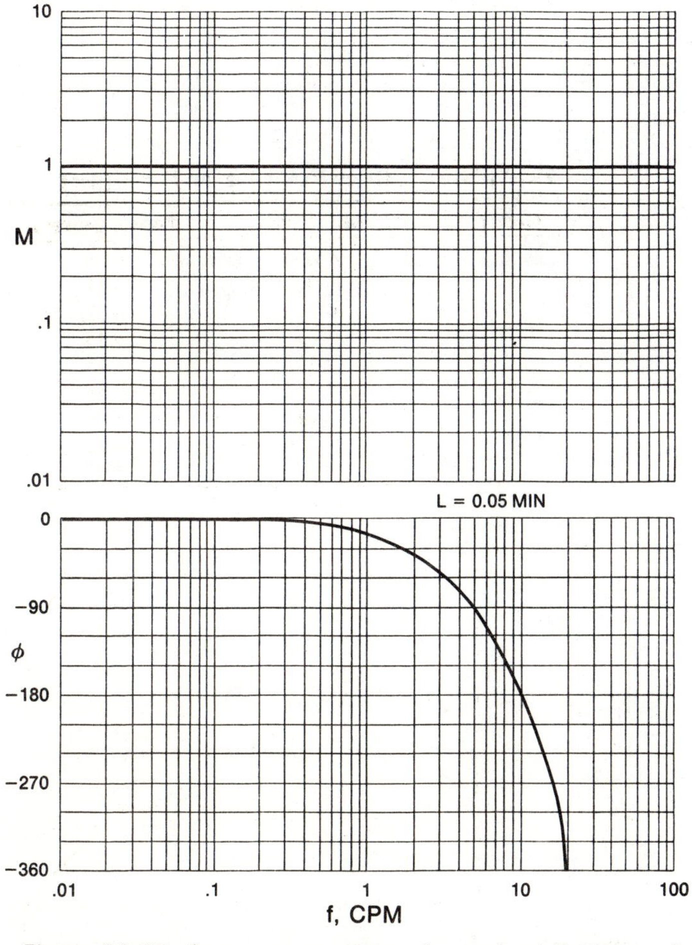

Figure 3A-35. Frequency response for a dead time lag of 0.05 min.

Chapter 4

1. $f_{c2} = 1/(2\pi T_2) = 0.159/0.1 = 1.59$ cpm
   $k_2 = 1.0$
   (See Figure 4A-4 for plot.)

2. $k_3 = 10$; phase angle = 0
   (See Figure 4A-4 for plot.)

3. magnitude equals 1 at all frequencies
   phase angle = $\Phi = -360f(0.15) = -54f$ (degrees)
   (See Figure 4A-4 for plot.)

4. $f_{c5} = 1/(2\pi T_5) = 0.159/0.05 = 3.18$ cpm; $k_5 = 0.12$

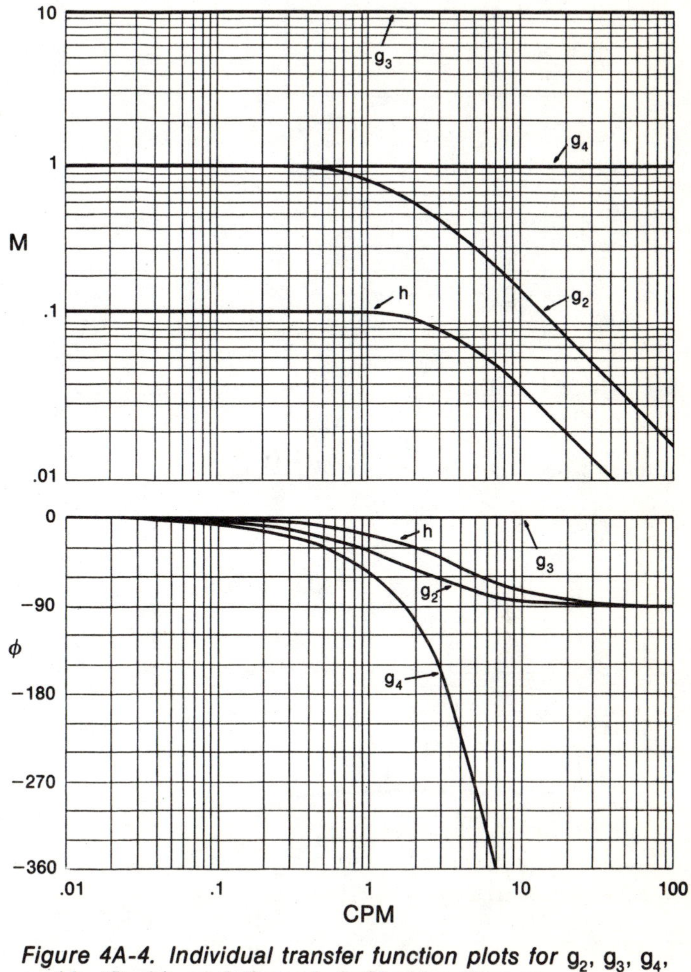

Figure 4A-4. Individual transfer function plots for g_2, g_3, g_4, and h. (Problems 1 through 4, Chapter 4)

5, 6, 7. The answers to Problems 5, 6, and 7 demonstrate a step-by-step procedure for combining individual components to obtain the process transfer function. As the student becomes more at home with the technique, the procedure can be completed in one step. As a further test of method and technique, the reader should combine $g_2g_3g_4h$ in one step using the Bode plot obtained in Problems 1 through 4. The result should be the same as shown in Figure 4A-7.

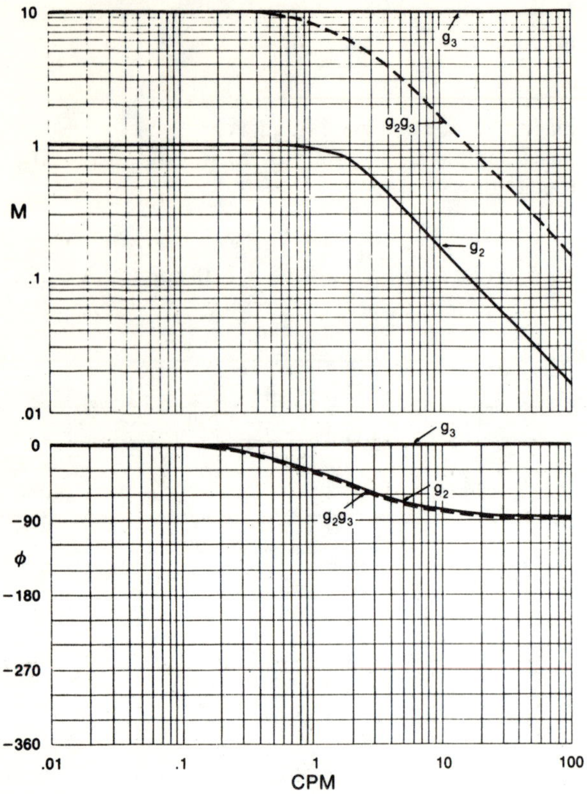

Figure 4A-5. Transfer function plot for g_2g_3.

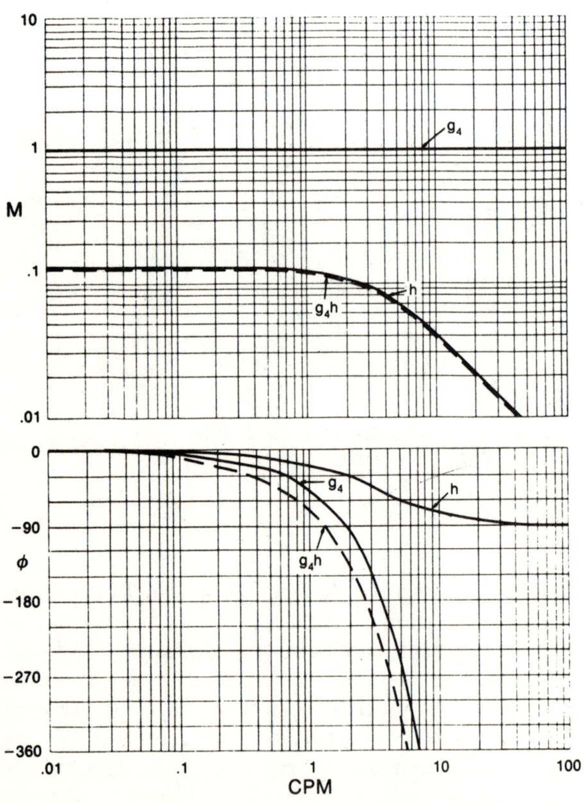

Figure 4A-6. Transfer function plot for g_4h.

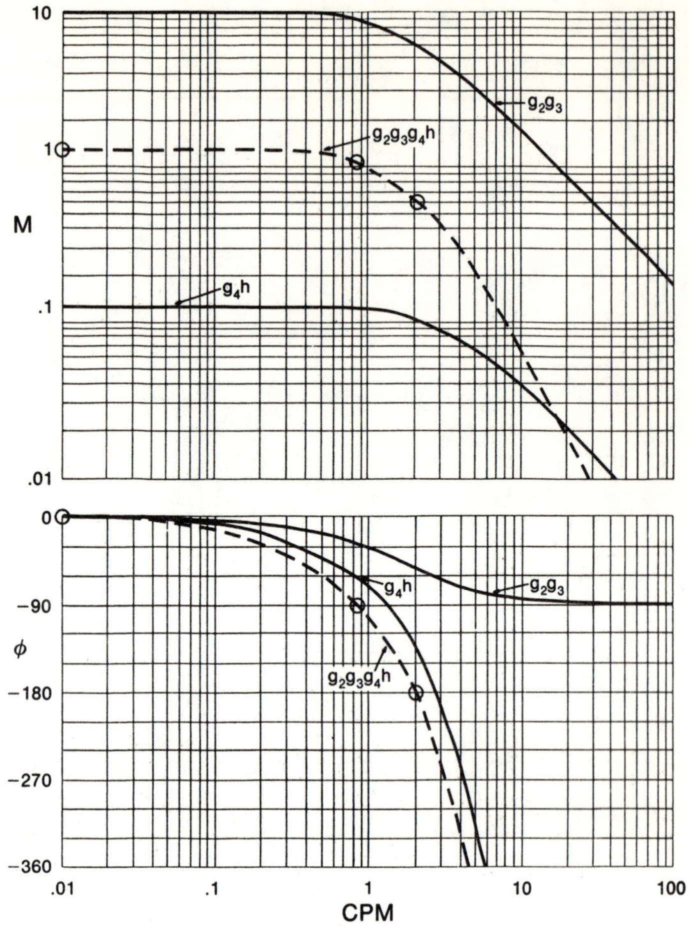

Figure 4A-7. Transfer function plot for $g_2g_3g_4h$. Note: Plotted points will be referred to in Chapter 5.

8. process

Chapter 5

1. direct
2. reverse
3. direct
4. odd
5. 180°
6. constant

7. "process," controller

8. 1, −180°

9. neglected

10. zero, "process"

11. −180°

12. 0.25

13. natural

14. 0.4 min

15. "process," 180

16. quarter

17. increase

18. 1.8

Chapter 6

1. analog

2. transfer function

3. transient

4. overdamped, large

5. critical

6. underdamped, small

7. $f_n = 1/(2\pi T_n)$

8. actual, critical

9. output/input = $1/(T_n{}^2 s^2 + 2\,\zeta\, T_n s + 1)$

10. peak frequency magnitude/zero frequency magnitude $\cong 1/(2\zeta)$

    From the Figure in Problem 10,

    peak frequency magnitude = 2.4
    zero frequency magnitude = 1.0

    $\zeta = 1/[2\,(2.4)] = 0.21$

    From the Figure in Problem 10,

    peak frequency = 0.97 cpm

    Assume $f_n = f_p = 0.97$ cpn

    zero frequency magnitude = 1.0

11. From Figure 6-6 for $\zeta = 0.21, Y/Z = 1.5$.

12. From Figure 6-6 for $\zeta = 0.21$,
    decay ratio = 0.25
    first overpeak = maximum deviation − final value
    final value = (zero frequency magnitude) (step input)
    = (1.0) (1.0) = 1.0
    first overpeak = 1.5 − 1.0 = 0.5
    second overpeak = (first overpeak) (decay ratio)
    = (0.5) (0.25) = 0.125
    third overpeak = (second overpeak) (decay ratio)
    = (0.125) (0.25) = 0.03125

13. Assume $f = f_n = f_p = 0.97$ cpm.
    period = $1/f = 1/0.97 = 1.03$ min.

14. From Figure 6-6 for $\zeta = 0.21$, $f_n X = 0.51$; from the solution to Problem 13, $f = f_n$ and $f_n = 1/\text{period}$; thus, $X = 0.51(\text{period}) = 0.51\,(1.03) = 0.52$ min.

15. Plot the first peak at $y = 1.5$ and $t = 0.52$. The second peak is $y = 1.125$ at $t = 0.52 + 1.03$. The third peak is $y = 1.03$ at $t = 0.52 + 2\,(1.03)$. Plot mirror images of first, second, and third peaks at appropriate times. (Example $y = 0.5$ at $t = 0.52$.) Sketch exponential envelope as shown in Figure 6A-15. First undershoot hits exponential envelope at $t = X + 1/2\,(\text{period}) = 0.52 + 0.515 = 1.04$ min. Second undershoot hits exponential envelope at $t = X + 1.5(\text{period}) = 0.52 + 1.545 = 2.07$ min. Sketch initial rise as a lazy S and

remaining curve as a damped sine wave. The solid line of Figure 6A-15 is the author's sketch. Plotted points within the squares represent exact solution calculated from Equation (6-23).

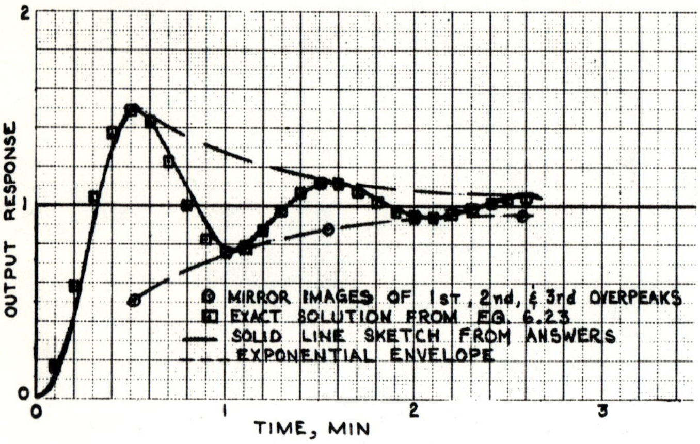

Figure 6A-15. Calculated time response.

Chapter 7

1. transfer functions

2. changes

3. no

4. closed loop transfer function = open loop transfer function/(1 + open loop transfer function)

5. $b/e = g_1g_2g_3h$
   $b/r = g_1g_2g_3h/(1 + g_1g_2g_3h)$

6. $f_{c2} = 1/\left[2\pi(0.4)\right] \cong 0.4$ cpm
   $f_{c3} = 1/\left[2\pi(1.2)\right] \cong 0.13$ cpm
   $f_{ch} = 1/\left[2\pi(0.1)\right] \cong 1.6$ cpm
   The process frequency response is shown in Figure 7A-8.

7. $P_G \cong 0.11$ at $\Phi = -180°$
   $G_U = 1/P_G = 1/0.11 \cong 9.1$
   $G_O = G_U/2 = 9.1/2 \cong 4.6$

8. process, controller
   The open loop frequency response is shown in Figure 7A-8.

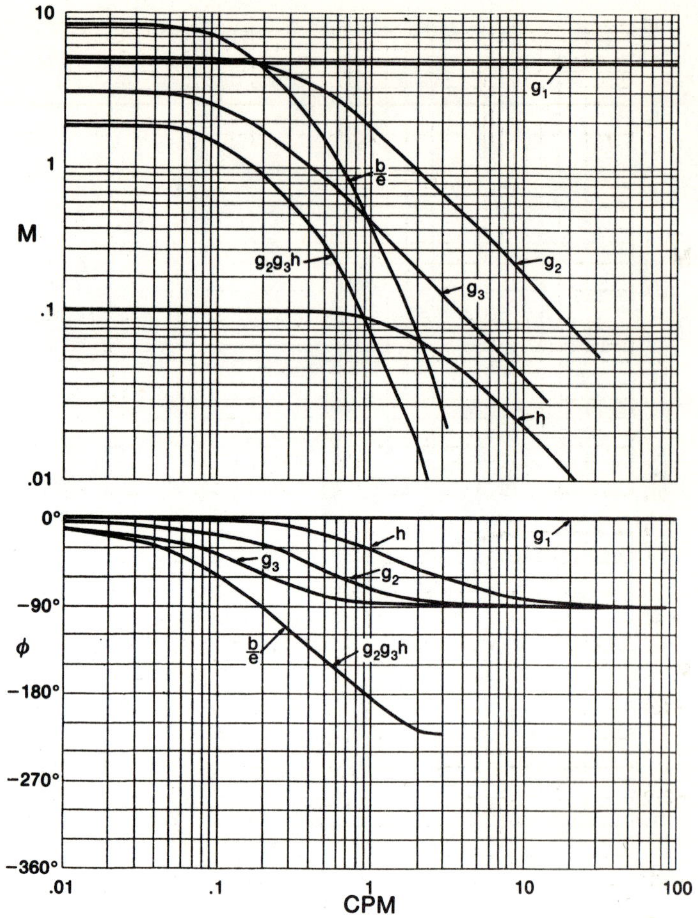

Figure 7A-8. Open loop frequency response, b/e.

9. Calculated data are shown in the table below. See Figure 7A-9 for the closed loop frequency response plot.

Frequency cpm	Open Loop Gain	Open Loop Phase Angle	Closed Loop Gain	Closed Loop Phase Angle
0.01	8.3	−7°	0.89	−1°
0.04	8.3	−29°	0.90	−3°
0.07	7.6	−45°	0.91	−5°
0.1	6.8	−58°	0.92	−7°
0.2	4.2	−90°	0.97	−14°
0.3	2.8	−116°	1.11	−21°
0.4	2.0	−134°	1.35	−30°
0.5	1.4	−145°	1.7	−40°
0.6	1.1	−157°	2.5	−68°
0.7	0.86	−163°	3.0	−115°
0.8	0.66	−171°	1.8	−155°
0.9	0.50	−180°	1.0	−180°
1.0	0.42	−186°	.75	−190°

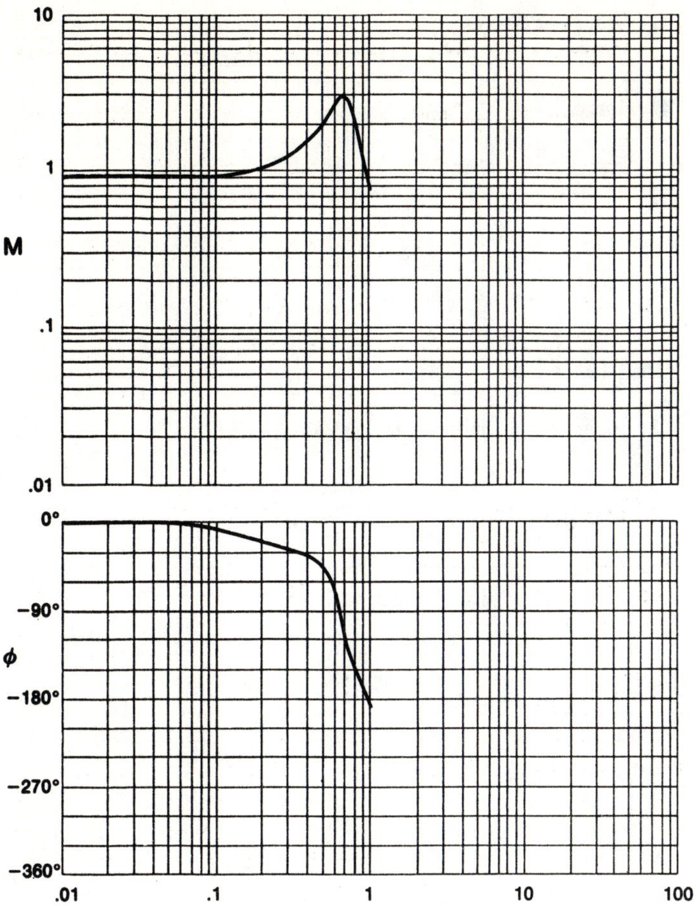

Figure 7A-9. Closed loop frequency response, b/r.

10. From Figure 7A-9 (or the table in Problem 9),

zero frequency magnitude = 0.89
peak frequency magnitude = 3.0
peak frequency = 0.7 cpm

Using Equation (6-34), the value of ζ is

$$\zeta = 0.89/[2(3)] = 0.15$$

Final value may be calculated as follows:

$b = 0.89r = (0.89)(10°) = 8.9°$F
absolute value = 140° + 8.9° = 148.9°F

From Figure 6-6 for $\zeta = 0.15$, $Y/Z \cong 1.62$. Thus,

maximum deviation = (1.62) (8.9°) = 14.4°F
absolute value = 140° + 14.4° = 154.4°F

From Figure 6-6 for $\zeta = 0.15$, decay ratio = 0.38, using the difference between maximum deviation and final value overpeaks may be determined.

14.4° − 8.9° = 5.5°
second overpeak = (0.38) (5.5°) = 2.1°
absolute value = 140° + 8.9° + 2.1° = 151°F
third overpeak = (0.38) (2.1°) = 0.8°
absolute value = 140° + 8.9° + 0.8° = 149.7°F

Assuming $f = f_n = f_p = 0.7$ cpm and reading a value of $f_n X = 0.51$ from Figure 6-6 allows period and time of initial overpeak to be calculated.

period = $1/f_p = 1/0.7 = 1.43$ min

$X = 0.51/f_n = (0.51)(\text{period}) = (0.51)(1.43) \cong 0.73$ min

first overpeak = X = 0.73 min
first undershoot = 0.73 + 0.5 (1.43) ≅ 1.45 min
second overpeak = 0.73 + 1.0 (1.43) = 2.16 min
second undershoot = 0.73 + 1.5 (1.43) ≅ 2.87 min
third overpeak = 0.73 + 2.0 (1.43) = 3.59 min
third undershoot = 0.73 + 2.5 (1.43) = 4.31 min

Using techniques demonstrated in Chapters 6 and 7, a time response may be plotted. The result is shown in Figure 7A-10.

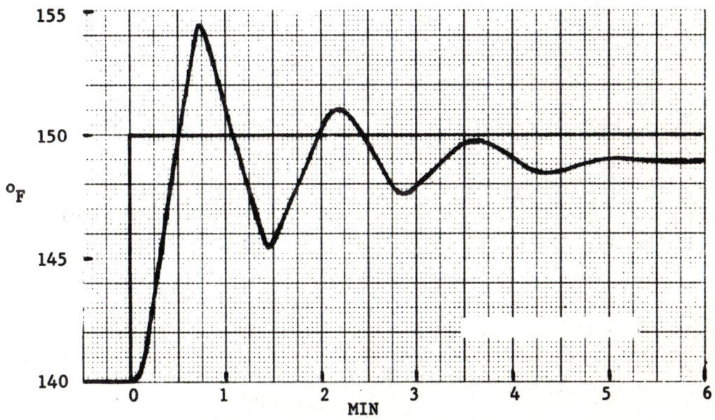

Figure 7A-10. Plot of calculated time response.

11. time response

12. mathematical, graphical

13. approximate

14. offset

15. one

Chapter 8

1. $b/U_4 = -\left[1/(g_1g_2g_3)\right]\left\{g_1g_2g_3g_4h/(1+g_1g_2g_3g_4h)\right\}$

2. $f_{c3} = 1/(2\pi T_3) = 1/\left[2\pi(0.4)\right] = 0.4$ cpm
See Figure 8A-2 for plot of b/U_4.

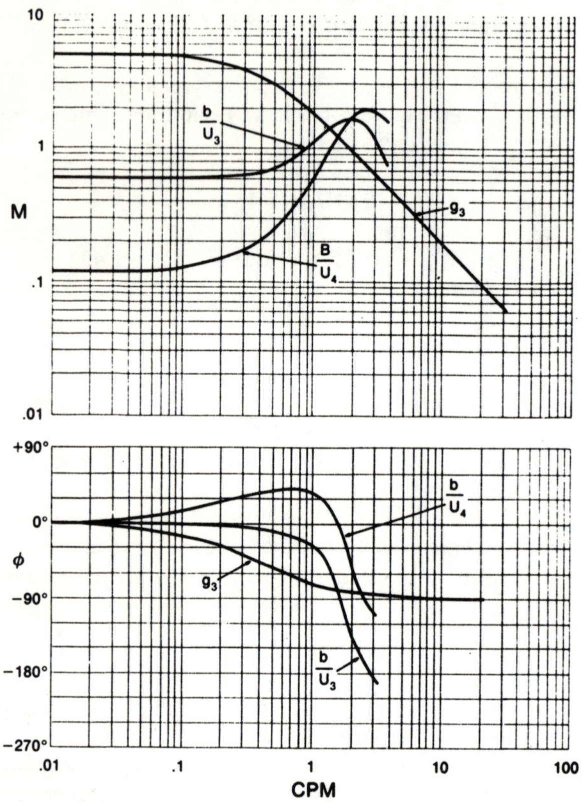

Figure 8A-2. Closed loop frequency response for b/U_4.

3. $b/r = g_1g_2g_3g_4h/(1+g_1g_2g_3g_4h)$
$b/U_2 = (1/g_1)(b/r)$
$b/U_3 = \left[1/(g_1g_2)\right](b/r)$
$b/U_4 = \left[1/(g_1g_2g_3)\right](b/r)$

4. offset

5. steady-state, zero frequency gain

6. forward, feedback

7. resonant

8. no

9. increase, decrease, increase

10. filters or delays

11. large

12. controller

13. small

Chapter 9

1. measured variable, reference or set point

2. high gain

3. relay valve, power amplifier

4. reduce

5. vary, output

6.

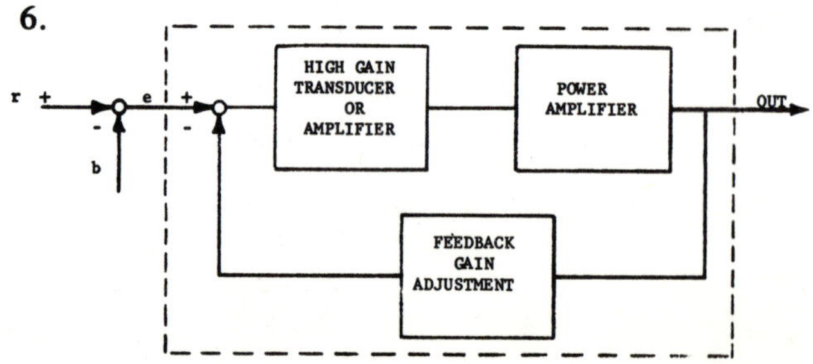

Figure 9A-6. Block diagram for a proportional controller circuit.

7. **Equivalent Proportional Controller Values**

Case	Gain	Prop. Band, %	Sensitivity, psi/in.
a	(5)	20	15
b	2/3	(150)	2
c	3 1/3	30	(10)

8. Input Scale Factor = 100%/16 ma
(output/input) multiplier = 100%/PB%
Output Scale Factor = 4 volts/100%
Output, volts = (100%/16 ma) (100%/40%)
(4 volts/100%) 4 ma = 2.5 volts

9. high

10. proportional

11. $\Delta P_0 = 3(1) + (3/1.5)\int_0^t dt = 3 + 2t$

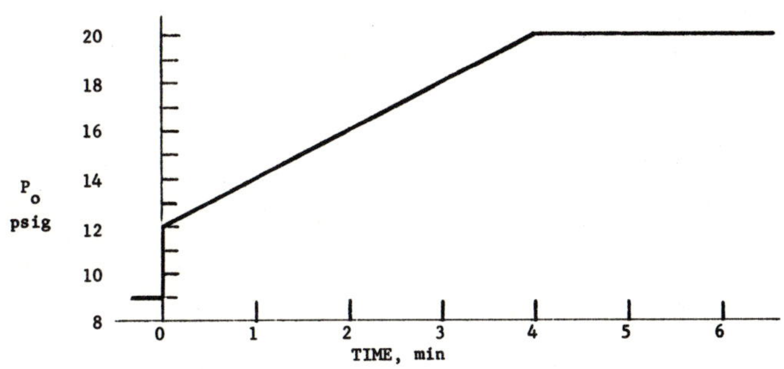

Figure 9A-11. Open loop time response for a proportional plus reset controller.

12. repeat

13. reset corner $= 1/(2\pi T_r) = 1/[2\pi(0.5)] = 0.32$ cpm
high frequency gain = 3
See Figure 9A-13 for plot.

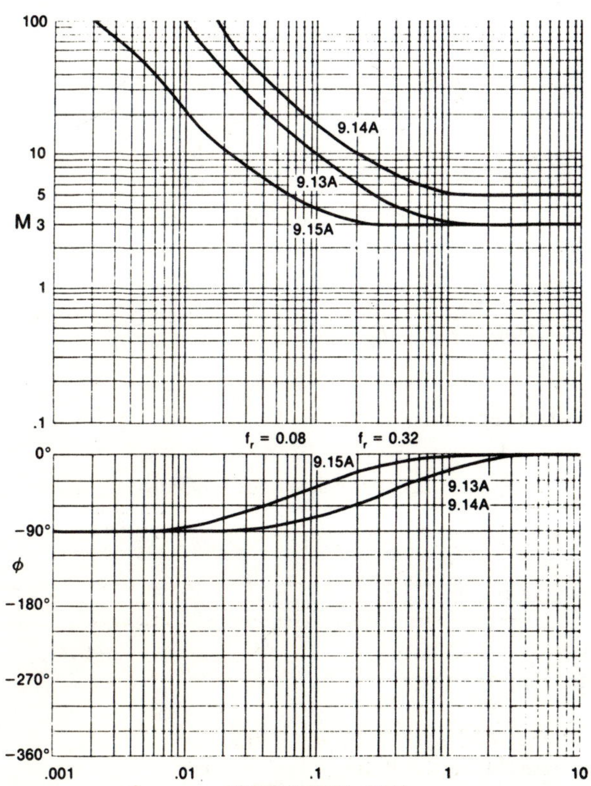

Figure 9A-13. Frequency responses for a proportional plus reset controller.

14. See Figure 9A-13.

15. reset corner $= 1/[2\pi(2)] = 0.08$ cpm
See Figure 9A-13.

16. $G_U = 1/2.5 = 0.4$
$P_U = 1/12 = 0.083$ min

17. $G_0 = 0.45\, G_U = 0.45\,(0.4) = 0.18$
$T_r = P_U/1.2 = 0.083/1.2 = 0.07$ min

18. offset

19. increasing

20. $I = 1/T_r = 1/0.07 = 14.3$ repeats/min

21. initial overpeaking, period

22. $P_o = 4(0.5t) + 4(0.4)(0.5)$
$P_o = 2t + 0.8$
See Figure 9A-22 for sketch.

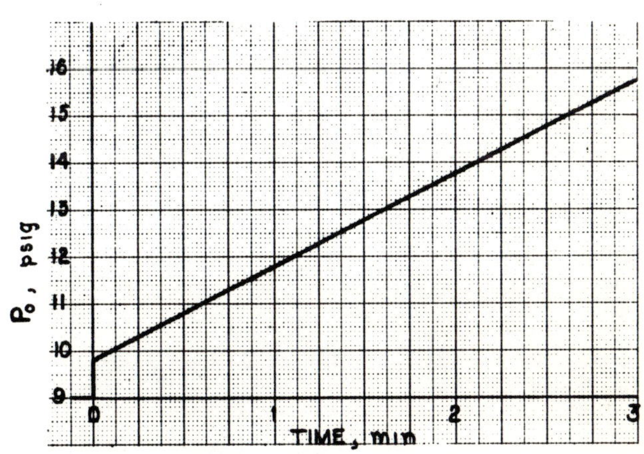

Figure 9A-22. Open loop time response for a proportional plus rate controller.

23. proportional

24. offset

25. rate corner $= 1/(2\pi T_d) = 1/[2\pi(0.5)] = 0.32$ cpm
zero frequency gain = 4
See Figure 9A-25 for plot.

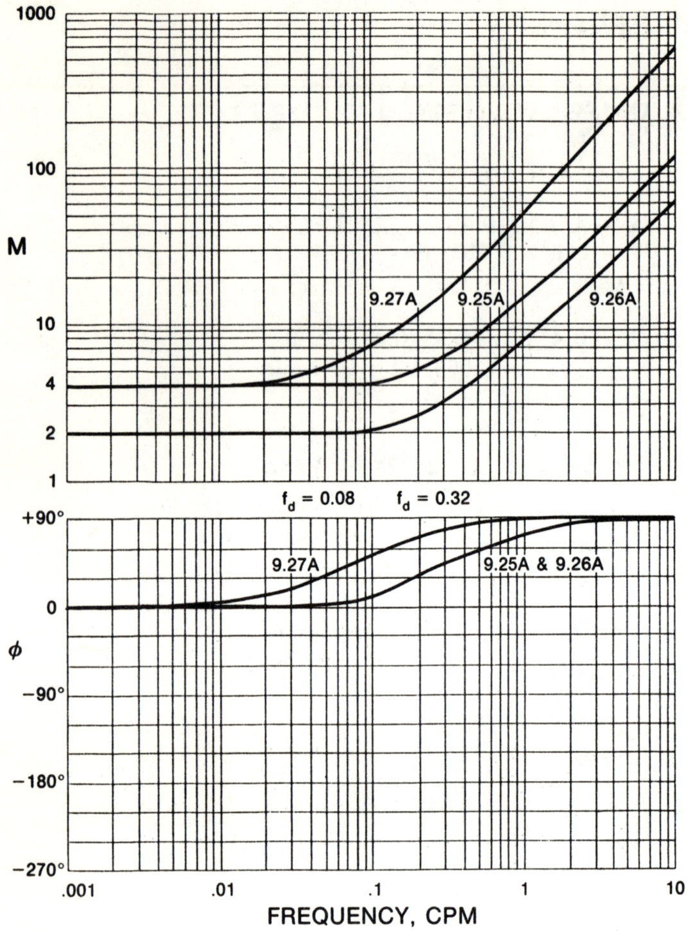

Figure 9A-25. Frequency responses for a proportional plus rate controller.

26. See Figure 9A-25.

27. reset corner $= 1/(2\pi T_d) = 1/\left[2\pi(2.0)\right] = 0.08$ cpm
See Figure 9A-25.

28. dead time lag

29. two

30. noise

31. $G_U = 1/P_G = 1/1.6 = 0.625$
$G_O = 0.6\,G_U = 0.6\,(0.625) = 0.375$
$P_U = 1/f_{-180} = 1/0.7 = 1.43$ min
$T_d = P_U/8 = 1.43/8 = 0.18$ min

32. period, initial overpeaking

33. offset

34. offset, period, initial overpeaking

35. maximum, minimum

36. G_U (or ultimate gain)

37. P_U (or ultimate period)

38. $G_O = 0.6\,G_U = 0.6\,(3) = 1.8$
$T_r = P_U/2 = 2/2 = 1.0$ min
$T_d = P_U/8 = 2/8 = 0.25$ min

39. one quarter, high

Chapter 10

1. true
2. percent
3. precision
4. reproducibility
5. constant sensitivity
6. operating point
7. span, zero
8. one, zero
9. increases
10. decreased